Livestock Diseases and Management

This book series discusses the various infectious diseases affecting the livestock, principle of the disease control, and specific disease management. It discusses the existing strategies to control infectious disease includes animal management programs, vaccination, targeted antimicrobial use, and food hygiene.

Despite public health and veterinary public health improvement within the last century, animal populations remain vulnerable to health threats caused by infectious diseases.

It reviews the current understanding of the zoonotic, emerging, and transboundary animal infections in relation to their transmission, epidemiology, clinical and pathological effects, diagnosis and treatment.

It also examines the recent advancements in the veterinary diagnostics including the existing capabilities, constraints, opportunities, and future potentials. In addition, it elaborates on the conventional and recombinant vaccines that are used in the veterinary medicines and the molecular approaches that have led to the development of new vaccines in recent years. A volume focusing on the various water- and food-borne diseases and its impact on the domestic animals is also a part of this series.

The book series examines the emergence of antimicrobial resistance in livestock, ongoing global surveillance, and monitoring program, its impact on the animal-human interface and strategies for combating resistance.

More information about this series at https://link.springer.com/bookseries/16404

Yashpal Singh Malik • Arockiasamy Arun Prince Milton • Sandeep Ghatak • Souvik Ghosh

Role of Birds in Transmitting Zoonotic Pathogens

Springer

Yashpal Singh Malik
College of Animal Biotechnology
Guru Angad Dev Veterinary and Animal Sciences University
Ludhiana, Punjab, India

Arockiasamy Arun Prince Milton
Division of Animal Health
ICAR Research Complex for NEH Region
Umiam, Meghalaya, India

Sandeep Ghatak
Division of Animal Health
ICAR Research Complex for NEH Region
Umiam, Meghalaya, India

Souvik Ghosh
Ross University School of Veterinary Med
Basseterre, Saint Kitts and Nevis

ISSN 2662-4346 ISSN 2662-4354 (electronic)
Livestock Diseases and Management
ISBN 978-981-16-4553-2 ISBN 978-981-16-4554-9 (eBook)
https://doi.org/10.1007/978-981-16-4554-9

This Springer imprint is published by the registered company Springer Nature Singapore Pte Ltd.
The registered company address is: 152 Beach Road, #21-01/04 Gateway East, Singapore 189721, Singapore

Preface

Zoonotic diseases pose a serious threat to global health and economy. Domestic and wild birds play crucial roles in transmission and spread of important zoonotic pathogens, with significant implications for human and avian health. Although zoonotic diseases have been extensively studied, information on various aspects of avian zoonotic pathogens has not been revisited or revised to a great extent. Currently, numerous research articles/news are available on several avian pathogens. This book is a comprehensive and updated compilation of important zoonotic diseases that are transmitted by domestic and wild birds, and consists of 21 chapters that meticulously describe the (1) aetiology and evolution, (2) complex epidemiology, such as migration pathways in the context of disease transmission, (3) pathogenesis, (4) clinical signs and necropsy findings, (5) diagnostics including the latest molecular assays, and (6) preventative and control strategies, with an emphasis on therapeutics and prophylaxis, of important zoonotic pathogens (bacterial, fungal, parasitic, and viral) of avian origin in humans and birds. Each chapter is aptly supported by interactive tables and figures and features an updated reference section at the end for readers who want to obtain further details of each topic. This book aims to create awareness and enlighten students of veterinary and human medicine on the role of birds in zoonoses and would serve as a useful reference for working veterinarians, human doctors, and public health experts. Such a resource is essential for the research community to understand the latest knowledge and trends in this field so that it can be utilized for improving the counteractions.

Considering the importance of avian migration in the spread of human infections including the zoonotic ones, in the first chapter on "Adaptation and Evolution of Bird's Migration", we have discussed the basics of avian migrations along with the role of birds in the dissemination of zoonotic infections to humans. The second chapter on "Migratory Birds and Public Health Risks" highlights the significance of migratory birds on public health. While humans would have to continue to share the ecosphere with their avian counterparts, a coordinated One Health approach is essential to address the challenges posed by migratory nature of the avian species, which is the focus of this chapter.

Chapters 3–8 describe the aetiology and evolution, epidemiology, disease transmission, pathogenesis, clinical signs and necropsy findings, diagnostics including the latest molecular assays, and preventative and control strategies, with an emphasis on therapeutics and prophylaxis, of important viral infections, including Japanese encephalitis, West Nile, avian influenza, Newcastle disease and other avian paramyxoviruses, Usutu virus, and some neglected birds associated viral diseases like western equine encephalomyelitis, eastern equine encephalomyelitis, Saint Louis encephalitis, Murray Valley encephalitis, Mayaro fever, Sindbis fever, Crimean-Congo haemorrhagic fever, and Kyasanur forest disease.

Chapters 9–16 highlight the important bacterial pathogens including clostridial infections (avian botulism), avian campylobacteriosis, avian chlamydiosis (psittacosis, ornithosis), avian colibacillosis (*E. coli*), avian erysipelas, avian mycoplasmosis, avian salmonellosis, and avian tuberculosis with coverage of discussions on their aetiology, hosts, pathogenesis, epidemiology, diagnosis, treatment, prevention, and control.

Other than viral and bacterial infections, parasitic infestations have also gained prime importance. Chapter 17 discusses cryptosporidiosis that is amongst the most common parasitic infections in domestic as well as wild birds around the world. In Chap. 18, another protozoan problem "Giardiasis" is elaborated that is a leading cause of diarrhoea in humans, transmitted mainly through faecal-oral route and distributed throughout the world. *Giardia* spp. also cause disease in pet and wild birds, which may act as an asymptomatic mechanical carrier of Giardia cysts to humans and other mammals.

Ticks are considered as the second most potential source of vector-borne diseases in humans. Migratory birds are acclaimed as long-distance transporters of ticks and have been accounted for carrying different human pathogens such as tick-borne encephalitis virus, Crimean Congo haemorrhagic fever virus, *Anaplasma marginale, Babesia divergens, Anaplasma phagocytophilum, Ehrlichia,* and *Borrelia burgdorferi*. These important problems have been deliberated in Chap. 19.

Fungal diseases are consistently ignored regardless of their disturbing consequence on human well-being. It is assessed that fungal infections kill more than 1.5 million individuals each year. Recent global estimates have found 3,000,000 cases of chronic pulmonary aspergillosis in a year. Chapter 20 deals with aspergillosis and Chap. 21 highlights the work on sporadic fungal infections including Dermatophytosis (Favus), Dactylariosis, Histoplasmosis, Cryptococcosis, Candidiasis, Rhodotoruliasis, Mucormycosis (*Mucor* spp., *Absidia* spp., and *Rhizopus* spp.).

We believe that owing to the in-depth knowledge of important aspects, the present book will be an excellent source of information for readers from diverse areas. The informative chapters could be useful for veterinary professionals, clinicians, public health experts, researchers, students/scholars, animal producers, faculty and students with interest in migratory birds, diseases, epidemiology, zoonoses and management of diseases and epidemics, and biomedicine experts and pave the way towards designing and adapting effective approaches from clinics to the laboratory for countering important diseases of birds.

We, the authors, would like to express our gratitude to all the peer reviewers whose able expertise and rigorous review of the contents presented here helped us to further improve the manuscript to reach publication stages. We convey our special thanks to Dr. Mohd Ikram Ansari (Research Associate, ICAR-NF project) for his irreplaceable inputs during the compilation of the book chapters. The authors are grateful to Springer Nature for accepting this book proposal and extend their special thanks to Dr. Bhavik Sawhney, Associate Editor-Biomedicine, and Mr. Lenold Esithor, Springer Nature, for providing all the editorial help and high cooperation while processing the manuscripts for its successful publishing.

Ludhiana, Punjab, India — Yashpal Singh Malik
Umiam, Meghalaya, India — Arockiasamy Arun Prince Milton
Umiam, Meghalaya, India — Sandeep Ghatak
Basseterre, Saint Kitts and Nevis — Souvik Ghosh

Contents

About the Authors

Yashpal Singh Malik is serving as Dean of the College of Animal Biotechnology at Guru Angad Dev Veterinary and Animal Sciences University (GADVASU), Ludhiana, Punjab (India), and previously he served as ICAR-National Fellow at Indian Veterinary Research Institute, Izatnagar, India. He works in viral disease epidemiology, virus-host interactions, microbial biodiversity, characterization, and diagnosis of pathogens. Prof. Malik has acquired advanced training in Molecular Virology from the University of Minnesota, USA; Division of Virology, University of Ottawa, Ontario, Canada; and Wuhan Institute of Virology, Wuhan, China. He is a recipient of several prestigious national, state, and academy awards and honours including the ICAR-Jawaharlal Nehru Award. He has supervised 5 Ph.D. and 17 M.V.Sc. students. He has authored 5 books, 25 book chapters, and published 217 scientific research and review articles. Dr. Malik has been the Editor-in-Chief of the *Journal of Immunology and Immunopathology*. He has also edited special issues of Springer's journal *VirusDisease* and Bentham's *Journal of Current Drug Metabolism* on emerging thematic areas. Prof. Malik is editing Springer's book series on Livestock Diseases and Management as "Series Editor".

Arockiasamy Arun Prince Milton is currently serving as a Scientist (Veterinary Public Health) in the Division of Animal and Fisheries Sciences, ICAR Research Complex for NEH region, Umiam, Meghalaya, under the Indian council of Agricultural Research (ICAR). He has fine experience in carrying out research programmes related to zoonoses and food safety. Dr. Milton has published over 45 scientific research articles in the field of zoonoses, food microbiology, and epidemiology of important animal diseases in various peer-reviewed, indexed international and national journals. He has also published ten book chapters in books published by reputed publishers like Springer, Taylor and Francis, etc. He has also presented various research papers in national and international conferences. He is a recipient of a number of prestigious awards including Mahendra Pal Zoonoses Award, Sir. F.M. Burnett Memorial Award, Dr. C.M. Singh National Award of Excellence, Young Researcher Award, etc. Currently, he is studying rodent and foodborne zoonoses, and developing diagnostics for important foodborne pathogens.

Sandeep Ghatak is currently serving as Principal Scientist (Veterinary Public Health) in the Division of Animal and Fisheries Sciences, ICAR Research Complex for NEH Region, Umiam, Meghalaya, under the Indian council of Agricultural Research (ICAR). He is experienced in research on zoonoses and antimicrobial resistance. Dr. Ghatak has published over 40 scientific research articles in the field of zoonoses, food microbiology, antimicrobial resistance, and epidemiology of important animal diseases in various peer-reviewed, international and national journals. He has also published six book chapters in books published by reputed international publishers. He has also presented various research papers and invited lectures in national and international conferences. He received postdoctoral training in the United States Department of Agriculture. Currently he is involved in research on genomic epidemiology of animal and foodborne diseases and antimicrobial resistance.

Souvik Ghosh, *BVSc & AH (Gold Medalist), MVSc, PhD*, is a professor of infectious diseases, director of the One Health Center for Zoonoses and Tropical Veterinary Medicine, and course director of DVM Virology at the Ross University School of Veterinary Medicine (RUSVM), an AVMA accredited vet school in St. Kitts and Nevis, West Indies. Dr. Ghosh's research focuses on molecular epidemiology, genetic diversity, and interspecies transmission of animal and human viruses. He has published 99 research papers and review articles in peer-reviewed biomedical journals with over 2400 citations so far. Dr. Ghosh serves as the section editor of *Archives of Medical Science* and *Frontiers in Microbiology*, review editor of *Frontiers in Veterinary Sciences*, and on the editorial board of *Pathogens* (MDPI) and *Epidemiologia* (MDPI). He is a member of the ICTV *Picobirnaviridae* study group, distinguished member of the World Society for Virology, and fellow of Indian Virological Society. He was involved in research capacity building and training in tropical countries. Because of his contributions to teaching, he received twice the CARE award from RUSVM.

Part I
Introduction

Chapter 1
Adaptation and Evolution of Bird Migration

Abstract Migration in birds is a fascinating biological phenomenon. Migration denotes the annual predictable return movements among breeding and nonbreeding populations of birds. Over millennia, the migration in birds evolved with and by various adaptive mechanisms enabling them to survive in the harshest of the conditions. Migration helps in population differentiation by exposing various groups to different environmental conditions or restricting the chances for a genetic exchange through selection by hybridization or assortative mating. The bird migration routes, also known as flyways, are influenced by various ecological factors. There are broadly eight primary flyways worldwide which are followed for long-distance migration of birds often across continents and large expanses of land and water. Avian migration also impacts human health. In the course of annual migration, birds disseminate many bacterial, viral, fungal, and parasitic diseases or disease-causing agents that affect humans. Birds may act as natural reservoirs (viz. psittacosis, Newcastle disease, avian flu, etc.), may act as asymptomatic carriers (salmonellosis and mite diseases), may disseminate various arthropod vectors of diseases (eastern equine encephalitis, western equine encephalitis, St. Louis encephalitis, etc.) and might create new reservoirs for infections on through fecal pollution of the environment (histoplasmosis, cryptococcosis). Considering the importance of avian migration in spreading of human infections including the zoonotic ones, in this chapter, we look into the basics of avian migrations along with the role of birds in dissemination of zoonotic infections to humans.

Keywords Bird migration · Population differentiation · Flyways · Human health · Arthropod vectors · Zoonotic infections

1.1 Introduction

Among the animal taxa, birds are the most mobile creature. Their mobility has significantly affected different parts of ecology and evolution (Boyle 2017). Migration is a term that denotes the annual predictable return movements among breeding and nonbreeding populations. Regular bird migration comprises movements of all

Y. S. Malik et al., *Role of Birds in Transmitting Zoonotic Pathogens*, Livestock Diseases and Management, https://doi.org/10.1007/978-981-16-4554-9_1

the individuals to long distances in a population to distinct areas of breeding and wintering (Terrill and Able 1988) through the movements which are naturally controlled (Berthold 1991). The majority of the bird migration studies are mainly based on the obligate, native, long-distance movements of large-scale ecological studies (Dingle 2008; Somveille et al. 2013) to field-intensive and population studies (Ydenberg et al. 2002; Stanley et al. 2012).

In the last few decades, many of the bird species have altered their migratory behavior. The most common changes noticed are the timing of relocation and the distance traveled during migration, the two of which are generally linked with climate changes (Newton 2008; Palacín et al. 2017). However, other reasons are not well known, especially for species found in the human-modified environments. Several different animals attempt seasonal migration among breeding and nonbreeding areas. These seasonal migrations in birds are due to many ecological benefits, like providing a favorable niche, allowing the migratory birds to escape competition and to escape parasites and predators (Milner-Gulland et al. 2011). In seasonal migration, birds need a group of various adaptions to survive in a different environment and take long-distance movements (Milner-Gulland et al. 2011). Thus, migration helps in population differentiation by exposing various groups to different environmental conditions or restricting the chances for a genetic exchange through selection by hybridization or assortative mating (Turbek et al. 2018).

1.2 Adaptation and Evolution of Bird Migration

Migration is a process that requires adaptation of different qualities like orientation, morphological adaptations of the locomotory framework, and metabolism (Berthold 2001; Hansson and Akesson 2014). Many studies recommend that major components of the migration, for example, the direction of migration or capacity to migrate, are genetically controlled (Berthold 1991; Pulido 2007; Liedvogel et al. 2011). Nevertheless, migratory behavior can unexpectedly change in a short time. This is shown in the case of blackcaps (*Sylvia atricapilla*) where a particular route of migration emerged in the twentieth century achieving frequencies of around 10% in 30 generations of the population (Helbig et al. 1994). Besides, it is related to adjustments in wing morphology (Rolshausen et al. 2009) that constructs a basis for selection and standing allelic variations being the main driving factor of the change.

Most of the researchers agree that birds have acquired most of their capabilities from their ancestors for migration as the migratory movements are a common and evolutionary feature among ancestral animals (Berthold 1999; Piersma et al. 2005). However, those explaining the evolution of migration does not explain the deep origin; instead, they explain capacities of migration of different lineages (Rappole et al. 2003). Three most essential questions in the bird migration are: (1) Whether migration evolved from wintering grounds to breeding grounds or vice versa? (2) Does the long-distance migrants have their phylogenetic origin in tropical environments or at low latitudes which generally have Palearctic region as their breeding ground? (3) Whether evolution is a characteristic of specific phylogenetic groups or the adaptive changes in a particular individual?

The population which is dispersed in seasonal breeding areas, and individuals among them having a high genetic predisposition of migration, has a selective advantage when migrating to less severe climate during the nonbreeding period. The term dispersal and migration should be clearly understood as the dispersal can be used in different linguistic ways (Salewski and Bruderer 2007). The summary of evolutionary theories of the historical development of bird migration highlights autumn departure from various severe conditions in northern ancestral homes (Dixon 1892), while further range expansion for the reproduction, considered the southern equatorial ancestral home (Dixon 1897). Thus, dispersal movements are regarded as an integral part of the evolution of bird migration (Gauthreaux 1982).

Diverse migratory birds have continuously evolved with a low level of genome-wide variation in a related population (Bensch et al. 2009; Lundberg et al. 2013). At the point when populaces contrasting in migratory behavior come into auxiliary contact, hybrid zones with intermediate and mixed migratory phenotypes can be formed (Delmore and Irwin 2014). Such migratory partitions will, in general, spatially overlap with hybrid zones of different organisms (suture zones) developing because of population separation amid glacial vicariance and resulting recolonization of deglaciated territory (Hewitt 2000; Møller et al. 2011). The regular example of clinal variety in extra quantitative characteristics crosswise over migratory divides underpins this hypothesis (Bensch et al. 2002; Ruegg 2008) and recommends that distinctions in migratory traits, for the most part, developed in allopatry.

Consequently, migratory divides work as natural research facilities where the evolutionary procedures related to hybridization and early speciation can be examined. The development of migratory divides may have a significant effect on the ecology and evolution of species. Sympatric reproducing populaces using diverse migratory routes can be presented to various selection pressures, for example, parasite networks (von Rönn et al. 2015), climate conditions (Newton 2007), or generally the distance required for morphological adaptations (Rolshausen et al. 2009; Alvarado et al. 2014). In the event of transgressive or intermediate hybrid phenotypes failure which may have the risk of high mortality or diminished condition, then disruptive selection can advance for reproductive isolation of parental populaces (Rolshausen et al. 2009) which inevitably can result in summed-up genome-wide differentiation (Shafer and Wolf 2013). The intermediate migratory phenotypes give rise to the selection, which has been supported indirectly in a different study (Helbig 1996; Delmore and Irwin 2014), but the pattern of selection may differ in different systems as a function of primary adaptive landscape (Irwin 2009). Furthermore, to the selection, assortative mating can altogether add to the stable support of the phenotypic dissimilarity against gene flow (Poelstra et al. 2014); however, proof here has likewise been mingled (Bearhop et al. 2005; Liedvogel et al. 2014). Further, in-depth and microevolutionary knowledge on mating and wellness segments connected with migration-related attributes is essential to comprehend the mode and quality of selection and the subsequent populace genetic concerns.

1.3 Route of Bird Migration

Migratory birds consistently perform migrations over the globe (Fig. 1.1), covering significant distances through long continuous flights (Klaassen et al. 2011; Åkesson et al. 2016). Bird migration routes generally develop in response to ecological factors like geology, accessibility of stopover destinations, favorable wind movements, and orientation signals (Alerstam et al. 2003). For better performance, the migration itself includes different adaptations to orientation, flight, timing, and fueling in the birds (Åkesson and Hedenström 2007). The geometry of worldwide routes pursued by birds during migration has been assessed for distances and courses, and for these, two different types of routes have been defined, i.e., orthodromes and loxodromes (Imboden and Imboden 1972; Gudmundsson and Alerstam 1998). The orthodrome route represents the big circular route, which is the shortest distance between the two points on the globe and needs constant changing of direction along the way. While the loxodrome (rhumb line) route is somewhat longer and is created, a consistent geographic course is kept all through the route, expecting a simple orientation mechanism. The two courses have been reflected in different studies assessing alternative compass routes in migratory birds (Muheim et al. 2003; Grönroos et al. 2010; Åkesson and Bianco 2016).

How the birds cross natural boundaries is the most exciting problem of bird migration. Aside from extensive waterways, highlands and arid zones, where no refueling is conceivable, may likewise be hindrances for temperate zone land fowls.

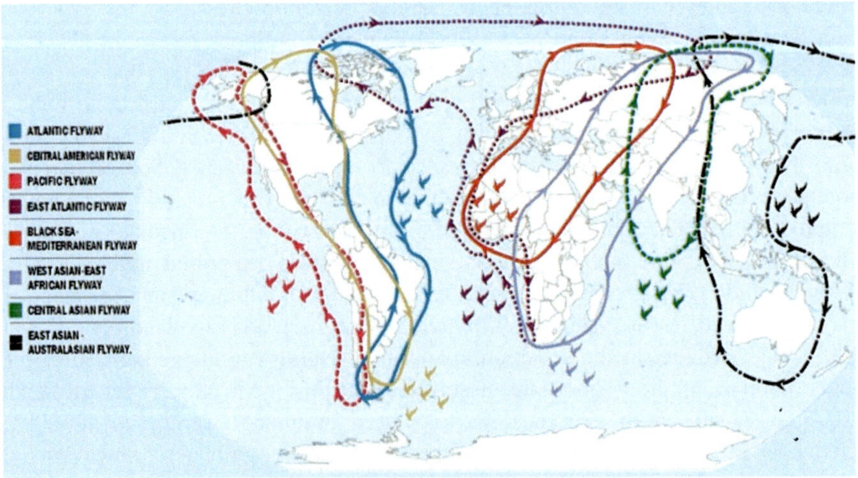

Fig. 1.1 The worldwide routes pursued by birds during migration are presented

Various studies have been done on the migration strategies of passerines crossing the Sahara (Bairlein 1988; Biebach et al. 2000; Biebach 1990; Biebach and Bauchinger 2003) and in the highlands and deserts of western Central Asia (Dolnik 1990; Bolshakov 2001). Studies in western Central Asia, together with moon-watching and catches at stopovers, recommended the speculation that nocturnal passerine migrants reproducing in Siberia and wintering in Africa abstain from crossing the deserts in autumn; instead, they make a reroute toward the north and northwest and fly north of the Caspian Sea (Bolshakov 2001, 2002). This theory is upheld by moon-watching information from the northwestern edge of the desert north of the Caspian Sea (Bulyuk and Chernetsov 2005). In light of cage experiments, a few compasses have been portrayed in birds that depend on data from the sun and the associated pattern of skylight polarization, stars, and the geomagnetic field (Able 1980; Wiltschko and Wiltschko 1995; Åkesson et al. 2014). The use of sun compass by which birds can gradually correct their apparent movement by a time-compensation mechanism, over the sky for the day generally equivalent to a move of 15° every hour of the sun azimuth for the horizon at high latitudes (Kramer 1952; Schmidt-Koenig 1990). While the star compass gives direction concerning the rotation center, i.e., geographic north of the night sky, without using a time-compensation mechanism (Emlen 1970), the geomagnetic field gives a universally accessible source of data, which might be used by migratory birds for compass orientation and route (Wiltschko and Wiltschko 1995, 2009).

The magnetic compass of birds is reliant on the angle of inclination by which the geomagnetic field lines cross the earth surface, and not on the polarity of the geomagnetic field (Wiltschko and Wiltschko 1972), giving a method to separate directions along a north-south axis driving toward the equator or poles. Young birds have an acquired ability to investigate alternative routes. However, they have to encounter a mix of characteristic compass data amid ontogeny to utilize the data for compass introduction (Able and Able 1996; Emlen 1975) and build up a populace-specific orientation (Weindler et al. 1996). Compasses may additionally be recalibrated amid movement (Cochran et al. 2004; Muheim et al. 2006; Åkesson et al. 2015). Despite significant aggregated information of compass components and calibration processes, we do not know precisely what and how compasses are utilized amid active flights of migration. However, the inquiry has been drawn nearer by foreseeing flight courses dependent on alternative compass mechanisms (Alerstam and Pettersson 1991).

1.4 Flyways and Stopovers

Mostly bird species undergo significant scale movement to protect them from unfavorable conditions or potentially use high-quality resources somewhere else. These movements can appear as migration (back and forth movements among breeding and wintering locations), partial migration when not all individuals in a populace are involved (Dean 2004). The large-scale movements of birds are a global phenomenon, and around 20% of bird species are migratory (Somveille et al. 2015). The three main migration system can be defined, the Nearctic–Neotropical system, the Palearctic–African system, and the Palearctic–Asian system (Rappole and Jones 2003). Nomadism is, for the most part, connected to semi-arid and arid conditions around the world (Dean 2004) with nomadism representing around 10% of all flying creature species in southern Africa (Dean 1997) and 26% of all flying creature species in Australia (Smith 2015).

A flyway is a flight way utilized in bird migration – flyways by and large range over landmasses and oceans. When going between their rearing and wintering grounds, feathered creatures do not pick their ways arbitrary. They pursue set courses that incorporate reasonable living spaces where they can stop to rest and refuel enroot. A wide range of animal groups bunch together to go along comparable courses, which have been inexactly part into eight primary flyways. (1) Atlantic Flyway, (2) Central Flyway, (3) Pacific Flyway, (4) East Atlantic Flyway, (5) Black Sea-Mediterranean Flyway, (6) West Asian-East African Flyway, (7) Central Asian Flyway, and (8) East Asian-Australasian Flyway.

Numerous migratory bird species having a wide geographic distribution are made out of various subpopulations following particular migratory routes. Thus, these flyways are delineated depending on mark-recapture information, while in numerous species like passerines, they are supported by the molecular data (Ruegg and Smith 2002; Clegg et al. 2003; Pérez-Tris et al. 2004), waterfowl (Scribner et al. 2001; Gay et al. 2004), and waders (Wennerberg 2001). The absence of genetic divergence among migratory flyways may recommend that the migratory behavior is commonly adaptable or the migration route has changed recently in specific populations. The different flyways are frequently deciphered as a heritage of the ice ages. Various subpopulations would have been bound to different refugia, which they held as wintering grounds following a postglacial range development (Ruegg and Smith 2002). These periods in allopatry would prompt some genetic divergence among subpopulations, then which may be disintegrated step by step through gene flow.

Long-distance migratory birds are under pressure to migrate rapidly. Stopovers need additional time than flying and are utilized by birds to refuel amid movement, yet the impact of fuel loads obtained at stopover destinations during the migration has not been measured or quantified (Gómez et al. 2017). As far as time is concerned, the maximum expense of migration is during stopovers rather than during times of flight (Hedenström and Alerstam 1997; Wikelski et al. 2003), and the birds depend on the time spent at stopover destinations to rest and refuel for the next leg of their journeys (Smith and McWilliams 2014). Optimal migration theory gives a system to examine stopover behavior and its outcomes by checking whether migrants are time or energy minimizers utilizing information on fueling rate, stopover length, fuel burdens, and potential flight ranges (Alerstam 2011). Individuals endeavoring to limit the total time spent on migration are relied upon to augment the measure of fuel they can obtain at every stopover in the briefest time conceivable. A key result of this procedure is that it expands the distances that can be flown between stopovers (Hedenström and Alerstam 1997; Weber and Houston 1997). Subsequently, the fuel loads (a measure of fat conveyed) of a period minimizer ought to be firmly connected to local conditions at stopover destinations just as to the conditions expected ahead because these conditions impact fueling rates (Hedenström and Alerstam 1997; Weber and Houston 1997). Besides, stopover lengths in time minimizers are relied upon to have been molded by or to react straightforwardly to experienced fueling conditions (Alerstam 2011; Hedenström and Alerstam 1997). Bigger takeoff fuel loads ought to take into consideration longer and quicker flights, large pace of migration since individuals securing adequate fuel in the briefest time conceivable should make less stopovers and have the option to take more straightforward routes to their destinations, including having the option to fly over physical obstructions or substantial zones of inadequate territory, for example, deserts or seas, as opposed to evading these areas (Alerstam 2001).

1.5 Significance in Disease Transmission

The birds can transmit infectious diseases to humans by several mechanisms. The infections caused by birds can be assembled into four groups. In group 1 infections, the birds act as natural repositories for the infection, which causes disease among them. The unhealthy birds at that point spread the infectious agent into nature, and people become infected as unintentional hosts. Instances of such infections are psittacosis, Newcastle sickness, avian flu, and yersiniosis. In group 2 and 3 infections, fowls are the characteristic reservoirs for the infection, yet do not turn out to be sick themselves. The infection from group 2 (for instance, salmonellosis and mite diseases) scatters from the colonized birds directly into the environment, while the agents of group 3, for instance, eastern equine encephalitis, western equine encephalitis, St. Louis encephalitis, and Japanese B encephalitis, are dispersed by the arthropod vectors and include humans as unintentional hosts. In group 4, the organism transmits through their fecal matter in the environment. Instances of infections of the last class are the fungal disease like histoplasmosis and cryptococcosis (Levison 2015).

The disease caused by fowls can be intense, chronic, latent, or asymptomatic. Infections that can be transported are the arboviruses (Eastern equine encephalitis, West Nile infection), Newcastle disease virus, Usutu virus, avian pox virus, duck plague virus, St. Louis encephalitis virus, and influenza A virus (Bengis et al. 2004; Hubalek 2004). Similarly, the causal agents of ornithosis, avian cholera, mycoplasmosis, avian tuberculosis, erysipelas, coxiellosis, campylobacteriosis, Lyme borreliosis, colibacillosis, cholera, salmonellosis, listeriosis, and yersiniosis can likewise be transferred. In the same manner, drug-resistant enteropathogens can also be dispersed (Bengis et al. 2004; Hubalek 2004; Chuma et al. 2000; Hudson et al. 2000; Rappole and Hubalek 2003) like fungi and endoparasites, for example, *Aspergillus* spp., *Candida* spp., *Haemoproteus* spp., *Leucocytozoon* spp., *Sarcocystis* spp., *Toxoplasma* spp., and *Cryptosporidium* spp. (Bengis et al. 2004; Hasle et al. 2004).

Mechanical carriers transmit either outer or interior microbial pathogens. Exterior pathogens, as fungal spores, are situated on the body of the fowl and can retain up to 12 days on the plumes of migratory swallows. Internal pathogens do not develop in the avian body but go through the digestive tract and are viable when discharged. Foot and mouth disease infection has been believed to be transmitted by mechanical carriers (Bengis et al. 2004; Hubalek 2004).

Migratory fowls can act as carriers of tainted hematophagous ectoparasites, which act as vectors for a few infections. In this way, the infected immature ixodid and argasid ticks are transported from one spot then onto the next and even from one continent to another (Hasle et al. 2004). Tick-borne pathogens can be viruses, for example, tick-borne encephalitis, Meaban, Hughes group, Crimean-Congo hemorrhagic fever virus, Bhanja, Great Island complex, Thogoto, and Dhori viruses; or the bacteria, for example, *Rickettsia* spp. and *Anaplasma phagocytophilum*; or then again protozoa-like *Babesia microti*. Further, bugs can be transmitted on migrating birds to long distances (Bengis et al. 2004; Hubalek 2004).

The method of transmission of all these microbes can either be direct or indirect. Direct transmission is brought about by the migratory fowl itself through close contact, contact by the inward breath of released respiratory droplets from sneezing or coughing, or by infectious feces. Indirect transmission happens using an arthropod, for example, a bug, mite, mosquito, sandfly, or tick, or a lifeless vehicle like water, soil, food, and so on. Likewise, the airborne infection spread by droplet nuclei, dust, and so forth is viewed as an indirect method of transmission (Hubalek 2004). The mode of transportation of pathogens by migratory birds relies upon the course of transmission (Hasle et al. 2004).

The agents in water-borne infections, for example, avian flu infection, *Chlamydia psittaci*, Newcastle sickness infection, *Escherichia coli, Enterococcus faecalis, Yersinia* spp., *Clostridium* spp., *Pasteurella multocida*, and *Candida* spp., are shed by tainted migratory fowls in excreta, in nasal release, and in respiratory exudate into water. In tick-borne contaminations, the infectious larval or nymphal tick is dropped into another geographic region amid relocation (Hubalek 2004).

References

Able KP (1980) Mechanisms of orientation, navigation and homing. In: Gauthreaux S Jr (ed) Animal migration, orientation and navigation. Springer, Berlin, pp 283–373

Able KP, Able MA (1996) The flexible migratory orientation system of the Savannah sparrow (*Passerculus sandwichensis*). J Exp Biol 199:3–8

Åkesson S, Bianco G (2016) Assessing vector navigation in long distance migrating birds. Behav Ecol 27:865–875

Åkesson S, Hedenström A (2007) How migrants get there: migratory performance and orientation. Bioscience 57:123–133

Åkesson S, Boström J, Liedvogel M, Muheim R (2014) Animal navigation. Animal movement across scales. Oxford University Press, Oxford, pp 151–178

Åkesson S, Odin C, Hegedüs R, Ilieva M, Sjöholm C, Farkas A, Horváth G (2015) Testing avian compass calibration: comparative experiments with diurnal and nocturnal passerine migrants in South Sweden. Biol Open 4:35–47

Åkesson S, Bianco G, Hedenström A (2016) Negotiating an ecological barrier: crossing the Sahara in relation to winds by common swifts. Phil Trans R Soc B 371:20150393

Alerstam T (2001) Detours in bird migration. J Theor Biol 209:319–331

Alerstam T (2011) Optimal bird migration revisited. J Ornithol 152:5–23

Alerstam T, Pettersson SG (1991) Orientation along great circles by migrating birds using a sun compass. J Theor Biol 152:191–202

Alerstam T, Hedenström A, Åkesson S (2003) Long-distance migration: evolution and determinants. Oikos 103:247–260

Alvarado AH, Fuller TL, Smith TB (2014) Integrative tracking methods elucidate the evolutionary dynamics of a migratory divide. Ecol Evol 4:3456–3469

Bairlein F (1988) How do migratory songbirds cross the Sahara? (a review). Trends Ecol Evol 3: 191–194

Bearhop S, Fiedler W, Furness RW et al (2005) Assortative mating as a mechanism for rapid evolution of a migratory divide. Science 310:502–504

Bengis RG, Leighton FA, Fisher JR, Artois M, Mörner T, Tate CM (2004) The role of wildlife in emerging and re-emerging zoonoses. Rev Sci Tech 23(2):497–511

Bensch S, Akesson S, Irwin DE (2002) The use of AFLP to find an informative SNP: genetic differences across a migratory divide in willow warblers. Mol Ecol 11:2359–2366

Bensch S, Grahn M, Muller N, Gay L, Akesson S (2009) Genetic, morphological, and feather isotope variation of migratory willow warblers show gradual divergence in a ring. Mol Ecol 18: 3087–3096

Berthold P (1991) Genetic control of migratory behavior in birds. Trends Ecol Evol 6:254–257

Berthold P (1999) A comprehensive theory for the evolution, control and adaptability of avian migration. Ostrich 70:1–11

Berthold P (2001) Bird migration: a general survey, 2nd edn. Oxford University Press, New York

Biebach H (1990) Strategies of trans-Sahara migrants. In: Gwinner E (ed) Bird migration. Springer, Berlin, pp 352–367

Biebach H, Bauchinger U (2003) Energetic savings by organ adjustment during long migratory flights in Garden Warblers (*Sylvia borin*). In: Berthold P, Gwinner E, Sonnenschein E (eds) Avian migration. Springer, Berlin, pp 269–280

Biebach H, Biebach I, Friedrich W, Heine G, Partecke J, Schmidl D (2000) Strategies of passerine migration across the Mediterranean Sea and the Sahara Desert: a radar study. Ibis 142:623–634

Bolshakov CV (2001) Results of the large-scale study of nocturnal bird migration in the arid and mountainous zone of western Central Asia ("Asia" programme). In: Kurochkin EN, Rakhimov II (eds) Achievements and problems of ornithology of northern Eurasia on the border between the centuries. Magarif, Kazan, pp 372–393. (in Russian)

Bolshakov CV (2002) The Palaearctic–African bird migration system: the role of desert and highland barrier of western Asia. Ardea 90:515–523

Boyle WA (2017) Altitudinal bird migration in North America. Auk Ornithol Adv 134(2):443–465

Bulyuk VN, Chernetsov N (2005) Why fewer Siberian–African passerines cross the deserts of western Central Asia in autumn than during return migration in spring? Alauda 73:256–257

Chuma T, Hashimoto S, Okamoto K (2000) Detection of thermophilic *Campylobacter* from sparrows by multiplex PCR: the role of sparrows as a source of contamination of broilers with *Campylobacter*. J Vet Med Sci 62:1291–1295

Clegg SM, Kelly JF, Kimura M, Smith TB (2003) Combining genetic markers and stable isotopes to reveal population connectivity and migration patterns in a Neotropical migrant, Wilson's warbler (*Wilsonia pusilla*). Mol Ecol 12:819–830

Cochran WW, Mouritsen H, Wikelski M (2004) Migrating songbirds recalibrate their magnetic compass daily from twilight cues. Science 304:405–408

Dean WRJ (1997) The distribution and biology of nomadic birds in the Karoo, South Africa. J Biogeogr 24:769–779. https://doi.org/10.1046/j.1365-2699.1997.00163.x

Dean WRJ (2004) Nomadic desert birds. Springer, Berlin

Delmore KE, Irwin DE (2014) Hybrid songbirds employ intermediate routes in a migratory divide. Ecol Lett 17:1211–1218

Dingle H (2008) Bird migration in the southern hemisphere: a review comparing continents. Emu 108:341–359

Dixon C (1892) The migration of birds. Windsor House, London

Dixon C (1897) The migration of birds. Horace Cox, Windsor House, London

Dolnik VR (1990) Bird migration across arid and mountainous regions of Middle Asia and Kazakhstan. In: Gwinner E (ed) Bird migration. Springer, Berlin, pp 368–386

Emlen ST (1970) Celestial rotation: its importance in the development of migratory orientation. Science 170:1198–1202

Emlen ST (1975) Migration: orientation and navigation. In: Farner DS, King JR (eds) Avian biology, vol 5. Academic Press, New York, pp 129–219

Gauthreaux SA (1982) The ecology and evolution of avian migration systems. In: Farner DS, King JR, Parkes KC (eds) Avian biology. Academic Press, New York, pp 93–168

Gay L, Defos Du Rau P, Mondain-Monval J, Crochet P (2004) Phylogeography of a game species: the red-crested pochard (*Netta rufina*) and consequences for its management. Mol Ecol 13: 1035–1045

Gómez C, Bayly NJ, Norris DR, Mackenzie SA, Rosenberg KV, Taylor PD, Hobson KA, Cadena CD (2017) Fuel loads acquired at a stopover site influence the pace of intercontinental migration in a boreal songbird. Sci Rep 7(2017):3405

Grönroos J, Muheim R, Åkesson S (2010) Orientation and autumn migration routes of juvenile sharp-tailed sandpipers at a staging site in Alaska. J Exp Biol 213:1829–1835

Gudmundsson GA, Alerstam T (1998) Optimal map projections for analysing long-distance migration routes. J Avian Biol 29:597–605

Hansson L-A, Akesson S (2014) Animal movement across scales. Oxford University Press, Oxford

Hasle G, Mehl R, Bjune G, Leinaas HR (2004) Transport of ticks by migratory birds in Norway. In: Proc. third African acarology symposium, 11–15 January, Giza. www.reiseklinikken.no/bstract_Cairo.htm. Accessed 29 Nov 2008

Hedenström A, Alerstam T (1997) Optimum fuel loads in migratory birds: distinguishing between time and energy minimization. J Theor Biol 189:227–234

Helbig AJ (1996) Genetic basis, mode of inheritance and evolutionary changes of migratory directions in palearctic warblers (Aves: Sylviidae). J Exp Biol 199:49–55

Helbig AJ, Berthold P, Mohr G, Querner U (1994) Inheritance of a novel migratory direction in Central European blackcaps. Naturwissenschaften 81:184–186

Hewitt GM (2000) The genetic legacy of the Quaternary ice ages. Nature 405:907–913

Hubalek Z (2004) An annotated checklist of microorganisms associated with migratory birds. J Wildlife Dis 40(4):639–659

Hudson CR, Quist C, Lee MD, Keyes K, Dodson SV, Morales C, Sanchez S, White DG, Maurer JJ (2000) Genetic relatedness of Salmonella isolates from nondomestic bird in south eastern Unites States. J Clin Microbiol 38:1860–1865

Imboden C, Imboden D (1972) Formel für Orthodrome und Loxodrome bei der Berechnung von Richtung und Distanz zwischen Beringungs-und Wiederfundort. Vogelwarte 26:336–346

Irwin DE (2009) Speciation: new migratory direction provides route toward divergence. Curr Biol 19:R1111–R1113

Klaassen RHG, Alerstam T, Carlsson P, Fox JW, Lindstrom A (2011) Great flights by great snipes: long and fast non-stop migration over benign habitats. Biol Lett 7:833–835

Kramer G (1952) Experiments on bird orientation. Ibis 94:265–285

Levison ME (2015) Diseases transmitted by birds. Microbiol Spec 3(4). https://doi.org/10.1128/microbiolspec.IOL5-0004-2015

Liedvogel M, Akesson S, Bensch S (2011) The genetics of migration on the move. Trends Ecol Evol 26:561–569
Liedvogel M, Larson KW, Lundberg M et al (2014) No evidence for assortative mating within a willow warbler migratory divide. Front Zool 11:1–9
Lundberg M, Boss J, Canback B et al (2013) Characterisation of a transcriptome to find sequence differences between two differentially migrating subspecies of the willow warbler *Phylloscopus trochilus*. BMC Genomics 14:330
Milner-Gulland EJ, Fryxell JM, Sinclair ARE (eds) (2011) Animal migration: a synthesis. Oxford University Press, New York
Møller AP, Garamszegi LZ, Peralta-Sanchez JM, Soler JJ (2011) Migratory divides and their consequences for dispersal, population size and parasite–host interactions. J Evol Biol 24: 1744–1755
Muheim R, Åkesson S, Alerstam T (2003) Compass orientation and possible migration routes of passerine birds at high arctic latitudes. Oikos 103:341–349
Muheim R, Phillips JB, Åkesson S (2006) Polarized light cues underlie compass calibration in migratory songbirds. Science 313:837–839
Newton I (2007) Weather-related mass-mortality events in migrants. Ibis 149:453–467
Newton I (2008) The migration ecology of birds. Academic, Amsterdam
Palacín C, Alonso JC, Martín CA, Alonso JA (2017) Changes in bird-migration patterns associated with human-induced mortality. Conserv Biol 31(1):106–115
Pérez-Tris J, Bensch S, Carbonell R, Helbig A, Tellería J (2004) Historical diversification of migration patterns in a passerine bird. Evolution 58:1819–1832
Piersma T, Pérez-Tris J, Mouritsen H, Bauchinger U, Bairlein F (2005) Is there a 'migratory syndrome' common to all migrant birds? Ann N Y Acad Sci 1046:282–293
Poelstra JW, Vijay N, Bossu C et al (2014) The genomic landscape underlying phenotypic integrity in the face of gene flow in crows. Science 344:1410–1414
Pulido F (2007) The genetics and evolution of avian migration. Bioscience 57:165–174
Rappole JH, Hubalek Z (2003) Migratory birds and West Nile virus. J Appl Microbiol 94 (Suppl):47S–58S
Rappole JH, Jones P (2003) Evolution of old and new world migration systems. Ardea 90:525–537
Rappole JH, Helm B, Ramos MA (2003) An integrative framework for understanding the origin and evolution of avian migration. J Avian Biol 34:124–128
Rolshausen G, Segelbacher G, Hobson KA, Schaefer HM (2009) Contemporary evolution of reproductive isolation and phenotypic divergence in sympatry along a migratory divide. Curr Biol 19:2097–2101
Ruegg K (2008) Genetic, morphological, and ecological characterization of a hybrid zone that spans a migratory divide. Evolution 62:452–466
Ruegg KC, Smith TB (2002) Not as the crow flies: a historical explanation for circuitous migration in Swainson's thrush (*Catharus ustulatus*). Proc R Soc B 269:1375–1381
Salewski V, Bruderer B (2007) The evolution of bird migration—a synthesis. Naturwissenschaften 94:268–279
Schmidt-Koenig K (1990) The sun compass. Experientia 46:336–342
Scribner KT, Petersen MR, Fields RL, Talbot SL, Pearce JM, Chesser RK et al (2001) Sex-biased gene flow in spectacled eiders (Anatidae): inferences from molecular markers with contrasting modes of inheritance. Evolution 55:2105–2115
Shafer ABA, Wolf JBW (2013) Widespread evidence for incipient ecological speciation: a meta-analysis of isolation-by-ecology. Ecol Lett 16:940–950
Smith JE (2015) Effects of environmental variation on the composition and dynamics of an arid-adapted Australian bird community. Pacif Cons Biol 21:74–86. https://doi.org/10.1071/PC14905
Smith AD, McWilliams SR (2014) What to do when stopping over: behavioural decisions of a migrating songbird during stopover are dictated by initial change in their body condition and mediated by key environmental conditions. Behav Ecol 25:1423–1435

Somveille M, Manica A, Butchart SHM, Rodrigues ASL (2013) Mapping global diversity patterns for migratory birds. PLoS One 8:e70907. https://doi.org/10.1371/journal.pone.0070907
Somveille M, Rodrigues ASL, Manica A (2015) Why do birds migrate? A macroecological perspective. Glob Ecol Biogeogr 24:664–674
Stanley CQ, MacPherson M, Fraser KC, McKinnon EA, Stutchbury BJM (2012) Repeat tracking of individual songbirds reveals consistent migration timing but flexibility in route. PLoS One 7: e40688. https://doi.org/10.1371/journal.pone.0040688
Terrill SB, Able KP (1988) Bird migration terminology. Auk 105:205–206
Turbek SP, Scordato ESC, Safran RJ (2018) The role of seasonal migration in population divergence and reproductive isolation. Trends Ecol Evol 33(3):164–175
von Rönn JAC, Harrod C, Bensch S, Wolf JBW (2015) Transcontinental migratory connectivity predicts parasite prevalence in breeding populations of the European barn swallow. J Evol Biol 28:535–546
Weber TP, Houston AI (1997) A general model for time-minimizing avian migration. J Theor Biol 185:447–458
Weindler P, Wiltschko R, Wiltschko W (1996) Magnetic information affects the stellar orientation of young bird migrants. Nature 383:158–160
Wennerberg L (2001) Breeding origin and migration pattern of dunlin (*Calidris alpina*) revealed by mitochondrial DNA analysis. Mol Ecol 10:1111–1120
Wikelski M et al (2003) Costs of migration in free-flying songbirds. Nature 423:704
Wiltschko W, Wiltschko R (1972) Magnetic compass of European robins. Science 176:62–64
Wiltschko R, Wiltschko W (1995) Magnetic orientation in animals. Springer, Berlin
Wiltschko R, Wiltschko W (2009) Avian navigation. Auk 126:717–743
Ydenberg RC, Butler RW, Lank DB, Guglielmo CG, Lemon M, Wolf N (2002) Trade-offs, condition dependence and stopover site selection by migrating sandpipers. J Avian Biol 33: 47–55

Chapter 2
Migratory Birds and Public Health Risks

Abstract Migration of avian species is a special characteristic enabling them great mobility to cover distances over a relatively short period of time. Coupled with their abundance in almost everywhere on earth, birds serve as potential disseminators of pathogens, many of which affect humans causing zoonotic infections. Migration in birds takes forms ranging from short-distance to local and seasonal movements to long-distance migration across continents. As birds coevolved and coexist with humans, over millennia they shared pathogen of humans and vice versa. Available evidences indicate that birds play important roles in maintenance and dissemination of many zoonotic pathogens such as influenza viruses, arboviral agents, tick-borne pathogens, *Salmonella* serovars, *Escherichia coli*, *Campylobacter jejuni*, *Chlamydia psittaci*, and *Mycobacterium avium*. Due to increasing industrialization, globalization, climatic changes, intensification of commercial farming, urbanization, destruction of habitats, and other factors, the traditional flyways of birds are being disturbed and therefore the interaction between humans and birds have become more complex opening new opportunities for transmission and perpetuation of zoonotic diseases. While humans would have to continue to share the ecosphere with their avian counterparts, a coordinated One Health approach is essential to address the challenges posed by migratory nature of the avian species.

Keywords Bird migration · Zoonosis · Human infection · One health · Flyway · Zoonotic infection

2.1 Introduction: Migration

Birds are one of the most abundant, evolved, and mobile life forms on earth. They are found in all continents and have adapted well to exist in every possible environment on earth. One of the most striking features of birds are a unique mobility feature called "migration." Though the term might be used for other species too, in birds migration usually denotes seasonal movement which occurs annually between breeding place and another nonbreeding area (usually wintering place). Ornithologists believe that migration is a survival strategy to optimize in the best of two very

Y. S. Malik et al., *Role of Birds in Transmitting Zoonotic Pathogens*, Livestock Diseases and Management, https://doi.org/10.1007/978-981-16-4554-9_2

different environments. When one particular environment turns out to be inhospitable or costly in terms of survival and reproduction, birds migrate to other place which is more suitable for survival and reproduction. Since the seasonal changes generally dictate suitability of a place for survival, the event of migration in birds is also seasonal and annually repeated.

The distances involved in migration varies and may involve very large distance over thousands of kilometers depending on the avian species involved and their natural habitat. In addition, one important feature of avian migration is "homing," which means that following movement to a new environment, the concerned avian species are able to come to its original habitat. This particular feature is also observed in many other organisms. For "homing" to happen successfully, it is imperative that migrating birds are able to navigate during the journey so that they can return back to the starting point of the journey at a suitable time. Many studies proved the ability of navigation in birds though the exact mechanism behind successful navigation still remains elusive. Whatever be the underlying cause, ability to navigate gave rise to more or less set path of movement for birds during annual migration, thus creating a pattern of movement. This path developed over millennia are known as flyways for the birds.

Armed with the ability to navigate over long distance and unique adaptability, birds act as a bridge between distant geographical locations which are otherwise separated by hydrological, geological, or climatic barriers. Thus, birds due to their migrating nature can connect two completely different environments which would otherwise have not been possible, and this connectivity has their implication too for the environment, ecology, and the residing flora and fauna at these locations. For this current discussion, it is important to note that birds are infected by many human pathogens and vice versa. Due to their ability to connect distant place, it is possible that migrating birds may act as efficient disseminator of many diseases that are of great public health importance.

2.2 Types of Migration

Avian migration has intrigued ornithologists since ages, and there is a great deal of literatures available on the subject. Contrary to common perception of migration involving very long-distance flights, as in popular imagination, avian migration can be of various types depending on the pattern involved (Table 2.1).

Avian migration can also be classified according to geographic regions involved or the flying routes of the birds. According to such classification, there are eight major migrations systems worldwide. These system are Palearctic–African, Nearctic–Neotropical, Palearctic–South Asian, Himalayan–South Asian, Intra-African, Intra-South American, Tropical–Subtropical Intra-Asian, and Intra-Australian (Newton 2010; Rappole 2013).

Table 2.1 Various types of avian migration and associated species of birds

Type	Avian species example
Short distance	White-throated bee-eater of Africa
Long distance	Bar-tailed godwit of Alaska, snowy egret of Africa
Stepwise	Seen in long-distance migrant, e.g., snowy egret in tropical Africa
Facultative	Cardueline finches, snowy owl
Partial	Song sparrow, European robin
Altitudinal	Hummingbirds of Andes, white-throated robin of Mexico
Differential	Dark-eyed Junco, song sparrow
Nomadic	Waterfowls, e.g., pink-eared ducks, cootsia
Wandering	Pelagic birds like albatrosses, petrels
Local and seasonal	Tropical birds—Northern flickers of Guatemala, masked Tityra of Costa Rica

Data source: (Bairlein and Wiltschko 2017; Cox 2010; Newton 2010; Rappole 2013)

2.3 Diseases Associated with Avian Migration

As birds routinely migrate over large geographical distances and harbors many pathogens of zoonotic importance to humans, it is conceivable that they may aid in spread of various infectious agents worldwide. Various studies have been done on the subject, and a number of diseases are to have been facilitated by the migrating birds (Altizer et al. 2011; Causey and Edwards 2008; Cohen et al. 2015; Comstedt et al. 2006; Contreras et al. 2016; Foti et al. 2011; Hasle 2013; Jourdain et al. 2007a, b; Najdenski et al. 2018; Palomar et al. 2012; Rappole 2013; Reed et al. 2003; Waldenström et al. 2007; Yin et al. 2020). Not only the movements of the birds that have attracted attention of researchers, environmental contamination by pathogens during "stopovers" have also been under scrutiny for possible human health risks (Altizer et al. 2011; Antilles et al. 2015; Contreras et al. 2016; Haag-wackernagel and Moch 2004; Hubálek 2004; Jourdain et al. 2007a, b; Najdenski et al. 2018).

Among many diseases that have been studied, one of the most studied zoonosis associated with migratory birds is avian influenza. Avian influenza is a zoonotic virus infection of birds and humans caused by *Influenza virus* A under the family *Orthomyxoviridae*. The disease is highly infectious in nature and has a global distribution. In humans, influenza causes the proverbial flu-like symptoms, though in birds the infection may go unnoticed, especially in wild birds. The natural transmission cycle of the disease involves circulation of the virus among wild and migratory birds without apparent outbreak of disease in birds. From wild birds, the disease is transmitted to peridomestic and commercial avian species and is introduced in the human population of the area though other mammals like pigs also play important role in recombination and emergence of novel influenza viruses. The disease has caused a number of pandemics causing millions of deaths in humans. While there are many types and subtypes of influenza viruses, based on their pathogenic potential, the viruses are categorized into LPAI (lowly pathogenic avian influenza) and HPAI (highly pathogenic avian influenza) types (Hlinak et al. 2006; Li et al. 2019; OIE 2018; Sendor et al. 2020; Wu et al. 2020).

Wild birds are considered natural reservoirs of *Influenza virus*. During migration of wild birds for wintering and breeding, these birds come in contact with other species of birds and contaminate the water bodies in and around their roosting sites. Several studies have documented this potential spread. *Influenza virus* causing avian influenza has been isolated from more than 100 different species of birds from various parts of the world. Among these, waterfowls have been identified as high-risk hosts, and in many cases of outbreak of avian influenza, direct or indirect contact with waterfowls has been observed. Though spread of avian influenza is dependent on a large number of ecological and population factors, the role of avian migration in dissemination of *Influenza virus* has been discussed convincingly (Altizer et al. 2011; Causey and Edwards 2008; Contreras et al. 2016; Hubálek 2004; Jourdain et al. 2007a, b; Meri 2014; Najdenski et al. 2018; OIE 2018; Rappole 2013; Reed et al. 2003; Sendor et al. 2020; Tsiodras et al. 2008).

West Nile fever is an arboviral zoonosis which is emerging in many parts of the world. The disease is caused by West Nile virus (WNV) belonging to the family *Flaviviridae*. The virus is naturally maintained in a bird-mosquito-bird cycle. The mosquito species involved in the transmission of the virus is of *Culex* group of which *C. pipiens* is a prominent vector. In the United States, WNV has been isolated from more than 300 bird species including passerine birds and American robins. Humans acquire the infection from infected mosquito bites. Available evidences indicate that humans do not develop sufficient viremia during the course of the disease to transmit the disease to mosquitoes and therefore birds play a major role. In avian species, the disease is usually inapparent and migration of birds over short and long distances may introduce the virus in new territories. Studies from other parts of the world including European countries also indicated a similar epidemiological feature with avian species playing a major yet under-recognized role in dissemination of West Nile fever (David and Abraham 2016; Jourdain et al. 2007a, b; Petersen et al. 2013; Sule et al. 2018; Tyson-Pello and Olsen 2020). Recent reports from Germany suggested that human epidemic of WNV started following an epizootic in birds and horses, thus indicating the inherent role of birds in dissemination of the disease (Ziegler et al. 2020). Further, recent studies indicated that the virus is evolving to overwinter inside roosts of peridomestic birds such as crows instead of traditional wild bird cycle indicating movement of the source of infection closer to human habitations (Montecino-Latorre and Barker 2018).

Wild birds and their migrating nature contribute to dissemination of a number of other pathogens too. It has been reported that Swanison's thrush that annually migrate thousands of kilometers are capable carrying Lyme disease and WNV (Altizer et al. 2011). Migratory birds overwintering in Central and South America have been reported to be chronically infested with *Amblyomma* ticks which in turn carry rickettsial pathogens especially *R. paerkeri* and *R. monacensis*. The results indicated that these apparently healthy migrating flocks of birds can introduce insect vectors as well as other pathogens in new areas along their migration routes (Cohen et al. 2015). Similar to *Amblyomma* ticks, migratory birds are also important carriers of Ixodid and Argasid ticks which are vectors for Lyme borreliosis. Studies have documented the potential of migratory birds in spreading the ticks and *Borrelia* spp. in naive population and territory (Humair 2002; Meri 2014; Steele et al. 2013; Tsiodras et al. 2008). In fact, migratory birds are now considered an important reservoir of borreliosis (Comstedt et al. 2006). Tick-borne encephalitis virus and *Ixodes* ticks were recovered in Sweden from migratory birds such as Thrush, Robin, and Redstart (Waldenström et al. 2007). Migratory birds and ticks have a close ecological relation allowing for dispersal of tick species over long distances. With dispersal of tick species, many tick-borne pathogens of zoonotic importance also spread to newer regions, and often these migrating birds serve as a mobile reservoir for tick-borne diseases (Hamer et al. 2012a, b; Hasle 2013; Palomar et al. 2012; Waldenström et al. 2007; Yang et al. 2015).

In addition to viral and tick-borne pathogens, a number of researchers also documented potential dispersal of bacterial pathogens by migrating birds. In 2004, Hubalek drafted an annotated list of pathogens associated with migratory birds in which bacteria belonging to five different families were identified (Hubálek 2004). However, since then many studies appeared which underlined the role of migratory birds in dissemination of zoonotic bacteria including *Salmonella* serovars, *Escherichia coli*, *Campylobacter jejuni*, *Campylobacter coli*, *Chlamydia psittaci*, *Mycobacterium avium*, *Klebsiella pneumoniae*, *Yersinia enterocolitica*, *Staphylococcus aureus*, *Listeria* spp., *Acinetobacter baumannii*, and *Pseudomonas aeruginosa*, among others (Antilles et al. 2015; Foti et al. 2011; Hamer et al. 2012a, b; Meri 2014; Najdenski et al. 2018; Rappole 2013; Reed et al. 2003; Steele et al. 2013; Tsiodras et al. 2008).

2.4 Public Health Concerns

Humans and birds have coexisted over thousands of years. As they coevolved, they shared their pathogens too. However, due to increasing industrialization, globalization, climatic changes, intensification of agriculture, urbanization, destruction of habitats of birds and other species, and other factors, the delicate balance has been disrupted to a great extent. Disturbances in traditional fly ways of migrating birds is such a phenomenon. As a result, the interaction between birds and humans has become more complex and unpredictable which in turn lead to emergence of new zoonotic pathogens in humans. These pathogens and their natural avian hosts were in a stable balance over many thousand years but now have been perturbed by anthropogenic influences. While efforts must be taken toward minimization of health impacts due to introduction, maintenance, and dissemination of zoonotic pathogens by migrating birds through surveillance and research, it is also imperative to take necessary steps for conservation of the avian species and their habitats. As this requires well-coordinated plan with multiple stake holders, only "One Health" approach which aggregates expertise and resources from multiple disciplines should be the rational way ahead (Altizer et al. 2011; Bairlein and Wiltschko 2017; Berthold

1996; Chung et al. 2018; Cox 2010; Cunningham et al. 2017; Klaassen et al. 2012; Meri 2014; Newton 2010; Rappole 2013; Sehgal 2010; Tsiodras et al. 2008; Tyson-Pello and Olsen 2020).

References

Altizer S, Bartel R, Han BA (2011) Animal migration and infectious disease risk. Science 331:296–302. https://doi.org/10.1126/science.1194694

Antilles N, Sanglas A, Cerdà-Cuéllar M (2015) Free-living waterfowl as a source of zoonotic bacteria in a dense wild bird population area in northeastern Spain. Transbound Emerg Dis 62: 516–521. https://doi.org/10.1111/tbed.12169

Bairlein F, Wiltschko W (2017) Bird migration. J Comp Physiol A Neuroethol Sens Neural Behav Physiol 203(6–7):381–382. https://doi.org/10.1007/s00359-017-1187-3

Berthold P (1996) Control of bird migration, 1st edn. Chapman & Hall, London

Causey D, Edwards SV (2008) Ecology of avian influenza virus in birds. J Infect Dis 197:S29–S33. https://doi.org/10.1086/524991

Chung DM, Ferree E, Simon DM, Yeh PJ (2018) Patterns of bird-bacteria associations. EcoHealth 15:627–641. https://doi.org/10.1007/s10393-018-1342-5

Cohen EB, Auckland LD, Marra PP, Hamer SA (2015) Avian migrants facilitate invasions of Neotropical ticks and tick-borne pathogens into the United States. Appl Environ Microbiol 81: 8366–8378. https://doi.org/10.1128/aem.02656-15

Comstedt P, Bergström S, Olsen B, Garpmo U, Marjavaara L, Mejlon H, Barbour AG, Bunikis J (2006) Migratory birds as reservoirs of Lyme borreliosis. Emerg Infect Dis 12:1087–1095

Contreras A, Gomez-Martin A, Paterna A, Tatay-Dualde J, Prats-Van Der Ham M, Corrales JC, De La Fe C, Sanchez A (2016) Epidemiological role of birds in the transmission and maintenance of zoonoses. Rev Sci Tech 35:845–862. https://doi.org/10.20506/rst.35.3.2574

Cox GW (2010) Bird migration and global change, 1st edn. Island Press, Washington

Cunningham AA, Daszak P, Wood JLN (2017) One health, emerging infectious diseases and wildlife: two decades of progress. Philos Trans R Soc Lond B 372:4

David S, Abraham AM (2016) Epidemiological and clinical aspects on West Nile virus, a globally emerging pathogen. Infect Dis 48:571–586. https://doi.org/10.3109/23744235.2016.1164890

Foti M, Rinaldo D, Guercio A, Giacopello C, Aleo A, De Leo F, Fisichella V, Mammina C (2011) Pathogenic microorganisms carried by migratory birds passing through the territory of the island of Ustica, Sicily (Italy). Avian Pathol 40:405–409. https://doi.org/10.1080/03079457.2011.588940

Haag-wackernagel D, Moch H (2004) Health hazards posed by feral pigeons. J Infect 48:307–313. https://doi.org/10.1016/j.jinf.2003.11.001

Hamer SA, Goldberg TL, Kitron UD, Brawn JD, Anderson TK, Loss SR, Walker ED, Hamer GL (2012a) Wild birds and urban ecology of ticks and tick-borne pathogens, Chicago, Illinois, USA, 2005–2010. Emerg Infect Dis 18:1589–1595

Hamer SA, Lehrer E, Magle SB (2012b) Wild birds as sentinels for multiple zoonotic pathogens along an urban to rural gradient in greater Chicago, Illinois. Zoonoses Public Health 59:355–364. https://doi.org/10.1111/j.1863-2378.2012.01462.x

Hasle G (2013) Transport of ixodid ticks and tick-borne pathogens by migratory birds. Front Cell Infect Microbiol 3:1–6. https://doi.org/10.3389/fcimb.2013.00048

Hlinak A, Mühle RU, Werner O, Globig A, Starick E, Schirrmeier H, Hoffmann B, Engelhardt A, Hübner D, Conraths FJ, Wallschläger D, Kruckenberg H, Müller T (2006) A virological survey in migrating waders and other waterfowl in one of the most important resting sites of Germany. J Vet Med Ser B Infect Dis Vet Public Heal 53:105–110. https://doi.org/10.1111/j.1439-0450.2006.00935.x

Hubálek Z (2004) An annotated checklist of pathogenic microorganisms associated with migratory birds. J Wildl Dis 40:639–659

Humair PF (2002) Birds and Borrelia. Int J Med Microbiol 291:70–74. https://doi.org/10.1016/S1438-4221(02)80015-7

Jourdain E, Gauthier-clerc M, Bicout DJ, Sabatier P (2007a) Bird migration routes and risk for pathogen dispersion into western Mediterranean wetlands. Emerg Infect Dis 13:365–372

Jourdain E, Toussaint Y, Leblond A, Bicout DJ, Sabatier P, Gauthier-Clerc M (2007b) Bird species potentially involved in introduction, amplification, and spread of West Nile virus in a Mediterranean wetland, the Camargue (Southern France). Vector Borne Zoonotic Dis 7:15–33. https://doi.org/10.1089/vbz.2006.0543

Klaassen M, Hoye BJ, Nolet BA, Buttemer WA (2012) Ecophysiology of avian migration in the face of current global hazards. Philos Trans R Soc B Biol Sci 367:1719–1732. https://doi.org/10.1098/rstb.2012.0008

Li Y-T, Linster M, Mendenhall IH, Su YCF, Smith GJD (2019) Avian influenza viruses in humans: lessons from past outbreaks. Br Med Bull 132:81–95. https://doi.org/10.1093/bmb/ldz036

Meri S (2014) Birds as carriers of human disease. Duodecim 130:1287–1293

Montecino-Latorre D, Barker CM (2018) Overwintering of West Nile virus in a bird community with a communal crow roost. Sci Rep 8:6088. https://doi.org/10.1038/s41598-018-24133-4

Najdenski H, Dimova T, Zaharieva MM, Nikolov B, Petrova-Dinkova G, Dalakchieva S, Popov K, Hristova-Nikolova I, Zehtindjiev P, Peev S, Trifonova-Hristova A, Carniel E, Panferova YA, Tokarevich NK (2018) Migratory birds along the Mediterranean—Black Sea Flyway as carriers of zoonotic pathogens. Can J Microbiol 64:915–924. https://doi.org/10.1139/cjm-2017-0763

Newton I (2010) Bird migration, 1st edn. Harper Collins Publishers, London

OIE (2018) Avian influenza (infection with avian influenza viruses). In: OIE terrestrial manual. OIE, Paris, pp 821–843

Palomar AM, Santibáñez P, Mazuelas D, Roncero L, Santibáñez S, Portillo AA, Oteo JAJA, Santibanez P, Mazuelas D, Roncero L, Santibanez S, Portillo AA, Oteo JAJA (2012) Role of birds in dispersal of etiologic agents of tick-borne zoonoses, Spain, 2009. Emerg Infect Dis 18: 1188–1191. https://doi.org/10.3201/eid1807.111777

Petersen LR, Brault AC, Nasci RS (2013) West Nile virus: review of the literature. JAMA 310:308–315. https://doi.org/10.1001/jama.2013.8042

Rappole JH (2013) The avian migrant: the biology of bird migration, 1st edn. Columbia University Press, New York

Reed KD, Meece JK, Henkel JS, Shukla SK, Reed KD, Shukla SK (2003) Birds, migration and emerging zoonoses: West Nile virus, Lyme disease, influenza A and enteropathogens. Clin Med Res 1:5–12

Sehgal RNM (2010) Deforestation and avian infectious diseases. J Exp Biol 213:955–960. https://doi.org/10.1242/jeb.037663

Sendor AB, Weerasuriya D, Sapra A (2020) Avian influenza. Treasure Island (FL)

Steele CM, Brown RN, Botzler RG (2013) Prevalences of zoonotic bacteria among seabirds in rehabilitation centers along the Pacific Coast of California and Washington, USA. J Wildl Dis 41:735–744. https://doi.org/10.7589/0090-3558-41.4.735

Sule WF, Oluwayelu DO, Hernández-Triana LM, Fooks AR, Venter M, Johnson N (2018) Epidemiology and ecology of West Nile virus in sub-Saharan Africa. Parasit Vectors 11:414. https://doi.org/10.1186/s13071-018-2998-y

Tsiodras S, Kelesidis T, Kelesidis I, Bauchinger U, Falagas ME (2008) Human infections associated with wild birds. J Infect 56:83–98. https://doi.org/10.1016/j.jinf.2007.11.001

Tyson-Pello SJ, Olsen GH (2020) Emerging diseases of avian wildlife. Vet Clin North Am Exot Anim Pract 23:383–395. https://doi.org/10.1016/j.cvex.2020.01.002

Waldenström J, Lundkvist Å, Falk KI, Garpmo U, Bergström S, Lindegren G, Sjöstedt A, Mejlon H, Fransson T, Haemig PD, Olsen B (2007) Migrating birds and tickborne encephalitis virus. Emerg Infect Dis 13:1215–1218. https://doi.org/10.1016/S0140-6736(54)92781-6

Wu X, Xiao L, Li L (2020) Research progress on human infection with avian influenza H7N9. Front Med 14:8–20. https://doi.org/10.1007/s11684-020-0739-z

Yang J, Liu Z, Niu Q, Tian Z, Liu J, Guan G, Liu G, Luo J, Wang X, Yin H (2015) Tick-borne zoonotic pathogens in birds in Guangxi, Southwest China. Parasit Vectors 8:1–3. https://doi.org/10.1186/s13071-015-1249-8

Yin S, de Knegt HJ, de Jong MCM, Si Y, Prins HHT, Huang ZYX, de Boer WF (2020) Effects of migration network configuration and migration synchrony on infection prevalence in geese. J Theor Biol 502:110315. https://doi.org/10.1016/j.jtbi.2020.110315

Ziegler U, Santos PD, Groschup MH, Hattendorf C, Eiden M, Höper D, Eisermann P, Keller M, Michel F, Klopfleisch R, Müller K, Werner D, Kampen H, Beer M, Frank C, Lachmann R, Tews BA, Wylezich C, Rinder M, Lachmann L, Grünewald T, Szentiks CA, Sieg M, Schmidt-Chanasit J, Cadar D, Lühken R (2020) West Nile virus epidemic in Germany triggered by epizootic emergence, 2019. Viruses 12(4):448. https://doi.org/10.3390/v12040448

Part II
Viral Infections

Chapter 3
Japanese Encephalitis

Check for updates

Abstract Japanese encephalitis (JE), an important mosquito-transmitted viral encephalitis, is caused by Japanese encephalitis virus (JEV). Globally, JE is a foremost cause of encephalitis with about 70,000 human cases per year, leading to 10,000–15,000 deaths and causing psychiatric or neurological sequelae in around 30–50% of the survivors. The enzootic cycle of JE is sustained among mosquitoes and wild bird reservoirs, mainly large ardeid (*Ardeidae* family) water birds like pond herons (*Ardeola grayii*) and cattle egret (*Bubulcus ibis*). Both domestic and feral pigs serve as the main amplifying hosts. Although humans and horses are the dead-end hosts, they can develop the fatal disease. Approximately, 99% of human JE patients are asymptomatic, but disease in symptomatic patients can be devastating with even mortality of up to 30%. The World Health Organization (WHO) advocates the integration of JE vaccine into the national vaccination programs in endemic countries. Accomplishing high vaccination coverage in combination with continuous surveillance and vector control measures is the cornerstone to prevent and control JE.

Keywords Japanese encephalitis · Mosquito-borne · *Flavivirus* · Water birds · Pigs · Zoonotic

3.1 Introduction

Japanese encephalitis (JE), one of the prominent viral encephalitis, is a mosquito-transmitted flaviviral disease. It is well prevalent across Asia, the northern tip of Australia, and the western Pacific region (Nath et al. 2020). In Asia, JE is the leading cause of mosquito-transmitted virus infection. Globally, more than three billion people live in JE risk areas, where 75% of affected individuals are children (Choe et al. 2020). Two-thirds of the population at risk are in India and China only, and 95% JE cases are reported from these two countries (Rustagi et al. 2019). Despite the accessibility of different effective vaccines, JE is still a severe clinical disease (Auerswald et al. 2020). In 2002, the estimated DALY (disability-adjusted life year) due to JE was 709,000 globally; however, it was considered as an

Y. S. Malik et al., *Role of Birds in Transmitting Zoonotic Pathogens*, Livestock Diseases and Management, https://doi.org/10.1007/978-981-16-4554-9_3

underestimation as many countries have not yet implemented the surveillance system effectively (Campbell et al. 2011; Garjito et al. 2018). Epidemic outbreaks of JE are observed in subtropical and temperate areas, comprising China, Northern India, Nepal, Northern Thailand, Bhutan, Burma, Bangladesh, Pakistan, Northern Vietnam, Korea, Afghanistan, Taiwan, Madagascar, Southern Russia, and Oriental Africa (Wang and Liang 2015; Garjito et al. 2018). Endemic cases are reported mainly in Southern Asian and Australasia including Southern Thailand, Southern India, Malaysia, Vietnam, Philippines, Cambodia, Indonesia, Sri Lanka, Brunei Darussalam, Lao PDR, Papua New Guinea, Timor Leste, and Australia (Saxena et al. 2011; Wang and Liang 2015; Garjito et al. 2018). With humans as incidentals hosts, JE has complex ecological cycles involving water birds, pigs, and mosquitoes (Ricklin et al. 2016). The first report of JE was from horses and humans in 1871. Japan reported a severe epidemic in 1924, where a filterable agent was obtained from the brain of human being, and notable epidemics were reported every 10 years affecting about 6000 people (Miyake 1964). The disease has been replicated in monkeys by inoculating intracerebrally in 1934. In 1935, the virus was isolated from the human brain for the first time in Tokyo, Japan, and the Nakayama (prototype) strain was established. In 1937, the virus was also recovered from a sick horse's brain. In the early 1930s itself, mosquito transmission was suspected and JEV was also isolated from *Culex tritaeniorhynchus* (Mitamura et al. 1936). In 1959, the role of water birds and pigs in the eco-epidemiology of JE was determined (Buescher and Schere 1959). The name Japanese B encephalitis was given to differentiate it from type A encephalitis a.k.a. "summer epidemics of Von Economo's encephalitis lethargica." Of late, the term B was not often used, and now it is simply referred as Japanese encephalitis (Misra and Kalita 2010).

3.2 Epidemiology

3.2.1 Causative Agent

JEV fits to the genus *Flavivirus* under family *Flaviviridae*. JEV is the prototype virus of the JE serocomplex, which contains nine others antigenically- and genetically-related viruses. JEV is a positive-sense and single-stranded virus with 11-kb-length RNA genome unpolyadenylation at the $3'$ end and a cap at $5'$ end. The genome carries a long single open reading frame (ORF) which encodes a polyprotein flanked by 50 and 30 nontranslated regions. The polyprotein of JEV is cleaved by different viral and host proteases into three structural (capsid, C; envelope, E; and membrane, M) and seven nonstructural (NS) (NS1, NS2A, NS2B, NS3, NS4A, NS4B, and NS5) proteins (Unni et al. 2011; Nath et al. 2020). E gene-based phylogenetic analysis categorizes JEV into five (G-I to G-V) genotypes, with at least 12% differences in nucleotide between them and different subtypes, that cocirculate in most endemic areas. However, all five genotypes belong to a single serotype. The current vaccine is based on genotype III, as it was the most detected strain. However, genotype I is becoming dominant displacing GIII throughout East Asia in recent years. GI and GIII from humans have been extensively studied and characterized (Erra et al. 2013; Bharucha et al. 2018; Barzon and Palù 2018).

3.2.2 Hosts

The enzootic cycle of JE is sustained among mosquitoes and wild bird reservoirs, mainly large ardeid (*Ardeidae* family) water birds like pond herons (*Ardeola grayii*) and cattle egret (*Bubulcus ibis*) (van den Hurk et al. 2003; Miller et al. 2012; Mansfield et al. 2017). Both domestic and wild pigs play as amplifying hosts. Some other domestic animal hosts are cows, goats, horses, dogs, and chicken. Flying foxes, snakes, frogs, and ducks are the sylvatic hosts as the mosquito vector also feeds on all of them. As all these creatures cannot develop adequate viremia to pass on the virus to feeding mosquitoes, they are considered as dead-end hosts. Although humans and horses are the dead-end hosts, they can develop the fatal disease. Factors which favor pigs to act as the main amplifying host is their high fecundity and a rapid turnover of population, leading to constant generation of an immunologically naive population. Besides *Culex tritaeniorhynchus*, an important JE mosquito vector prefers to feed on pigs (Miller et al. 2012; Ricklin et al. 2016; Mansfield et al. 2017). As JE is an occupational disease, people working in waterlogged areas and paddy fields and living near pigs are at greater risk of contracting JEV and develop the disease (Hegde and Gore 2017).

3.2.3 Transmission

The principal mode of transmission of JEV is by mosquito bite, particularly by *Culex* species, that too mostly by *Culex tritaeniorhynchus.* This mosquito species is ornithophilic and so the natural ecological cycle encompasses mosquitoes and birds. However, mammalophilic nature of *Cx. tritaeniorhynchus* has also been demonstrated, and hence, it acts as a bridging vector that transmits the virus to livestock and humans (Guo et al. 2014; Mansfield et al. 2017). Irrigated paddy fields attract migratory waders (reservoirs) and also offer a breeding ground for vector mosquitoes and thus helps in maintaining the sylvatic cycle (Jeffries and Walker 2015; Barzon and Palù 2018). Epizootic transmission cycle is established involving pigs (amplifiers). Human infections are mostly from spillover from above sylvatic/enzootic and epizootic cycle. Depending upon the mosquito activities and densities, and diverse temporal transmission patterns, human JE infections occur throughout the year in tropical and subtropical zones, with the peak during the rainy season and summer in temperate regions (Mansfield et al. 2017; Barzon and Palù 2018). *Cx. tritaeniorhynchus* can breed in low lying areas like fallow rice fields and also in ponds, wells, and ditches of the urban environment, thus bringing the virus into

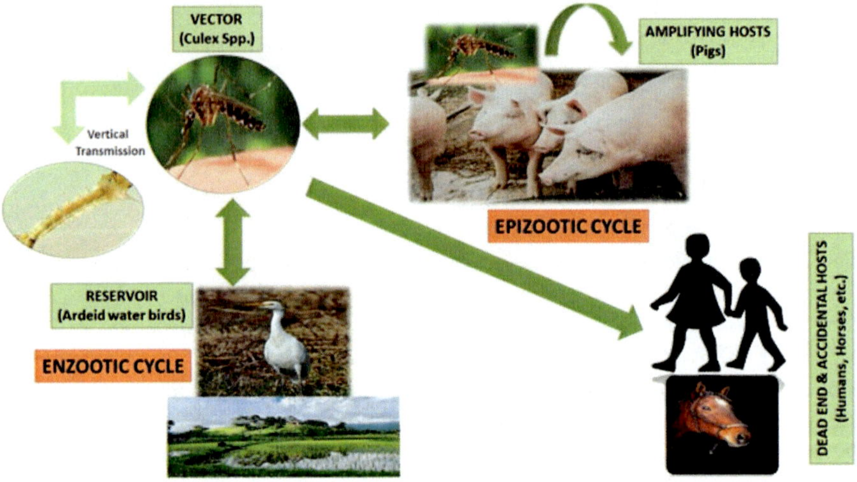

Fig. 3.1 Schematic representation of the JEV transmission cycle

contact with both humans and livestock (Reuben et al. 1994). The proximity of pigs to wild birds with the presence of *Culex* mosquitoes increases the risk of JEV transmission to humans. Figure 3.1 schematically depicts the transmission cycle of JEV.

Between pigs, the vector-free transmission (through oronasal excretions) has also been recently described under experimental conditions (Ricklin et al. 2016). Unlike tropical regions, where JE is endemic throughout the year, in temperate regions JE cases occur only in the summer season. Hence, the re-emergence of JEV every year would require either overwintering in anonymous hosts or reintroduction by migrating birds (Rosen 1986). In Hokkaido, the re-emergence of JE in pigs at the same locations indicates the presence of overwintering phenomenon. It is also supported by recent molecular data from temperate regions showing the phylogenetic relationship between strains isolated from same sampling location (Schuh et al. 2013, 2014). Overwintering in vertebrate hosts such as cold-blooded species, bats, or in invertebrates such as ticks and mosquitoes have been proposed by researchers. *Culex* mosquito can locally overwinter and can transmit the virus after experimental hibernation. Experimental vertical transmission of JEV in mosquitoes has also been demonstrated (Ricklin et al. 2016).

3.3 Role of Birds

Wild birds of family *Ardeidae* (i.e., egrets, herons, black-crowned night herons, and bitterns) play an important role as a reservoir and amplifying hosts in the epidemiology of JEV. By their seasonal migration, they transmit JEV over longer distances (Kamimura 1998). Besides, these birds are ubiquitous and regularly share their habitats with mosquitoes and humans. On natural and experimental infections, several bird species have shown JEV viremia, and naturally picked up anti-JEV antibodies have been documented in a range of avian species (Nemeth et al. 2012). In Asia, although poultry is a foremost component of livestock farming, their role in JEV transmission is not significant. Serosurveillance studies have described that they can get infected (Kalaiyarasu et al. 2016); however, detection of low magnitude viremia in adult ducks and chickens (Dhanda et al. 1977) implies that they are not likely to be the part of the disease epidemiology. However, this necessitates revision in light of the results of recent studies, where high viremia has been demonstrated in chicks and ducklings following experimental infection. Viremia levels decreased swiftly in older birds, and juvenile ones did not present overt morbidity. These data indicate that if ducks and chickens infected below 3 days of age, they can act as a JEV source for feeding mosquitoes (Cleton et al. 2014; Mansfield et al. 2017).

3.4 Disease

3.4.1 Pathogenesis

As JEV is a neurotropic virus, it can gain access into the central nervous system (CNS) and multiply in several types of cells like microglia, microvascular endothelial cells, neural precursors, and developing neurons. Viral infection elicits an inflammatory response with perivascular infiltration of mainly T cells, activation of endothelial cells, macrophage infiltration, microglia activation, and vascular damage, which alters the blood-brain barrier (BBB), impedes proliferation and differentiation of neural cells, and induces death of neuronal cells (Lannes et al. 2017; Myint et al. 2014). Viral replication and entry into CNS have to be controlled by host immune response by producing neutralizing antibodies to the JEV; otherwise, there is increased chance of neuroinvasive infection and death (Burke et al. 1985). Although JEV-exposed people develop lifelong adaptive immunity, it gets waned in elderly people (Solomon 2004). Production of JEV-neutralizing antibodies following vaccination corresponds with protection, and 50% plaque reduction neutralization titer of 1:10 have been described to be protective in humans. JEV in humans is probably protected by cell-mediated immunity (van Gessel et al. 2011; Turtle et al. 2016). Observations from an animal (mouse) model study indicated that the effector CD8 T cells can protect against infection by stopping the breach in the BBB and neural damage, as CD8 T cells deficit mice were highly susceptible to neuroinvasion and died (Jain et al. 2017). Another mouse model study revealed that T cells play an important part in immunopathology and neuronal damage (Larena et al. 2011).

3.4.2 Clinical Signs/Clinical Profile

Approximately, 99% of JE patients remains asymptomatic, but disease in symptomatic patients can be devastating with mortality of approximate 30% patients. Following a 4–14 days incubation period, symptomatic individuals can present with chills, high fever, confusion, and myalgia. Children suffer from gastrointestinal illnesses (vomiting and cramp) initially, and 75% will have seizures. Fifty percent survivors will suffer from persistent neurological complaints, and this is significant in children as it may result in a lifetime disability. Elderly is also at more risk of infection (Burchard et al. 2009; Batchelor and Peterson 2015). JE infection in the course of the first and second trimesters of pregnancy can result in spontaneous abortion or fetal death. Recurrence of symptoms after a year of the acute encephalitic episode has been noticed in some children, and virus can also be recovered from the PBMC during latent phase (Susarla et al. 2020).

In pigs, generally, the infection is of asymptomatic nature except for some reports of pyrexia, anorexia, and reproductive illness in large farms (Gresham 2003). The manifestations of reproductive illness are stillbirth, congenital deformity, and poor reproductive performance. However, the observations from outbreaks in Japan have observed increased abortion in pigs with periodic recurrence (Takashima et al. 1988). In the horse, most JEV infections seem to be subclinical. But earlier reports from Japan indicate that there have been regular epidemics involving horses in thousands (Sugiura and Shimada 1999). Since 2000, only two instances of infection in equines have been reported from Hong Kong and Japan. In both cases, the animal presented with pyrexia and neurological problems that included ataxia. One among them died within 3 days and the other was euthanized on welfare grounds (Lam et al. 2005; Yamanaka et al. 2006). In cattle, a similar clinical picture is described with rare clinical disease presented with loss of appetite, depression, circling movements, and walking abnormalities (Katayama et al. 2013; Kako et al. 2014).

3.4.3 Pathology/Lesions

Typical laboratory findings include modest leukocytosis with a left shift, hyponatremia, and mild anemia. Examination of cerebrospinal fluid (CSF) includes lymphocyte predominance, moderate pleocytosis, normal CSF to plasma glucose ratio, and slightly elevated protein. Important electroencephalogram observation is scattered slow waves with "periodic lateralized epileptiform discharges (PLEDS)." Magnetic resonance imaging (MRI) findings exhibited white matter edema, atypical signals in the basal ganglia, thalamus, cerebellum, pons, midbrain, and spinal cord. Most commonly demonstrated abnormality is lesions in the thalamus (David et al. 2010).

JE presents polymorphic pathological lesions involving several anatomical areas of the brain. Diffuse edema and tonsillar herniation are noticed in the brain. On gross examination, small, round-shaped necrolytic lesions seen in the diencephalic nuclei and the superficial cortical ribbon are characteristic to JE. Other features are diffuse microglial reaction, microglial nodule formation, and neuronophagia. Variable grades of meningitis with perivascular lymphocytic infiltration and polymorph reaction are seen on histological observation. Another characteristic microscopic feature is the presence of small, round, and confluent necrolytic areas in the gray matter and deep nuclei. These necrolytic areas carry polymorphic infiltration and lipid-containing macrophages. Consequently, these necrolytic areas turn into pale, moderately acellular zones with loss of myelin and axons. Astroglial reaction will be absent. These necrolytic areas, although confined to gray matter, may occasionally also found in fiber tracts of white matter. Rarely, focal hemorrhagic lesions may be noticed in the brainstem and thalamus, generally after 7–10 days. These lesions represent bystander immune and allergic reaction (Ishii et al. 1977; King et al. 2007; Susarla et al. 2020).

3.5 Disease Management

3.5.1 Diagnosis

JEV diagnosis is usually achieved through serological assays and virus detection by molecular methods in the viremic phase. But virus detection is challenging owing to short time viremia. Spotting of JEV-specific IgM and IgG antibodies are usually preferred and effective methods of confirming infection. Serum surveillance exposes the viral presence in specific areas in the lack of clinical disease (Yang et al. 2007; Mansfield et al. 2017). Table 3.1 presents the diagnostic options for detection of JEV.

3.6 Prevention and Control

Given the paucity of effective antiviral therapy, vaccination is the most important preventive intervention to prevent JE in the susceptible population (Yun and Lee 2014). The WHO advocates the integration of JE vaccine into the national vaccination program in endemic countries. Accomplishing high vaccination coverage in combination with continuous surveillance is the cornerstone to prevent and control JE (Choe et al. 2020). Presently four types of JE vaccines are licensed and available for human use: killed-inactivated mouse brain-derived vaccine; killed-inactivated cell culture-derived vaccines; live-attenuated cell culture-derived vaccine SA14-14-2; and live-attenuated cell culture-derived chimeric vaccines (Beasley et al. 2008; Halstead and Thomas 2011; Wilder-Smith and Halstead 2010). However, the production of mouse brain-derived vaccine was stopped in 2006 due to certain limitations and drawbacks (Fischer et al. 2010). Additionally, plasmid DNA-based

Table 3.1 Comprehensive diagnostic options for diagnosis of JEV

Diagnostic methods	Techniques	References
Virus isolation	• Suitable brain samples are cortex or thalamus, corpus striatum, and blood and spinal cord • Primary cell cultures used for virus isolation are hamster or porcine kidney cells and chicken embryos • Established cell lines like BHK, Vero, and C6/36 (*Aedes albopictus*) can also be used • In vivo virus isolation can be performed in 2–4-day-old mice (intracerebral inoculation). JEV-positive mice exhibit neurological signs and die within 14 days. Brains are removed and processed for isolation of the virus	Mansfield et al. (2017)
Serological methods	• Serological assays available are enzyme-linked immunosorbent assay (ELISA), hemagglutination inhibition tests (HI), virus neutralization tests (VNT), and complement fixation tests (CFT) • Plaque reduction neutralization test (PRNT) is the most specific assay with minimal cross-reactivity. It is the gold standard test for JE diagnosis • IgM ELISAs are most usually deployed for JE diagnosis to detect recent infection. IgM is detectable in CSF and serum within 7 days after onset of disease • Commercially available ELISAs are JE detect IgM capture ELISA (InBios, USA) and the Japanese encephalitis-dengue IgM combo ELISA (Panbio, Australia) • Indirect immunofluorescence tests (IIFT) are also commercially available, which detect either anti-JEV IgM or IgG (Euroimmun, Germany) • Published ELISAs are available for screening pig serum samples. A DIVA ELISA is also available for veterinary application	Lian et al. (2002), Konishi et al. (2006), Yang et al. (2006), WHO (2006), Litzba et al. (2010), OIE (2012), Do et al. (2015)
Molecular assays	• Several RT-PCR and real-time RT-PCR methods to detect JEV from human samples are available in the literature and OIE manual • A RT-LAMP method is also described for use in the resource-limited	Tanaka (1993), Chung et al. (1996), Jan et al. (2000), Lian et al. (2002), Parida et al. (2006), Gao et al. (2013), Marston et al. (2013), Li et al. (2014), Do et al. (2015), Dhanze et al. (2015)

(continued)

Table 3.1 (continued)

Diagnostic methods	Techniques	References
	settings • Whole-genome sequencing techniques can provide more insights on genetic identity and helps in identifying the geographic origin of a specific JEV strain • A study compared the diagnostic efficiency of three (RT-PCR, real-time RT-PCR, and RT-LAMP) assays for the detection of JEV in pig serum samples. Results suggested RT-LAMP and real-time RT-PCR as more specific and sensitive assays	

vaccines (Kaur and Vrati 2003), recombinant protein-based vaccines (Mason et al. 1989; Seif et al. 1995, 1996; Verma et al. 2009), and poxvirus-based vaccines (Konishi et al. 1991; Mason et al. 1991; Jan et al. 1993; Kaur and Vrati 2003) have also been tried in animal models with variable efficacies. Pig vaccination could also be an important strategy to mitigate JEV in humans and to avoid reproductive losses in pigs. Vaccines for both pigs and horses are commercially available in some Asian countries (Igarashi 2002; Nah et al. 2015). Studies are required to investigate the cost-benefit ratio of pig vaccination in comparison with breeding performance and public health (Ruget et al. 2017). Vector suppression measures and prevention of mosquito bites in human and animal houses using mesh screening and other repellants during peak vector activity period are crucial. Preventing mosquitoes feeding from the pigs effectively breaks the epizootic JEV cycle (Zanin et al. 2003).

Aggressive vaccination programs and vector control measures will prevent the occurrence of JE and protect the population dwelling in risk areas. Establishing a sensitive and strong surveillance system for humans, pigs, and mosquitoes for JE infection in all the endemic countries should be addressed in future.

References

Auerswald H, Ruget AS, Ladreyt H, In S, Mao S, Sorn S, Tum S, Duong V, Dussart P, Cappelle J, Chevalier V (2020) Serological evidence for Japanese encephalitis and West Nile virus infections in domestic birds in Cambodia. Front Vet Sci 7:15. https://doi.org/10.3389/fvets.2020.00015. eCollection 2020

Barzon L, Palù G (2018) Recent developments in vaccines and biological therapies against Japanese encephalitis virus. Expert Opin Biol Ther 18(8):851–864

Batchelor P, Peterson K (2015) Japanese encephalitis: a review of clinical and vaccine availability in Asia. Trop Dis Travel Med Vaccines (TDTMV) 1:11

Beasley DW, Lewthwaite P, Solomon T (2008) Current use and development of vaccines for Japanese encephalitis. Expert Opin Biol Ther 8:95–106

Bharucha T, Sengvilaipaseuth O, Vongsouvath M, Vongsouvath M, Davong V, Panyanouvong P et al (2018) Development of an improved RT-qPCR assay for detection of *Japanese encephalitis virus* (JEV) RNA including a systematic review and comprehensive comparison with published methods. PLoS One 13(3):e0194412. https://doi.org/10.1371/journal.pone.0194412

Buescher EL, Schere WF (1959) Ecological studies of Japanese encephaltiis in Japan. IX. Epidemiological correlations and conclusions. Am J Trop Med Hyg 8:719–722

Burchard G, Caumes E, Connor B, Freedman D, Jelinek T, Jong E et al (2009) Expert opinion on vaccination of travelers against Japanese encephalitis. J Travel Med 16:204–216

Burke DS, Lorsomrudee W, Leake CJ et al (1985) Fatal outcome in Japanese encephalitis. Am J Trop Med Hyg 34:1203–1210

Campbell G, Hills S, Fischer M, Jacobson J, Hoke C, Hombach J, Marfin A, Solomon T, Tsai T, Tsui V, Ginsburg A (2011) Estimated global incidence of Japanese encephalitis. Bull World Health Organ 89:766–774. https://doi.org/10.2471/BLT.10.085233

Choe YJ, Jee Y, Takashima Y, Lee JK (2020) Japanese encephalitis in the Western Pacific Region: implication from the Republic of Korea. Vaccine. pii: S0264-410X (20) 30284-X. https://doi.org/10.1016/j.vaccine.2020.02.061

Chung Y-J, Nam J-H, Ban S-J, Cho H-W (1996) Antigenic and genetic analysis of Japanese encephalitis viruses isolated from Korea. Am J Trop Med Hyg 55:91–97

Cleton NB, Bosco-Lauth A, Page MJ, Bowen RA (2014) Age-related susceptibility to Japanese encephalitis virus in domestic ducklings and chicks. Am J Trop Med Hyg 90:242–246. https://doi.org/10.4269/ajtmh.13-0161

David WV, Alan B, Tom S (2010) Flaviviruses (yellow fever, dengue, dengue hemorrhagic fever, Japanese encephalitis, West Nile encephalitis, St. Louis encephalitis, tick-borne encephalitis). In: Mandell LG, Bennett EJ, Dolin R (eds) Principles and practice of infectious diseases, 7th edn. Churchill Livingstone Elsevier, Philadelphia, p 2133e51

Dhanda V, Banerjee K, Deshmukh PK, Ilkal MA (1977) Experimental viraemia and transmission of Japanese encephalitis virus by mosquitoes in domestic ducks. Indian J Med Res 66:881–888

Dhanze H, Bhilegaonkar KN, Ravi Kumar GV, Thomas P, Chethan Kumar HB, Suman Kumar M, Rawat S, Kerketta P, Rawool DB, Kumar A (2015) Comparative evaluation of nucleic acid-based assays for detection of Japanese encephalitis virus in swine blood samples. Arch Virol 160:1259–1266

Do LP, Bui TM, Hasebe F, Morita K, Phan NT (2015) Molecular epidemiology of Japanese encephalitis in northern Vietnam, 1964–2011: genotype replacement. Virol J 12:51

Erra EO, Askling HH, Yoksan S et al (2013) Cross-protective capacity of Japanese encephalitis (JE) vaccines against circulating heterologous JE virus genotypes. Clin Infect Dis 56:267–270

Fischer M, Lindsey N, Staples JE, Hills S, Centers for Disease C, Prevention (2010) Japanese encephalitis vaccines: recommendations of the Advisory Committee on Immunization Practices (ACIP). MMWR Recomm Rep 59:1–27

Gao X, Lui H, Wang H, Fu S, Guo Z, Liang G (2013) Southernmost Asia is the source of Japanese encephalitis virus (genotype 1) diversity from which the viruses disperse and evolve throughout Asia. PLoS Negl Trop Dis 19:e2459

Garjito TA, Widiarti, Anggraeni YM, Alfiah S, Satoto T, Farchanny A et al (2018) Japanese encephalitis in Indonesia: an update on epidemiology and transmission ecology. Acta Trop 187:2

Gresham A (2003) Infectious reproductive disease in pigs. In Pract 25:466–473

Guo XX, Li CX, Wang G, Zheng Z, Dong YD, Zhang YM, Xing D, Zhao TY (2014) Host feeding patterns of mosquitoes in a rural malaria-endemic region in Hainan Island, China. J Am Mosq Control Assoc 30:309–311

Halstead SB, Thomas SJ (2011) New Japanese encephalitis vaccines: alternatives to production in mouse brain. Expert Rev Vaccines 10:355–364

Hegde NR, Gore MM (2017) Japanese encephalitis vaccines: immunogenicity, protective efficacy, effectiveness, and impact on the burden of disease. Hum Vaccin Immunother 13(6):1320–1337. https://doi.org/10.1080/21645515.2017.1285472

Igarashi A (2002) Control of Japanese encephalitis in Japan: immunization of humans and animals, and vector control. Curr Top Microbiol Immunol 267:139–152

Ishii T, Matsushita M, Hamada S (1977) Characteristic residual neuropathologic features of Japanese encephalitis. Acta Neuropathol 38:181–186

Jain N, Oswal N, Chawla AS et al (2017) CD8 T cells protect adult naive mice from JEV-induced morbidity via lytic function. PLoS Negl Trop Dis 11:e0005329

Jan LR, Yang CS, Henchal LS, Sumiyoshi H, Summers PL, Dubois DR et al (1993) Increased immunogenicity and protective efficacy in outbred and inbred mice by strategic carboxyl-terminal truncation of Japanese encephalitis virus envelope glycoprotein. Am J Trop Med Hyg 48:412–423

Jan L-R, Yueh Y-Y, Wu Y-C, Horng C-B, Wang G-R (2000) Genetic variation of Japanese encephalitis virus in Taiwan. Am J Trop Med Hyg 62:446–452

Jeffries CL, Walker T (2015) The potential use of Wolbachia-based mosquito biocontrol strategies for Japanese encephalitis. PLoS Negl Trop Dis 9:e0003576

Kako N, Suzuki S, Sugie N, Kato T, Yanase T, Yamakawa M, Shirafuji H (2014) Japanese encephalitis in a 114-month-old cow: pathological investigation of the affected cow and genetic characterization of Japanese encephalitis virus isolate. BMC Vet Res 10:63

Kalaiyarasu S, Mishra N, Khetan RK, Singh VP (2016) Serological evidence of widespread West Nile virus and Japanese encephalitis virus infection in native domestic ducks (Anas platyrhyncus var domesticus) in Kuttanad region Kerala, India. Comp Immunol Microbiol Infect Dis 48:61–68

Kamimura K (1998) Studies on the population dynamics of the principal vector mosquito of Japanese encephalitis. Med Entomol Zool 49:181–185

Katayama T, Saito S, Horiuchi S, Maruta T, Kato T, Yanase T, Yamakawa M, Shirafuji H (2013) Nonsuppurative encephalomyelitis in a calf in Japan and isolation of Japanese encephalitis virus genotype 1 from the affected calf. J Clin Microbiol 51:3448–3453

Kaur R, Vrati S (2003) Development of a recombinant vaccine against Japanese encephalitis. J Neurovirol 9:421–431

King NJ, Getts DR, Getts MT et al (2007) Immunopathology of flavivirus infections. Immunol Cell Biol 85:33–42

Konishi E, Pincus S, Fonseca BA, Shope RE, Paoletti E, Mason PW (1991) Comparison of protective immunity elicited by recombinant vaccinia viruses that synthesize E or NS1 of Japanese encephalitis virus. Virology 185:401–410

Konishi E, Shoda M, Kondo T (2006) Analysis of yearly changes in levels of antibodies to Japanese encephalitis virus non-structural 1 protein in racehorses in central Japan showing high levels of natural virus activity still exist. Vaccine 24:516–524

Lam KHK, Ellis TM, Williams DT, Lunt RA, Daniels PW, Watkins KL, Riggs CM (2005) Japanese encephalitis in a racing thoroughbred gelding in Hong Kong. Vet Rec 157:168–173

Lannes N, Summerfield A, Filgueira L (2017) Regulation of inflammation in Japanese encephalitis. J Neuroinflammation 14:158

Larena M, Regner M, Lee E, Lobigs M (2011) Pivotal role of antibody and subsidiary contribution of CD8+ T cells to recovery from infection in a murine model of Japanese encephalitis. J Virol 85:5446–5455

Li MH, Fu SH, Chen WX, Wang HY, Cao YX, Liang GD (2014) Molecular characterization of full-length genome of Japanese encephalitis virus genotype V isolated from Tibet, China. Biomed Environ Sci 27:231–239

Lian W-C, Liau M-Y, Mao C-L (2002) Diagnosis and genetic analysis of Japanese encephalitis virus infected in horses. J Vet Med B Infect Dis Vet Public Health 49:361–365

Litzba N, Klade CS, Lederer S, Niedrig M (2010) Evaluation of serological diagnostic test systems assessing the immune response to Japanese encephalitis vaccination. PLoS Trop Negl Dis 4: e883

Mansfield KL, Hernandez-Triana LM, Banyard AC, Fooks AR, Johnson N (2017) Japanese encephalitis virus infection, diagnosis and control in domestic animals. Vet Microbiol 201: 85–92. https://doi.org/10.1016/j.vetmic.2017.01.014

Marston DA, McElhinney LM, Ellis RJ, Horton DL, Wise EL, Leech SL, David D, De Lamballerie X, Fooks AR (2013) Next generation sequencing of viral RNA genomes. BMC Genomics 14:444

Mason PW, Dalrymple JM, Gentry MK, McCown JM, Hoke CH, Burke DS et al (1989) Molecular characterization of a neutralizing domain of the Japanese encephalitis virus structural glycoprotein. J Gen Virol 70(Pt 8):2037–2049

Mason PW, Pincus S, Fournier MJ, Mason TL, Shope RE, Paoletti E (1991) Japanese encephalitis virus-vaccinia recombinants produce particulate forms of the structural membrane proteins and induce high levels of protection against lethal JEV infection. Virology 180:294–305

Miller RH, Masuoka P, Klein TA, Heung-Chul K, Todd S, John G (2012) Ecological niche modeling to estimate the distribution of Japanese encephalitis virus in Asia. PLoS Negl Trop Dis 6:e1678

Misra U, Kalita J (2010) Overview: Japanese encephalitis. Prog Neurobiol 91(2):108–120

Mitamura T, Kitoka M, Watanabe M et al (1936/1992) Study of Japanese encephalitis virus:animal experiments and mosquito transmission experiments, Kansai Lji 1, 260. Quoted by Vaughn DW, Hoke CH. The epidemiology of Japanese encephalitis: prospects for prevention. Epidemiol Rev 14:197–221

Miyake M (1964) The pathology of Japanese encephalitis. Bull World Health Organ 30:153–160

Myint KS, Kipar A, Jarman RG et al (2014) Neuropathogenesis of Japanese encephalitis in a primate model. PLoS Negl Trop Dis 8:e2980

Nah J-J, Yang D-K, Kim H-H, Song J-Y (2015) The present and future of veterinary vaccines for Japanese encephalitis in Korea. Clin Exp Vacc Res 4:130–136

Nath B, Vandna, Saini HM, Prasad M, Kumar S (2020) Evaluation of Japanese encephalitis virus E and NS1 proteins immunogenicity using a recombinant Newcastle disease virus in mice. Vaccine 38(7):1860–1868

Nemeth N, Bosco-Lauth A, Oesterle P, Kohler D, Bowen R (2012) North American birds as potential amplifying hosts of Japanese encephalitis virus. Am J Trop Med Hyg 87(4):760–767. PMID:22927494

OIE (2012) Manual of diagnostic tests and vaccines for terrestrial animals, 7th edn. World Health Organisation for Animal Health (OIE), Paris

Parida MM, Santhosh SR, Dash PK, Tripathi NK, Saxena P, Ambuj S, Sahni AK, Lakshmana Rao PV, Morita K (2006) Development and evaluation of reverse transcription-loop-mediated isothermal amplification assay for rapid and real-time detection of Japanese encephalitis virus. J Clin Microbiol 44:4172–4178

Reuben R, Tewart SC, Hiriayan H, Akiyama J (1994) Illustrated keys to species of culex (Culex) associated with Japanese encephalitis in Southeast Asia (Diptera: culicidae). Mosq Syst 26:75–96

Ricklin ME, GarciaNicolas O, Brechbuhl D, Python S, Zumkehr B, Nougairede A et al (2016) Vector-free transmission and persistence of Japanese encephalitis virus in pigs. Nat Commun 7: 10832

Rosen L (1986) The natural history of Japanese encephalitis virus. Annu Rev Microbiol 40:395–414

Ruget AS, Beck C, Gabassi A et al (2018) Japanese encephalitis circulation pattern in swine of Northern Vietnam and consequences for swine's vaccination recommendations. Transbound Emerg Dis 65(6):1485–1492

Rustagi R, Basu S, Garg S (2019) Japanese encephalitis: strategies for prevention and control in India. Indian J Med Spec 10:12–17

Saxena SK, Tiwari S, Saxena R, Mathur A, Nair MPN (2011) Japanese encephalitis: an emerging and spreading Arbovirosis. In: Růžek D (ed) Flavivirus encephalitis. Intech, pp 295–316. https://doi.org/10.5772/22145

Schuh AJ, Ward MJ, Brown AJ, Barrett AD (2013) Phylogeography of Japanese encephalitis virus: genotype is associated with climate. PLoS Negl Trop Dis 7:e2411

Schuh AJ, Ward MJ, Leigh Brown AJ, Barrett AD (2014) Dynamics of the emergence and establishment of a newly dominant genotype of Japanese encephalitis virus throughout Asia. J Virol 88:4522–4532

Seif SA, Morita K, Matsuo S, Hasebe F, Igarashi A (1995) Finer mapping of neutralizing epitope (s) on the C-terminal of Japanese encephalitis virus E-protein expressed in recombinant Escherichia coli system. Vaccine 13:1515–1521

Seif SA, Morita K, Igarashi A (1996) A 27 amino acid coding region of JE virus E protein expressed in E. coli as fusion protein with glutathione-S-transferase elicit neutralizing antibody in mice. Virus Res 43:91–96

Solomon T (2004) Flavivirus encephalitis. N Engl J Med 351:370–378

Sugiura T, Shimada K (1999) Seroepizootiological survey of Japanese encephalitis virus and Getah virus in regional horse race tracks from 1991 to 1997 in Japan. J Vet Med Sci 61:877–881

Susarla SK, Mhadevan A, Radotra B, Takao M, Wong KT. Flaviviruses 4: Japanese encephalitis. In: Chretien F, Wong KT, Sharer LR, Keohane C, Gray F, editors. Infections of the central nervous system: pathology and genetics. 1st ed. Wiley-Blackwell; 2020.p. 169–176

Takashima I, Watanabe T, Ouchi N, Hashimoto N (1988) Ecological studies of Japanese encephalitis virus in Hokkaido: interepidemic outbreaks of swine abortion and evidence for the virus to overwinter locally. Am J Trop Med Hyg 38:420–427

Tanaka M (1993) Rapid identification of flavivirus using the polymerase chain reaction. J Virol Methods 41:311–322

Turtle L, Bali T, Buxton G et al (2016) Human T cell responses to Japanese encephalitis virus in health and disease. J Exp Med 213:1331–1352

Unni SK, Ruzek D, Chhatbar C, Mishra R, Johri MK, Singh SK (2011) Japanese encephalitis virus: from genome to infectome. Microbes Infect 13:312–321

van den Hurk AF, Nisbet DJ, Hall RA, Kay BH, Mackenzie JS, Ritchie SA (2003) Vector competence of Australian mosquitoes (Diptera: culicidae) for Japanese encephalitis virus. J Med Entomol 40:82–90

van Gessel Y, Klade CS, Putnak R et al (2011) Correlation of protection against Japanese encephalitis virus and JE vaccine (IXIARO) induced neutralizing antibody titres. Vaccine 29: 5925–5931

Verma SK, Kumar S, Gupta N, Vedi S, Bhattacharya SM, Lakshmana Rao PV (2009) Bacterially expressed recombinant envelope protein domain III of Japanese encephalitis virus (rJEV-DIII) elicits Th1 type of immune response in BALB/c mice. Vaccine 27:6905–6909

Wang H, Liang G (2015) Epidemiology of Japanese encephalitis: past, present, and future prospects. Ther Clin Risk Manag. https://doi.org/10.2147/TCRM.S51168

Wilder-Smith A, Halstead SB (2010) Japanese encephalitis: update on vaccines and vaccine recommendations. Curr Opin Infect Dis 23:426–431

World Health Organisation (WHO) (2006) Japanese encephalitis vaccines. Wkly Epidemiol Rec 81:331–340

Yamanaka T, Tsujimura K, Kondo T, Yasuda W, Yasuda W, Okada A, Noda K, Okumura T, Matsumura T (2006) Isolation and genetic analysis of Japanese encephalitis virus from a diseased horse in Japan. J Vet Med Sci 68:293–295

Yang DK, Kim BH, Lim SI, Kwon JH, Lee KW, Choi CU, Kweon CH (2006) Development and evaluation of indirect ELISA for the detection of antibodies against Japanese encephalitis virus in swine. J Vet Sci 7:271–275

Yang DK, Kweon CH, Kim BW, Hwang IJ, Kang MI, So BJ, Cho KO (2007) The seroprevalence of Japanese encephalitis virus in goats raised in Korea. J Vet Sci 8:197–199

Yun SI, Lee YM (2014) Japanese encephalitis: the virus and vaccines. Hum Vaccin Immunother 10: 263–279

Zanin MP, Webster DE, Martin JL, Wessenlingh SL (2003) Japanese encephalitis vaccines: moving away from mouse brain. Expert Rev Vaccines 2:407–416

Chapter 4
West Nile

Abstract West Nile is a neurotropic problem caused by West Nile virus (WNV), a neurotropic arbovirus. The natural reservoir and biological vector of the WNV are birds and *Culex* sp. mosquito, respectively. WNV is naturally maintained in the bird–mosquito–bird transmission cycle. Horses and humans acquire the infection from mosquitoes, are incidental hosts with asymptomatic infection, or suffer mild to severe disease conditions. *Culex* mosquitoes may also play a role of overwintering host, as WNV is observed to be vertically transmitted to the progenies of females with high viral titers. Other important modes of WNV transmission to humans include organ transplantation, blood transfusion, transplacental and breast milk transmission from mother to child, and occupational exposure of laboratory personnel to the virus. WNV infection in humans presents with a wide clinical spectrum. Around 80% of infected individuals are asymptomatic, while about 20% experience a nonspecific pyretic illness known as West Nile fever (WNF), which can manifest in mild to severe clinical forms. A small subsection of infected people (<1%) develop a lethal neuroinvasive infection. To this point, no human vaccine is licensed or available against WNV; however, there are some vaccines licensed for veterinary use. As there is no licensed vaccine for humans till date, prevention of WNV infection mainly depends on mosquito control programs and screening of organ and blood donors.

Keywords West Nile · Arbovirus · *Culex* · Mosquito-borne · *Zoonotic* · Crows

4.1 Introduction

West Nile virus (WNV) is a neurotropic arbovirus belonging to the family *Flaviviridae*, which also includes other mosquito-transmitted viruses of public health importance such as yellow fever, Japanese encephalitis, and dengue viruses (Chianese et al. 2019). The virus is primarily sustained in transmission cycles between mosquitoes and birds, with horses and humans as the incidental hosts. Birds are the preferred and predominant reservoirs of WNV (Colpitts et al. 2012; Spickler and Anna Rovid 2013; Dahlin et al. 2016). WNV human infections are

Y. S. Malik et al., *Role of Birds in Transmitting Zoonotic Pathogens*, Livestock Diseases and Management, https://doi.org/10.1007/978-981-16-4554-9_4

usually asymptomatic except few cases showing symptoms like nausea, vomiting, rash, and fever. Severe disease may develop in less than 1% of patients with manifestations of disorientation, headache, paralysis, coma, and even death. People above the age of 50 experience a higher degree of neuroinvasive disorder (Huhn et al. 2003; Troupin and Colpitts 2016). WNV was first identified in 1937 in Uganda. Now, it can be found in Europe, West Asia, Australia, Africa, and North America. In North America, WNV made its first entry in 1999 via New York City triggering an encephalitis epidemic. Within 3 years, the disease was reported in 39 states with ~4000 morbidities and 300 deaths (Huhn et al. 2003). In 2014, the Centers for Disease Control and Prevention (CDC), United States, with its surveillance data support, reported WNV as the most frequent cause of neuroinvasive arthropod-borne viral disease in the United States. WNV disease was reported in 52 states of United States, with 50,830 cases in Washington, D.C. Out of that, 24,657 (48.5%) cases were categorized as neuroinvasive WNV disease, resulting in 2330 deaths (https://www.cdc.gov/westnile/statsmaps/cumMapsData.html). The European Centre for Disease Prevention and Control (ECDC) reports that WNV is perhaps in the new phase of spreading out, in comparison to the past. According to its recent 2018 data, there are 2083 human cases reported in the European Union and the European Economic Area (EU/EEA) countries (Vilibic-Cavlek et al. 2019a, b). Among the affected countries, Italy is the most affected followed by Serbia, Greece, Romania, and Hungary (Chianese et al. 2019). Since its discovery, WNV has spread to an extensive region of the world and is now regarded as the most significant etiological agent of viral encephalitis globally (Chancey et al. 2015). The rise in both average rainfall and temperatures may be the reason for this trend (Chianese et al. 2019). Due to the high endemic potential of the virus and expanded distribution of the vector, WNV has assumed the status of an important public health threat worldwide (Chianese et al. 2019).

4.2 Epidemiology

4.2.1 Causative Agent

WNV, a member of the Japanese encephalitis (JE) serocomplex, is under the genus *Flavivirus* and family *Flaviviridae*. The other related viruses of the JE serocomplex are St. Louis encephalitis virus, Japanese encephalitis virus, Murray Valley encephalitis virus, and Rocio virus. Like any other flaviviruses, it has a positive-sense single-stranded RNA genome of 11 kb approximate size, surrounded by an icosahedral nucleocapsid, encircled by a lipid bilayer envelope. The WNV genome has a single open reading frame and lacks the polyadenylation tail at the 3′ end (Chambers et al. 1990). The 5′ and 3′ noncoding regions of the WNV form stem-loop structures which are crucial for viral replication. WNV genome encodes a lone polyprotein which is co- and post-translationally cleaved into seven nonstructural (NS) proteins (NS1, NS2A, NS2B, NS3, NS4A, NS4B, and NS5) and three structural proteins: C (capsid), E (Envelope) and prM/M (pre-M/membrane) (Chancey et al. 2015; Chianese et al. 2019). WNV replicates in a vast array of cell types from an extensive range of vertebrate and invertebrate species that includes birds, mammals, insects, amphibians, and reptiles (Klenk et al. 2004; van der Meulen et al. 2005; Spickler and Anna Rovid 2013; Ahlers and Goodman 2018). WNV has got a capability of infecting abundant human cells comprising dendritic cells, B and T lymphocytes, hepatocytes, monocytes/macrophages, neurons, and endothelial cells (Rios et al. 2006). The virus is biologically diverse; up to nine lineages have been proposed with two major lineages differing by 25–30% in nucleotide homology (Rizzoli et al. 2015). However, only strains belonging to lineage 1 and 2 have been attributed to cause disease in humans, including neuroinvasive conditions. Lineage 1 is commonly distributed in Europe, North and Central America, Australia, Africa, Asia, and the Middle East. Further, there is a subclassification in lineage 1, i.e., 1a, 1b, and 1c. Lineage 1a has been most widely distributed in America (NY99 strain), Europe, Africa, and the Middle East and was accountable for most of the outbreaks in these regions. Lineage 2 was exclusively reported in Africa till 2004, and, since 2004, has been detected in Europe (David and Abraham 2016; Fall et al. 2017; Velasco et al. 2020).

4.2.2 Hosts

The natural reservoir and vector of the WNV are birds and *Culex* sp. mosquito, respectively. Humans and horses are the incidental hosts which get infected and acquire a mild to severe disease. Not only incidental, but they both are also dead-end hosts, as they cannot transmit the virus to any other mammalian or mosquito hosts during infection. Other reported incidental but asymptomatic hosts of WNV are cats, rabbits, squirrels, skunks, lake frogs, alligators, chipmunks, and some bats (Troupin and Colpitts 2016). These animals play an unknown role in the transmission cycle with a very low viremia titer revealing very less likelihood of becoming reservoirs (Hayes et al. 2005). However, the role of reptiles in the transmission of some other flaviviruses like St. Louis encephalitis virus and western and eastern equine encephalitis virus is well established (Dahlin et al. 2016). In laboratory conditions, some species like green iguanas (*Iguana iguana*), North American Bullfrogs (*Lithobates catesbeianus*), red-ear sliders (*Trachemys scripta elegans*), eastern garter snakes, and western fence-lizards (*Sceloporus occidentalis*) have developed noticeable viremia titers of WNV (Steinman et al. 2006; Reisen et al. 2007; Dahlin et al. 2016). And snakes have also died after mounting high WNV viremic titers (Steinman et al. 2006). In several countries, WNV has also been presented to be most pathogenic in crocodilians, producing high titer viremias (Marschang 2011). Despite the rising evidence that reptiles may certainly play a part in the ecoepidemiology of WNV, information of WNV in reptiles are mostly based on experimental infections or farmed entities rather than wild ones. Nevertheless, a recent study showed 41% WNV antibodies in wild Morelet's crocodiles (*Crocodylus moreletii*), specifying that studies on wild population can offer novel insights into the ecoepidemiology of WNV (Machain-Williams et al. 2013).

4.2.3 Transmission

WNV is disseminated in a bird–mosquito–bird transmission cycle. *Culex* mosquitoes are the most implicated vector involved in transmitting the virus to humans, while wild birds are considered as the primary reservoirs. However, WNV has been found in 65 mosquito species, 326 avian species, and >30 nonavian hosts in the United States (Petersen et al. 2013). WNV outbreaks have accompanied with increased populations of *Culex* mosquitoes during the rainy season in tropical regions and summers in temperate zones (Baqar et al. 1993). Numerous species of *Culex* that have been associated with transmission of WNV include *Culex modestus* in France; *Culex univittatus* in Israel, Egypt, and South Africa; and *Culex vishnui* complex, i.e., *Culex tritaenorhynchus* and *Culex vishnui* and *Culex pseudovishnuii*, in India and Pakistan; while *Culex restuans*, *Culex salinarius*, *Culex pipiens*, and *Culiseta melanura* are implicated in the United States (Dandawate et al. 1969; Umrigar and Pavri 1977; Hubalek and Halouzka 1999; Andreadis et al. 2001). WNV has also been noticed to a lesser extent in *Anopheles* and *Aedes* mosquitoes (Gubler 2007) and ixodid and argasid ticks (Lawrie et al. 2004). *Culex* mosquitoes may also play a role of overwintering host as WNV is observed to be vertically transmitted to the progenies of females having higher virus titer infections (Nelms et al. 2013). Other important modes of WNV transmission to humans are organ transplantation (Winston et al. 2014), blood transfusion (Stramer et al. 2005), transplacental (CDC 2002a), and breast milk transmission (CDC 2002b) from mother to child and occupational exposure in laboratory personnel (CDC 2002c). In 2002, 23 cases of transfusion-related transmission have been reported including 1 granulocyte transfusion (Pealer et al. 2003; Meny et al. 2011). Vector-borne transmission being the main transmission route, experimental trials conducted in 24 wild bird species (15 families) have shown that birds of families Laridae and Corvidae can develop high viremia titer and shed huge amounts of WNV in cloacal and oral secretions. This suggests that contact birds can contract infection through oral-oral or fecal-oral routes or through skin or feather picking (McLean et al. 2001; Komar et al. 2003a, b; Colpitts et al. 2012; Pérez-Ramírez et al. 2014). Figure 4.1 schematically depicts the WNV transmission cycle.

4.3 Role of Birds

Birds are both natural reservoirs and amplifying hosts of WNV, and over 100 species of birds are vulnerable to infection in North America. Birds are mostly capable of surviving the infection with the development of permanent immunity (Campbell

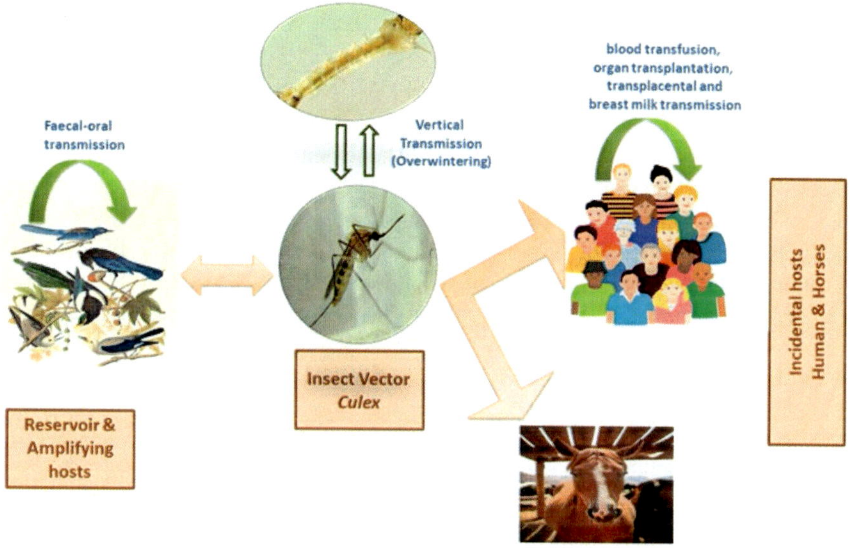

Fig. 4.1 Schematic representation of the WNV transmission cycle

et al. 2002; Hayes et al. 2005). Both migratory and sedentary birds develop sufficient viremia which allows them to infect the feeding mosquitoes. Thus, WNV is maintained in a well-established enzootic cycle involving birds as the main amplifying and reservoir hosts (Komar et al. 2003a, b). Infected birds develop higher viremic titers subsequently permitting transmission of the virus to ornithophilic mosquitoes. In North America, jays and crows (family *Corvidae*) usually die due to WNV infection and form the basis of dead-bird surveillance programs that help in tracking the spread of WNV (Campbell et al. 2002; Hayes et al. 2005). Viremic migratory birds can spread the virus to larger areas along its flyways such as from the breeding areas of Europe to wintering areas of Africa. Apart from migratory birds, local movement of resident birds may also aid in the spread of the virus (Rappole et al. 2000; Campbell et al. 2002; Calistri et al. 2010). Birds of the order *Passeriformes* act as amplifier hosts, as they develop adequate viremia (Komar et al. 2003a, b). Globally, birds of other species like the house sparrow, hooded crow (Africa, Asia, Europe), European starling, barn owl, hybrid falcon Japanese quail, cattle egret, mallard, rock pigeon, ring-necked pheasant, and common goose have all been associated in the transmission of WNV (Pérez-Ramírez et al. 2014).

4.4 Disease

4.4.1 Pathogenesis

Rodent models have offered significant insights into the pathogenesis of WNV. Subsequent to a bite from infected mosquitoes, WNV is placed in the skin and blood tissue. The virus then infects cutaneous Langerhans's cells and proceeds to the local draining lymph node. The migration of Langerhans's cells, draining lymph nodes accumulation, and commencement of lymph node shutdown due to WNV infection are dependent on interleukin-1b and can also happen in the absence of tumor necrosis factor. Multiplication of WNV in these tissues creates a low-level transient viremia, which drops with the arrival of anti-WNV IgM antibodies (Byrne et al. 2001; Busch et al. 2008). WNV further infects multiple vital organs like liver, spleen, kidney, and CNS. The virus was even spotted in the urine of an encephalitic patient 8 days following the onset of symptoms. WNV gains entry into the CNS either by direct infection and collapse of the blood-brain barrier (BBB) or through blood during high titer of infection. Without disrupting BBB, the virus enters into CNS as the infected microvascular endothelial cells of human brains allow the entry. E selectin and elevated vascular cell adhesion protein may help in the trafficking of virus-infected cells into the CNS through a Trojan horse mechanism and aid in the virus spread into the CNS. During experimental infection in mice, high viremia correlates with the severity of disease (Samuel and Diamond 2005; Tonry et al. 2005; Verma et al. 2009; Rossi et al. 2010). Host proteins like intercellular adhesion molecule (ICAM-1), death-associated protein-kinase related 2 (Drak2), matrix metalloproteinase 9 (MMP-9), and macrophage migration inhibitory factor (MIP) have been related with the alteration of BBB permeability. Sensors of innate immune like Toll-like receptor 3 (TLR3) could facilitate viral entry into the CNS by interceding the tumor necrosis factor-alpha (TNF-α) upregulation causing leakage of capillary and greater BBB permeability (Wang et al. 2004; Arjona et al. 2007; Dai et al. 2008; Wang et al. 2008; Rossi et al. 2010). The proinflammatory cytokines like macrophage migration inhibitory factor (MIF), monocyte chemoattractant protein 5 (MCP-5), interferon gamma-inducible protein 10 (IP-10), interferon-gamma (IFN-c), TNF-α, and monokine induced by gamma interferon (MIC) were upregulated in laboratory-infected mice, indicating that responses of the host to infection may contribute to nervous symptoms (Garcia-Tapia et al. 2007). The virus also spreads to olfactory bulbs and olfactory neurons (Monath et al. 1983). The movement of WNV in neurons is bidirectional, and the axonal transport eases the CNS entry of virus and acute limb paralysis. The viral infection in the CNS neurons causes degeneration, loss of architecture, and cell death succeeded by mononuclear cell infiltration at an advanced stage (Sampson et al. 2000; Samuel et al. 2007). The brain stem, spinal cord, and hippocampal injury have been detected in both diseased rodents and humans (David and Abraham 2016).

4.4.2 Clinical Signs/Clinical Profile

The human WNV infection presents a wide clinical spectrum. Around 80% of infected individuals are asymptomatic; about 20% experience a nonspecific pyretic illness denoted to as West Nile fever (WNF). This WNF can manifest from mild to severe, whereas a small subsection of infected people (<1%) develops a lethal neuroinvasive infection (Chancey et al. 2015; Luo and Wang 2018). The differential diagnosis is little complicated as WNF shows a dengue-like disease, frequently associated with headache, nausea, vomiting, malaise, chills myalgia, and lymphadenopathy and rarely presented with a rash on extremities and trunk (Hayes et al. 2005; Chianese et al. 2019). Symptoms may continue for a few days or weeks, subject to the age of the infected individual. Mild fever is usual in children; moderate to high fever accompanied by muscle pain and headache is frequently observed in young people. In elderly patients, more severe symptoms are common (Klee et al. 2004). Advancement to the serious neurological disorder may comprise encephalitis or meningitis, occasionally causing acute flaccid paralysis, analogous to that noticed with poliomyelitis (Colpitts et al. 2012). Patients manifesting neurological disease may experience altered mental status, stiff neck, fatigue, and movement complaints including Parkinsonism and shakes (Guillain-Barre syndrome). Older patients (>50 years) and immune-compromised people are at greater risk of developing the fatal neurological disease (Chianese et al. 2019).

American crows and yellow-billed magpies were the most affected bird species as their population size and demographic rates are substantially affected across the WNV-infected regions (Kilpatrick and Wheeler 2019). In some instances, it is worthy to note that pathogenicity of WNV and mortality are increasing among some wild birds which once were nonsusceptible amplifying hosts. The clinical disease development is due to the invasion of the virus to major organs like heart, liver, spleen, kidney, and CNS. In several cases, nonspecific signs like ataxia, dehydration, and anorexia appear on fifth or sixth day post-infection (dpi) (Gamino and Hofle 2013). The second clinical picture is among highly susceptible bird species like corvids; huge amount of virus is extensively distributed in vital organs, producing multiorgan failure and causing a rapid death without much development of clinical signs (Kipp et al. 2006; Shirafuji et al. 2008; Nemeth et al. 2011). The third clinical picture is the milder one with low-level virus replication leading to chronic infections. In certain cases, the infection becomes persistent and the virus can be spotted in tissues like skin, eye, brain, spleen, and kidney after several months of initial infection, and this has been demonstrated for Western scrub-jay (*Aphelocoma californica*), House sparrow, and House finch (*Haemorhous mexicanus*) surviving both experimental and natural infection (Reisen et al. 2006; Wheeler et al. 2012a). The virus persistence in birds might play a significant role in WNV overwintering and infection of mosquitoes in case of viremia recrudescence and host immune impairment; however, clear epidemiological consequences of persistence are still not clear (Wheeler et al. 2012b).

In WNV-infected horses, the clinical signs mainly include high fever, paresis, ataxia, tremors, and skin and muscle fasciculations (Ostlund et al. 2000, 2001). Some horses may recover spontaneously and some may need extensive supportive care like slinging to hold them upright during peak ataxia (Bosco-Lauth and Bowen 2019).

4.4.3 Pathology/Lesions

The most noticeable lesions in birds are gross brain hemorrhage, meningoencephalitis, splenomegaly, and necrotizing myocarditis. These lesions would alert pathologists to the likely occurrence of WNV (Steele et al. 2000). Microscopic lesions in CNS may be absent in highly susceptible bird species like crows, while pathological changes in other organs (vasculitis, nephritis and hepatitis, changes in striated muscle tissues of heart and skeletal muscle) may be seen with minimal inflammatory reaction (Weingartl et al. 2004; Gamino and Hofle 2013). In WNV neuroinvasive-diseased human patients, the histopathologic examination of CNS will show perivascular inflammation, neuronophagia, microglial nodules, and variable necrosis with neuronal loss with pathologic changes focused in the brainstem, deep gray matter nuclei, and anterior gray column of the spinal cord (Armah et al. 2007). The inflammation of spinal cord was reported in 17 of 23 individuals who died with WNV neuroinvasive disease. More prominent inflammation was observed in the anterior horns than in the posterior horns of nine patients. Gliosis, foci of demyelination, and intermittent perivascular infiltrate can be seen in patients with long clinical courses (Guarner et al. 2004).

4.5 Disease Management

4.5.1 Diagnosis

According to the diagnostic criteria of CDC and EU, WNV infection is described as confirmed or probable based on the clinical symptoms and laboratory reports. According to case definition of the EU, a human case of "WNF" is defined as any person with fever or with encephalitis or meningitis and at least one laboratory criterion for case confirmation, that is, isolation of WNV from blood or CSF, detection of WNV nucleic acid in blood or CSF, WNV-specific antibody response (IgM) in the CSF, WNV IgM high titer and detection of WNV IgG, and confirmation by neutralization (Table 4.1). Asymptomatic subjects with positive laboratory results are also considered as confirmed cases. The case is considered probable if only WNV-specific antibodies are present in serum. The laboratory reports must be interpreted only after considering the flavivirus vaccination status. In spite of all the efforts to implement surveillance procedures in countries with WNV circulation

Table 4.1 List of available laboratory diagnostic methods to detect WNV

Diagnostic method	Techniques involved	Samples required	Remarks	References
Viral isolation	Cell culture	Serum, CSF, urine tissues	• As WNV is a BSL-3 pathogen, it should be handled within a class II biological safety cabinet set in a BSL-3 facility • Virus isolation and propagation in mammalian cells (BHK-21 and Vero E6) or in mosquito cells (e.g., C6/36) • Characteristic cytopathic effect appears from 2 to 7 days	Brien et al. (2013), Barzon et al. (2014a, b)
Serological	IgM ELISA/ IFA	Serum, CSF	• IgM antibodies can be detected by day 4 after the start of symptoms • As IgM antibodies do not cross BBB, positive WNV IgM result (within 8 days of symptom onset) is a diagnostic criterion for infection	Prince et al. (2005), Niedrig et al. (2007)
	IgG ELISA/ IFA	Serum, CSF	• IgG antibodies can be detected by day 8 after the start of symptoms • A fourfold rise in WNV IgG antibody titer in paired (acute and convalescent) serum samples	Prince et al. (2005), Barzon et al. (2015), Niedrig et al. (2007), Ludolfs et al. (2007)
	Virus neutralization (VN)	Serum, CSF	• Positive results from immunofluorescence assay (IFA) and enzyme immunoassay (EIA) should be confirmed by VN assay due to the cross-reactivity issue of flaviviruses • VN assay is considered as the gold standard for diagnosis of WNV infection	Busch et al. (2008), Nelson et al. (2008) Calisher et al. (1989), Lustig et al. (2018)
	Antigen capture NS1 ELISA	Serum, plasma, CSF,	• NS1 protein is secreted at higher levels during acute infection, and it is demonstrable in the acute-phase sera of infected patients	Macdonald et al. (2005), Saxena et al. (2013), Ding et al. (2014)

(continued)

Table 4.1 (continued)

Diagnostic method	Techniques involved	Samples required	Remarks	References
Detection of viral genome (molecular method)	Real-time polymerase chain reaction (qRT-PCR	Serum, plasma, CSF, urine	• Due to short-lived viremia, molecular assays are of not much value; however, recent studies have detected virus genome in the urine as WNV is retained in the kidney for a longer period than in serum, plasma, or CSF	Busch et al. (2008), Barzon et al. (2013a), Lustig et al. (2016), Barzon et al. (2013a, b, 2014a, b, 2015)
Novel methods	Next-generation sequencing (NGS)	Serum, plasma, CSF	• In the latest study published from Tanzania, WNV was investigated by NGS and 2 out of 12 plasma samples turned positive	Williams et al. (2018)
	CRISPR-cas13 technology		• CRISPR-cas13 technology is used to detect RNAs of flaviviruses from the patient samples • It requires minimal equipment and is faster with comparable sensitivity to qRT-PCR	Myhrvold et al. (2018)

and developments in diagnostic tests, many WNV-infected patients are unnoticed clinically. As the molecular diagnosis of WNV is not reliable due to its short viremic phase, serological diagnosis is currently in much use. WNV diagnosis is further complicated by the recent emergence of Zika virus in Asia and America, as this virus significantly cross-reacts with other members of family *Flaviviridae* including WNV. Diagnosis is particularly challenging in patients from regions with cocirculation of flavivirus, and in travelers from dengue or Zika endemic countries traveling to WNV-infected areas or vice versa (Barzon et al. 2015; Lustig et al. 2018). In WNV-infected persons, the viremia starts 2–3 dpi and persists for 8–10 days. As the virus is RBC associated, viral load is more in whole blood than the plasma, particularly in the later stage of infection. Besides, the virus is also excreted in urine, and it can be spotted at a higher titer and for a lengthier period than in blood. During the later stage of infection, WNV is sequestered in peripheral organs or tissues, thus getting transmitted through organ donation or transplantation (Barzon et al. 2015).

4.6 Prevention and Control

No specific treatment regime is available for WNV disease, and clinical management is supportive. Extensive investigations through clinical and preclinical studies about the usage and repurposing of antiviral agents are in continuous progress (Diamond 2009; Beasley 2011; Lim and Shi 2013). Some of the promising agents against WNV in vitro are high-dose ribavirin, anti-WNV immunoglobulins, antisense gene-target compound, and interferon-α 2b. Nevertheless, clinical trials are still incomplete (Anderson and Rahal 2002; Petersen and Marfin 2002). Bearing in mind the therapeutic deficiencies for WNV, prevention of infection remains a priority. So far, no human vaccine is licensed or available; however, there are some vaccines licensed for veterinary use (De Filette et al. 2012). Fort Dodge Animal Health developed the very first vaccine by inactivating the virus with formalin (Rossi et al. 2010). Now, this vaccine, marketed as West Nile-Innovator™ in the United States, appears safe and effective without causing any adverse effects in vaccinated horses (Ng et al. 2003). Presently, there are four WNV vaccines on the market for veterinary use: three that encompass whole inactivated virus (WN Innovator™, Prestige® WNV, and Vetera™ WNV), and another one is Recombitek™ Equine WNV developed with a live chimeric virus containing the WNV prM/E in a canarypox backbone (Kaiser and Barrett 2019). While all four veterinary WNV vaccines are defensive in horses, all of them require two primary doses and a booster annually. Currently, six human vaccines have progressed into clinical trials. Out of the six vaccines, the two live attenuated vaccines have only elicited strong immunity following a single dose. None of the human vaccine candidates have advanced beyond phase II clinical trials (Kaiser and Barrett 2019). The most comprehensively studied human vaccine candidate, ChimeriVax-WN02, was shown to be effective and safe in multiple age groups, especially in the most vulnerable old population, and for this reason, the vaccine has been a prospective candidate for continued development (Dayan et al. 2012; Kaiser and Barrett 2019). As there is no licensed human vaccine till date, prevention of WNV infection mainly depends on mosquito control programs and screening of organ and blood donors. Mosquito control strategies mainly include the use of mosquito repellents (permethrin and *N*, *N*-diethyl-m-toluamide), mosquito nets, appropriate clothing, and activities limiting skin exposures (Marfin and Gubler 2001; Sampathkumar 2003). Another important phenomenon which helps in prediction and containment of the WNV infection is sentinel surveillance. Usually, trainers and owners of the horses closely monitor the health status of the horses and report them to veterinarians. This demonstrates the significance of horses as a sentinel animal indicating the WNV epizootic activity (Vilibic-Cavlek et al. 2019a, b). Further, some serological studies have shown that mules and donkeys have higher seroprevalence than horses. This feature makes them a better sentinel hosts for monitoring WNV in any given region (García-Bocanegra et al. 2012).

WNV is emerging as a global concern as the virus has the potential to expand into larger and new geographical areas and can cause severe neurological disease and large outbreaks. This is facilitated by human activities and altered environmental conditions supporting the expansion of the vector population. Creation of public awareness, national/international surveillance programs, sentinel surveillance in equine, and development of promising human vaccines is highly required. Ultimately, a continuous vector control program would be advantageous in controlling the spread of all mosquito-borne viruses.

References

Ahlers LRH, Goodman AG (2018) The immune responses of the animal hosts of West Nile virus: a comparison of insects, birds, and mammals. Front Cell Infect Microbiol 8:96

Anderson JF, Rahal JJ (2002) Efficacy of interferon alpha-2b and ribavirin against West Nile virus in vitro. Emerg Infect Dis 8:107–108

Andreadis TG, Anderson JF, Vossbrinck CR (2001) Mosquito surveillance for West Nile virus in Connecticut, 2000: isolation from Culex pipiens, Cx. restuans, Cx. salinarius, and Culiseta melanura. Emerg Infect Dis 7:670–674

Arjona A, Foellmer HG, Town T et al (2007) Abrogation of macrophage migration inhibitory factor decreases West Nile virus lethality by limiting viral neuroinvasion. J Clin Invest 117:3059–3066

Armah HB, Wang G, Omalu BI, Tesh RB, Gyure KA, Chute DJ, Smith RD, Dulai P, Vinters HV, Kleinschmidt-DeMasters BK, Wiley CA (2007) Systemic distribution of West Nile virus infection: postmortem immunohistochemical study of six cases. Brain Pathol 17:354–362

Baqar S, Hayes CG, Murphy JR et al (1993) Vertical transmission of West Nile virus by Culex and Aedes species mosquitoes. Am J Trop Med Hyg 48:757–762

Barzon L, Pacenti M, Franchin E, Pagni S, Martello T, Cattai M et al (2013a) Excretion of West Nile virus in urine during acute infection. J Infect Dis 208:1086–1092. https://doi.org/10.1093/infdis/jit290

Barzon L, Pacenti M, Palu G (2013b) West Nile virus and kidney disease. Expert Rev Anti-Infect Ther 11:479–487. https://doi.org/10.1586/eri.13.34

Barzon L, Pacenti M, Franchin E et al (2014a) Isolation of West Nile virus from urine samples of patients with acute infection. J Clin Microbiol 52:3411–3413

Barzon L, Pacenti M, Franchin E, Squarzon L, Sinigaglia A, Ulbert S et al (2014b) Isolation of West Nile virus from urine samples of patients with acute infection. J Clin Microbiol 52:3411–3413. https://doi.org/10.1128/JCM.01328-14

Barzon L, Pacenti M, Ulbert S, Palu G (2015) Latest developments and challenges in the diagnosis of human West Nile virus infection. Expert Rev Anti-Infect Ther 13:327–342. https://doi.org/10.1586/14787210.2015.1007044

Beasley DW (2011) Vaccines and immunotherapeutics for the prevention and treatment of infections with West Nile virus. Immunotherapy 3:269–285

Bosco-Lauth AM, Bowen RA (2019) West Nile virus: veterinary health and vaccine development. J Med Entomol 56(6):1463–1466

Brien JD, Lazear HM, Diamond MS (2013) Propagation, quantification, detection, and storage of West Nile virus. Curr Protoc Microbiol 31:15D.3.1–15D.3.18

Busch MP, Kleinman SH, Tobler LH et al (2008) Virus and antibody dynamics in acute West Nile virus infection. J Infect Dis 198:984–993

Byrne SN, Halliday GM, Johnston LJ et al (2001) Interleukin-1beta but not tumor necrosis factor is involved in West Nile virus-induced Langerhans cell migration from the skin in C57BL/6 mice. J Invest Dermatol 117:702–709

Calisher CH, Karabatsos N, Dalrymple JM, Shope RE, Porterfield JS, Westaway EG et al (1989) Antigenic relationships between flaviviruses as determined by cross-neutralization tests with polyclonal antisera. J Gen Virol 70(1):37–43

Calistri P, Giovannini A, Hubalek Z et al (2010) Epidemiology of West Nile in Europe and in the Mediterranean basin. Open Virol J 4:29–37

Campbell GL, Marfin AA, Lanciotti RS, Gubler DJ (2002) West Nile virus. Lancet Infect Dis 2: 519–529

Centers for Disease Control and Prevention (CDC) (2002a) Possible West Nile virus transmission to an infant through breastfeeding– Michigan, 2002. MMWR Morb Mortal Wkly Rep 51:877–878

Centers for Disease Control and Prevention (CDC) (2002b) Intrauterine West Nile virus infection—New York, 2002. MMWR Morb Mortal Wkly Rep 51:1135–1136

Centers for Disease Control and Prevention (CDC) (2002c) Laboratory acquired West Nile virus infections–United States, 2002. MMWR Morb Mortal Wkly Rep 51:1133–1135

Chambers TJ, Hahn CS, Galler R et al (1990) Flavivirus genome organization, expression, and replication. Annu Rev Microbiol 44:649–688

Chancey C, Grinev A, Volkova E, Rios M (2015) The global ecology and epidemiology of West Nile virus. Biomed Res Int 2015:376230

Chianese A, Stelitano D, Astorri R, Serretiello E, Rocca MTD, Melardo C, Vitiello M, Galdiero M, Franci G (2019) West Nile virus: an overview of current information. Transl Med Rep 3:8145

Colpitts TM, Conway MJ, Montgomery RR, Fikrig E (2012) West Nile virus: biology, transmission, and human infection. Clin Microbiol Rev 25(4):635–648. https://doi.org/10.1128/CMR.00045-12

Dahlin CR, Hughes DF, Meshaka WE Jr, Coleman C, Henning JD (2016) Wild snakes harbor West Nile virus. One Health 2:136–138. PMID:28616487

Dai J, Wang P, Bai F et al (2008) Icam-1 participates in the entry of West Nile virus into the central nervous system. J Virol 82:4164–4168

Dandawate CN, Rajagopalan PK, Pavri KM et al (1969) Virus isolations from mosquitoes collected in North Arcot district, Madras state, and Chittoor district, Andhra Pradesh between November 1955 and October 1957. Indian J Med Res 57:1420–1426

David S, Abraham AM (2016) Epidemiological and clinical aspects on West Nile virus, a globally emerging pathogen. Infect Dis 48:571–586. https://doi.org/10.3109/23744235.2016.1164890

Dayan GH, Bevilacqua J, Coleman D, Buldo A, Risi G (2012) Phase II, dose ranging study of the safety and immunogenicity of single dose West Nile vaccine in healthy adults $\geq$ 50 years of age. Vaccine 30:6656–6664

De Filette M, Ulbert S, Diamond M, Sanders NN (2012) Recent progress in West Nile virus diagnosis and vaccination. Vet Res 43:16

Diamond MS (2009) Progress on the development of therapeutics against West Nile virus. Antivir Res 83:214–227

Ding XX, Li XF, Deng YQ et al (2014) Development of a double antibody sandwich ELISA for West Nile virus detection using monoclonal antibodies against non-structural protein 1. PLoS One 9:e108623

Fall G, Di Paola N, Faye M et al (2017) Biological and phylogenetic characteristics of West African lineages of West Nile virus. PLoS Negl Trop Dis 11(11):e0006078. Published 2017 Nov 8. https://doi.org/10.1371/journal.pntd.0006078

Gamino V, Hofle U (2013) Pathology and tissue tropism of natural West Nile virus infection in birds: a review. Vet Res 44:39

García-Bocanegra I, Arenas-Montes A, Jaén-Téllez JA, Napp S, Fernández-Morente M, Arenas A (2012) Use of sentinel serosurveillance of mules and donkeys in the monitoring of West Nile virus infection. Vet J 194(2):262–264. https://doi.org/10.1016/j.tvjl.2012.04.017. PMID:22633828

Garcia-Tapia D, Hassett DE, Mitchell WJ et al (2007) West Nile virus encephalitis: sequential histopathological and immunological events in a murine model of infection. J Neurovirol 13: 130–138

Guarner J, Shieh WJ, Hunter S, Paddock CD, Morken T, Campbell GL, Marfin AA, Zaki SR (2004) Clinicopathologic study and laboratory diagnosis of 23 cases with West Nile virus encephalomyelitis. Hum Pathol 35:983–990

Gubler DJ (2007) The continuing spread of West Nile virus in the western hemisphere. Clin Infect Dis 45:1039–1046

Hayes EB, Komar N, Nasci RS, Montgomery S, O'Leary DR, Campbell GL (2005) Epidemiology and transmission dynamics of West Nile virus disease. CDC Emerg Infect Dis 11(8):1167–1173

Hubalek Z, Halouzka J (1999) West Nile fever—a reemerging mosquitoborne viral disease in Europe. Emerg Infect Dis 5:643–650

Huhn GD, Sejvar JJ, Montgomery SP, Dworkin MS (2003) West Nile virus in the United States: an update on an emerging infectious disease. Am Fam Physician 68(4):653–661

Kaiser JA, Barrett ADT (2019) Twenty years of progress toward West Nile virus vaccine development. Viruses 11:823

Kilpatrick MA, Wheeler SS (2019) Impact of West Nile virus on bird populations: limited lasting effects, evidence for recovery, and gaps in our understanding of impacts on ecosystems. J Med Entomol 56(6):1491–1497

Kipp AM, Lehman JA, Bowen RA, Fox PE, Stephens MR, Klenk K, Komar N, Bunning ML (2006) West Nile virus quantification in feces of experimentally infected American and fish crows. Am J Trop Med Hyg 75:688–690

Klee AL, Maidin B, Edwin B et al (2004) Long-term prognosis for clinical West Nile virus infection. Emerg Infect Dis 10:1405–1411

Klenk K, Snow J, Morgan K et al (2004) Alligators as West Nile virus amplifiers. Emerg Infect Dis 10(12):2150–2155. https://doi.org/10.3201/eid1012.040264

Komar N, Langevin S, Hinten S et al (2003a) Experimental infection of North American birds with the New York 1999 strain of West Nile virus. Emerg Infect Dis 9:311–322

Komar N, Langevin S, Hinten S, Nemeth N, Edwards E, Hettler D, Davis B, Bowen R, Bunning M (2003b) Experimental infection of North American birds with the New York 1999 strain of West Nile virus. Emerg Infect Dis 9:311–322. https://doi.org/10.3201/eid0903.020628

Lawrie CH, Uzcategui NY, Gould EA et al (2004) Ixodid and argasid tick species and West Nile virus. Emerg Infect Dis 10:653–657

Lim SP, Shi PY (2013) West Nile virus drug discovery. Viruses 5:2977–3006

Ludolfs D, Niedrig M, Paweska JT, Schmitz H (2007) Reverse ELISA for the detection of anti-West Nile virus IgG antibodies in humans. Eur J Clin Microbiol Infect Dis 26:467–473

Luo H, Wang T (2018) Recent advances in understanding West Nile virus host immunity and viral pathogenesis. F1000Res 7:338

Lustig Y, Mannasse B, Koren R, Katz-Likvornik S, Hindiyeh M, Mandelboim M et al (2016) Superiority of West Nile virus RNA detection in whole blood for diagnosis of acute infection. J Clin Microbiol 54:2294–2297. https://doi.org/10.1128/JCM.01283-16

Lustig Y, Sofer D, Bucris ED, Mendelson E (2018) Surveillance and diagnosis of West Nile virus in the face of flavivirus cross-reactivity. Front Microbiol 9:2421. https://doi.org/10.3389/fmicb.2018.02421

Macdonald J, Tonry J, Hall RA et al (2005) NS1 protein secretion during the acute phase of West Nile virus infection. J Virol 79:13924–13933

Machain-Williams C, Padilla-Paz SE, Weber M, Cetino-Trejo R, Juarez-Ordaz J, Lorono-Pino M et al (2013) Antibodies to West Nile virus in wild and farmed crocodiles in southeastern Mexico. J Wildl Dis 49(3):690–693

Marfin AA, Gubler DJ (2001) West Nile encephalitis: an emerging disease in the United States. Clin Infect Dis 33:1713–1719

Marschang RE (2011) Viruses infecting reptiles. Viruses 3:2087–2126

McLean RG, Ubico SR, Docherty DE, Hansen WR, Sileo L, McNamara TS (2001) West Nile virus transmission and ecology in birds. Ann N Y Acad Sci 951:54–57

Meny GM, Santos-Zabala L, Szallasi A et al (2011) West Nile virus infection transmitted by granulocyte transfusion. Blood 117:5778–5779

Monath TP, Cropp CB, Harrison AK (1983) Mode of entry of a neurotropic arbovirus into the central nervous system. Reinvestigation of an old controversy. Lab Investig J Tech Methods Pathol 48:399–410

Myhrvold C, Freije CA, Gootenberg JS, Abudayyeh OO, Metsky HC, Durbin AF et al (2018) Field-deployable viral diagnostics using CRISPRCas13. Science 360:444–448. https://doi.org/10.1126/science.aas8836

Nelms BM, Fechter-Leggett E, Carroll BD et al (2013) Experimental and natural vertical transmission of West Nile virus by California Culex (Diptera: Culicidae) mosquitoes. J Med Entomol 50: 371–378

Nelson S, Jost CA, Xu Q, Ess J, Martin JE, Oliphant T et al (2008) Maturation of West Nile virus modulates sensitivity to antibody mediated neutralization. PLoS Pathog 4:e1000060. https://doi.org/10.1371/journal.ppat.1000060

Nemeth NM, Thomsen BV, Spraker TR, Benson JM, Bosco-Lauth AM, Oesterle PT, Bright JM, Muth JP, Campbell TW, Gidlewski TL, Bowen RA (2011) Clinical and pathologic responses of American crows (*Corvus brachyrhynchos*) and fish crows (*Cossi fragus*) to experimental West Nile virus infection. Vet Pathol 48:1061–1074

Ng T, Hathaway D, Jennings N et al (2003) Equine vaccine for West Nile virus. Dev Biol (Basel) 114:221–227

Niedrig M, Donoso Mantke O, Altmann D, Zeller H (2007) First international diagnostic accuracy study for the serological detection of West Nile virus infection. BMC Infect Dis 7:72–76

Ostlund EN, Andresen JE, Andresen M (2000) West Nile encephalitis. Vet Clin North Am Pract 16: 427–441

Ostlund EN, Crom RL, Pedersen DD, Johnson DJ, Williams WO, Schmitt BJ (2001) Equine West Nile encephalitis, United States. Emerg Infect Dis 7:665–669

Pealer LN, Marfin AA, Petersen LR et al (2003) Transmission of West Nile virus through blood transfusion in the United States in 2002. N Engl J Med 349:1236–1245

Pérez-Ramírez E, Llorente F, Jiménez-Clavero MÁ (2014) Experimental infections of wild birds with West Nile virus. Viruses 6(2):752–781. Published 2014 Feb 13. https://doi.org/10.3390/v6020752

Petersen LR, Marfin AA (2002) West Nile virus: a primer for the clinician. Ann Intern Med 137: 173–179

Petersen LR, Brault AC, Nasci RS (2013) West Nile virus: review of the literature. JAMA 310:308

Prince HE, Tobler LH, Lape-Nixon M et al (2005) Development and persistence of West Nile virus-specific immunoglobulin M (IgM), IgA, and IgG in viremic blood donors. J Clin Microbiol 43: 4316–4320

Rappole JH, Derrickson SR, Hubalek Z (2000) Migratory birds and spread of West Nile virus in the Western hemisphere. Emerg Infect Dis 6:319–328

Reisen WK, Fang Y, Lothrop HD, Martinez VM, Wilson J, Oconnor P, Carney R, Cahoon-Young B, Shafii M, Brault AC (2006) Overwintering of West Nile virus in Southern California. J Med Entomol 43:344–355

Reisen WK, Brault AC, Martinez VM, Fang Y, Simmons K, Garcia S et al (2007) Ability of transstadially infected Ixodes pacificys (Acari: Ixodidae) to transmit West Nile virus to song sparrow or western fence lizards. J Med Entomol 44:320–327

Rios M, Zhang MJ, Grinev A et al (2006) Monocytes-macrophages are a potential target in human infection with West Nile virus through blood transfusion. Transfusion 46:659–667

Rizzoli A, Jimenez-Clavero MA, Barzon L, Cordioli P, Figuerola J, Koraka P, Martina B, Moreno A, Nowotny N, Pardigon N et al (2015) The challenge of West Nile virus in Europe: knowledge gaps and research priorities. Euro Surveill 20:21135. https://doi.org/10.2807/1560-7917.es2015.20.20.21135

Rossi SL, Ross TM, Evans JD (2010) West Nile virus. Clin Lab Med 30:47–65

Sampathkumar P (2003) West Nile virus: epidemiology, clinical presentation, diagnosis, and prevention. Mayo Clin Proc 78:1137–1143

Sampson BA, Ambrosi C, Charlot A et al (2000) The pathology of human West Nile virus infection. Hum Pathol 31:527–531

Samuel MA, Diamond MS (2005) Alpha/beta interferon protects against lethal West Nile virus infection by restricting cellular tropism and enhancing neuronal survival. J Virol 79:13350–13361

Samuel MA, Wang H, Siddharthan V et al (2007) Axonal transport mediates West Nile virus entry into the central nervous system and induces acute flaccid paralysis. Proc Natl Acad Sci U S A 104:17140–17145

Saxena D, Kumar JS, Parida M et al (2013) Development and evaluation of NS1 specific monoclonal antibody-based antigen capture ELISA and its implications in clinical diagnosis of West Nile virus infection. J Clin Virol 58:528–534

Shirafuji H, Kanehira K, Kubo M, Shibahara T, Kamio T (2008) Experimental West Nile virus infection in jungle crows (*Corvus macrorhynchos*). Am J Trop Med Hyg 78:838–842

Spickler, Anna Rovid (2013) West Nile virus infection. http://www.cfsph.iastate.edu/DiseaseInfo/factsheets.php

Steele KE, Linn MJ, Schoepp RJ et al (2000) Pathology of fatal West Nile virus infections in native and exotic birds during the 1999 outbreak in New York City. J Vet Pathol 37:208–224

Steinman A, Banet-Noach C, Grinfeld L, Grinfeld L, Aizenberg Z, Lahav D et al (2006) Experimental infection of common garter snakes (*Thamnophis sirtalis*) withWest Nile virus. Vector Borne Zoonotic Dis 6:361–368

Stramer SL, Fang CT, Foster GA et al (2005) West Nile virus among blood donors in the United States, 2003 and 2004. N Engl J Med 353:451–459

Tonry JH, Brown CB, Cropp CB et al (2005) West Nile virus detection in urine. Emerg Infect Dis 11:1294–1296

Troupin A, Colpitts TM (2016) Overview of West Nile virus transmission and epidemiology. Methods Mol Biol 1435:15–18. Epub 2016/05/18. PMID:27188546

Umrigar MD, Pavri KM (1977) Comparative biological studies on Indian strains of West Nile virus isolated from different sources. Indian J Med Res 65:596–602

van der Meulen KM, Pensaert MB, Nauwynck HJ (2005) West Nile virus in the vertebrate world. Arch Virol 150:637–657

Velasco M, Sánchez-Seco MP, Campelo C et al (2020) Imported human West Nile virus lineage 2 infection in Spain: neurological and gastrointestinal complications. Viruses 12(2):156. Published 2020 Jan 29. https://doi.org/10.3390/v12020156

Verma S, Lo Y, Chapagain M et al (2009) West Nile virus infection modulates human brain microvascular endothelial cells tight junction proteins and cell adhesion molecules: transmigration across the in vitro blood-brain barrier. Virology 385:425–433

Vilibic-Cavlek T, Savic V, Petrovic T et al (2019a) Emerging trends in the epidemiology of West Nile and Usutu virus infections in Southern Europe. Front Vet Sci 6:437. Published 2019 Dec 6. https://doi.org/10.3389/fvets.2019.00437

Vilibic-Cavlek T, Savic V, Sabadi D, Peric L, Barbic L, Klobucar A, Miklausic B, Tabain I, Santini M, Vucelja M, Dvorski E, Butigan T, Kolaric-Sviben G, Potocnik-Hunjadi T, Balenovic M, Bogdanic M, Andric Z, Stevanovic V, Capak K, Balicevic M, Listes E, Savini G (2019b) Prevalence and molecular epidemiology of West Nile and Usutu virus infections in Croatia in the 'One health' context, 2018. Transbound Emerg Dis 66(5):1946–1957. https://doi.org/10.1111/tbed.13225. PMID:31067011

Wang T, Town T, Alexopoulou L et al (2004) Toll-like receptor 3 mediates West Nile virus entry into the brain causing lethal encephalitis. Nat Med 10:1366–1373

Wang S, Welte T, McGargill M et al (2008) Drak2 contributes to West Nile virus entry into the brain and lethal encephalitis. J Immunol 181:2084–2091

Weingartl HM, Neufeld JL, Copps J, Marszal P (2004) Experimental West Nile virus infection in blue jays (*Cyanocitta cristata*) and crows (*Corvus brachyrhynchos*). Vet Pathol 41:362–370

Wheeler SS, Vineyard MP, Woods LW, Reisen WK (2012a) Dynamics of West Nile virus persistence in house sparrows (*Passer domesticus*). PLoS Negl Trop Dis 6:e1860

Wheeler SS, Langevin SA, Brault AC, Woods L, Carroll BD, Reisen WK (2012b) Detection of persistent West Nile virus RNA in experimentally and naturally infected avian hosts. Am JTrop Med Hyg:559–564

Williams SH, Cordey S, Bhuva N, Laubscher F, Hartley MA, Boillat-Blanco N et al (2018) Investigation of the plasma virome from cases of unexplained febrile illness in Tanzania from 2013 to 2014: a comparative analysis between unbiased and VirCapSeq-VERT high-throughput sequencing approaches. MSphere 3:e00311–e00318

Winston DJ, Vikram HR, Rabe IB et al (2014) Donor-derived West Nile virus infection in solid organ transplant recipients: report of four additional cases and review of clinical, diagnostic, and therapeutic features. Transplantation 97:881–889

Chapter 5
Avian Influenza

Abstract Avian influenza (AI), informally known as "avian flu" or "bird flu," causes devastating economic losses to poultry industries worldwide and is a well-recognized zoonotic threat to humans with pandemic potential. The causative agent of AI, the influenza A virus, is an eight-segmented single-stranded RNA virus fitting to the family *Orthomyxoviridae*. Mechanisms of genetic diversity of avian influenza viral strains include reassortment (antigenic shift) and mutations (antigenic drift), which may contribute to the emergence of highly virulent and/or zoonotic viruses. AI viruses are categorized as highly pathogenic (HPAI) or lowly pathogenic (LPAI) strains. HPAI strains are usually derived from LPAI in poultry flocks and cause severe disease in poultry. The natural reservoirs of avian influenza virus are mostly wild water birds of the orders *Anseriformes* (geese, ducks, and swans) and *Charadriiformes* (gulls, terns, shorebirds, and auks). AI viruses are chiefly transmitted by feco-oral and respiratory routes, and via water soiled with feces from infected birds. Migratory birds and international poultry trade play a vital role in expanding the geographical distribution of the virus. The majority of cases of human influenza from an avian source have been associated to direct or indirect interaction with poultry. Control and prevention strategies such as vaccination programs and biosecurity measures in poultry farms, routine surveillance in domesticated and wild birds, and constant monitoring of the evolutionary patterns of the virus are of utmost importance to combat the spread of avian influenza among birds and eventual transmission events to humans.

Keywords Avian flu · *Influenza virus A* · Water birds · Migratory birds · HPAI · LPAI

5.1 Introduction

Avian influenza, informally familiar as avian or bird flu, is an important, highly contagious disease with zoonotic potential, caused by species *influenza A virus*, a member of the genus *influenza virus A* and family *Orthomyxoviridae* (Sendor et al. 2020). Avian influenza viruses are widely distributed and have been reported in both

Y. S. Malik et al., *Role of Birds in Transmitting Zoonotic Pathogens*, Livestock Diseases and Management, https://doi.org/10.1007/978-981-16-4554-9_5

wild and domesticated birds, with wild avian species acting as the primary source of infection for poultry. Infected wild water birds are typically asymptomatic and serve as the natural reservoirs of the AI virus, while the disease can be severe in poultry birds, incurring devastating economic damages to the poultry industry (Paul et al. 2019). AI viruses have also been reported, albeit occasionally, in various mammals including humans. The first documented evidence for AI dates back to 1878, when researchers distinguished the disease (previously known as fowl plague) from other poultry diseases that cause high mortalities (Alexander and Brown 2009). The diverse nature of the AI virus is due to antigenic shift (genetic reassortment) and antigenic drift (mutations). These mechanisms of genetic diversity enable a greater chance for changes in the viral genome, thereby facilitating the emergence of new viruses. As a result, during twentieth–twenty-first centuries, four instances of influenza-driven pandemics were recorded in humans: H1N1 in 1918 and 2009, H2N2 in 1957, and H3N2 in 1968 (Su et al. 2015). In Asia, the AI strain H5N1 was first detected in domestic geese during the year 1996 in Southern China, and by 2000, the virus had extended its host range to domestic ducks that led to the genesis of outbreaks in 2003–2004 (Sims et al. 2005).

AI virus strains can be either highly pathogenic (HPAI) or lowly pathogenic (LPAI) based on their degree of pathogenicity in young chickens that are inoculated intravenously with the virus, or presence of certain features in the viral genome that have been associated with those of HPAI (Capua and Alexander 2007). Highly pathogenic avian influenza (HPAI) viruses were first reported at a small poultry farm in Scotland in 1959 (Dhingra et al. 2018). HPAI is a major cause for outbreaks with high case fatality rates in poultry (Capua and Alexander 2007; Paul et al. 2019). Outbreaks with HPAI strains have caused devastating monetary losses to the poultry industry and have had serious implications on international poultry trade (Ramos et al. 2017; Paul et al. 2019). Epidemics of HPAI strains that are derived from accumulation of mutations in LPAI (H5 and H7 subtypes) viruses likely occurred in 1983 in the United States, 1994 in Mexico, 1999–2000 in Italy, 2002 in Chile, 2003 in the Netherlands, 2008 in the UK, or due to direct entry of HPAI virus into a geographical location that occurred in 2007 in the UK (Cheung and Poon 2007; Capua and Alexander 2007; Bonfanti et al. 2014; Naguib et al. 2019; DEFRA 2007, 2008). Avian influenza caused by the subtypes H5 and H7 has been declared as a notifiable disease by the World Organisation for Animal Health (OIE) and are designated as lowly pathogenic notifiable avian influenza (LPNAI) and highly pathogenic notifiable avian influenza (HPNAI) (Gonzales et al. 2010). HPAI strain H5N1 and LPAI strain H7N9 virus strains are of zoonotic importance because of the disease severity among humans. It is reported that till date, subtypes H5, H6, H7, H9, and H10 have caused infections in humans crossing their particular avian species barrier (Short et al. 2015; El-Shesheny et al. 2018).

5.2 Epidemiology

5.2.1 Causative Agent

The causative agent of avian influenza, Influenza A virus (a member of the family *Orthomyxoviridae*), is a single-stranded, 8-segmented RNA viruses encoding 11 proteins (Naguib et al. 2019). Viral-encoded proteins include receptor-binding surface proteins such as hemagglutinin (HA) and neuraminidase (NA), integral membrane protein matrix protein 2 (M2); internal proteins such as polymerase basic protein 2 (PB2), polymerase basic protein1 (PB1), polymerase acidic protein (PA), nucleoprotein (NP), matrix protein 1 (M1), and nuclear export protein (NEP); and nonstructural proteins such as NS1 and PB1-F2. Based on HA and NA proteins, influenza A virus is categorized into at least 18 (H1-H18) and 11 (N1-N11) subtypes, among which H17-H18 and N10-N11 are reported only in bats (Sato et al. 2019). Influenza A virus has been responsible for four major pandemics in the history and has claimed millions of lives around the world (Table 5.1).

5.2.2 Hosts

Avian influenza virus infects a wide array of avian species that include domesticated and wild birds, with wild water birds (seagulls, dabbling ducks, and shorebirds) serving as the natural reservoir hosts. On the other hand, the virus has been occasionally reported in various mammals such as swine, equine, humans, felids, ferrets, dogs, civets, whales, mink, and seals (Sandrock and Kelly 2007). Except for the newly reported subtypes from bats, all other subtypes of AI virus (H1-H16 and N1-N9) are found to circulate among waterfowls and include the vast majority of the LPAI viruses (Fouchier and Munster 2009; Tong et al. 2013; Horman et al. 2018). HPAI viruses are not usually seen in wild birds (Stallknecht and Brown 2007). However, surveillance programs implemented for avian influenza are more focused on some avian species (especially in mallards) than others, and therefore, may not

Table 5.1 Past influenza pandemics (Data as per CDC)

Pandemics	Year	Strain	Estimated number of deaths	Origin of virus
Spanish flu	1918–1919	H1N1	50 million	Probably originated from the United States and then spread to Europe
Asian flu	1957–1958	H2N2	1.1 million	Southern China
Hong Kong flu	1968–1969	H3N2	1 million	Hong Kong
Swine flu	2009–2010	H1N1	151,700–575,400	Mexico

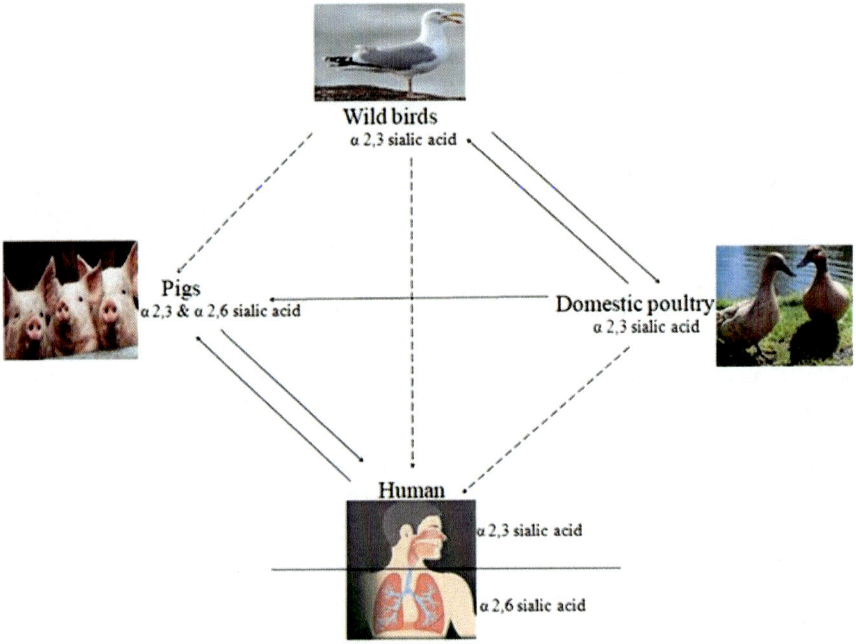

Fig. 5.1 Tissue tropism and receptor specificities

reveal the actual scenario of HPAI and LPAI strains circulating in various other species of wild birds (Olsen et al. 2006).

Influenza viruses are typically restricted to particular hosts, with receptor specificity and tissue tropism being the important restriction factors (Eriksson et al. 2018). The receptor tropism is determined by HA, which can bind to either α2,3-linked sialic acid (avian-derived HAs), or α2,6-linked sialic acid (human-derived HAs), or both. Sometimes, a change in tropism occurs, with swine being a significant "mixing vessel" because of their respiratory epithelium containing receptors with both α-2,3-linked and α-2,6-linked sialic acid moieties. The disease manifestation in pigs may range from acute respiratory disease to inapparent infection (Eriksson et al. 2018). This also facilitates generation of new influenza virus subtype/s due to genetic reassortment, which may result in emergence of new strains with pandemic potential. In human beings, α2,3-linked sialic acid is expressed at higher levels on nonciliated bronchiolar cells and alveolar type II cells of the lower respiratory tract, while α2,6-linked sialic acid is found at higher levels in the ciliated epithelium of the upper airways (Shinya et al. 2006). AI viruses generally attach to type II pneumocytes, nonciliated epithelial cells in the terminal bronchioles and alveoli, and alveolar macrophages of lower respiratory tract (i.e., 2,3-linked sialic acid), causing severe disease in humans (Shinya et al. 2006; Nicholls et al. 2007; Short et al. 2015). Tissue tropism and receptor specificities are illustrated in Fig. 5.1.

Infected pigs, housed in close vicinity to human settlement areas, can also transmit the virus to humans. Pigs usually do not contract H5N1 and H7N9 (highly pathogenic) strains, but there are in vivo and in vitro studies suggesting that H7N9 can replicate in the respiratory tract, cause pathological lesions, and get transmitted among pigs at low levels, emphasizing that pigs could turn out as an intermediate host for the H7N9 strain. Reports suggest that as the number of H7N2 infection cases increases, the possibility of H7N9 infection in pigs and their transmission in mammals may gain stability. LPAI subtypes H1 and H3 are endemic in pigs and infection with H4 and H9 is sporadic. H5N1 strains have been observed from pigs in a few countries (Jones et al. 2013; Xu et al. 2014; Yum et al. 2014; Balzli et al. 2016; Horman et al. 2018). Humans are infected with only a few subtypes of AI but with highly pathogenic magnitudes. The first outbreak of avian-derived H5N1 infection in humans was described in Hong Kong in the year 1997 with 18 confirmed cases constituting 6 deaths (Lai et al. 2016).

Later, a sporadic outbreak of H5N1 occurred in humans with nearly 900 cases which resulted in 450 deaths (Lai et al. 2016; Horman et al. 2018). H7N9 is an LPAI strain in chickens but is of specific concern in humans as they can cause a mortality rate of about 40% which is similar to that of H5N1 infections. Humans acquire H7N9 infection by direct contact with infected host species producing severe disease outcomes than any other virus in H7 family, and they get transmitted more efficiently than any other virus in the family (Imai et al. 2017; Farooqui et al. 2016; Horman et al. 2018). The first occurrence of avian-derived H7N9 infection in humans was described in China's Yangtze River Delta in March 2013, which by the year January 2018 caused 1566 cases throughout China with 613 deaths (Horman et al. 2018). H7N9 LPAI virus acquired some of the genes from H9N2 viruses, diversified ever since, and followed by emergence of regional lineages (Lam et al. 2013; Liu et al. 2014a, b).

5.2.3 *Transmission*

Avian influenza virus can be transmitted by feco-oral and respiratory routes, and also via water contaminated with feces of infected water birds like ducks, gulls, geese, and swans. In wild aquatic birds, feco-oral route appears to be a major route of transmission, as their feces contain large number of viruses (Sturm-Ramirez et al. 2004; Fouchier and Munster 2009; Hofle et al. 2012). Once an AI virus has entered a poultry farm, it can spread by both the feco-oral and aerosol routes, due to the close proximity. Flies can act as mechanical vectors, and fomites can also be significant in transmission (Nielsen et al. 2011; Scott et al. 2017). Birds with HPAI die shortly after infection (shed virus for ~2 days), while birds with LPAI may continue to shed the virus for weeks (both respiratory and cloacal shedding) (Villanueva-Cabezas et al. 2017).

During a H5N1 outbreak among domestic chicken in Hong Kong in the year 1997, it was hypothesized that the source of the virus was water contaminated with feces of migratory birds. Humans may have also acquired the H5N1 infection from birds during this outbreak, i.e., poultry cullers—9 out of 29 persons, 10% of workers in poultry markets showed seropositivity and most likely acquired the infection by a feco-oral route as evidenced in WHO reports (Thomas and Noppenberger 2007). Humans acquire infection by direct interaction with infected birds or their carcasses or exposure to secretions from infected birds, consumption of raw or parboiled poultry/wild birds or meat products, and during processing of poultry products (Malik 2009). Thomas and Noppenberger (2007) have suggested that the virus can also be transmitted from birds to humans by feco-oral route. Human-human transmission, i.e., among family members and healthcare workers, has been reported due to negligence of hygienic measures. HPAI H5N1 human infections have occurred following direct contact with dead or infected birds (Areechokchai et al. 2006; Dinh et al. 2006). HPAI H7N7 transmission to poultry workers in the Netherlands and H7N9 human infections in Hong Kong (Leung et al. 2017) and China (Ai et al. 2013; Liu et al. 2014a, b) are some other reports that are available. HPAI H5N1 are capable of human-human transmission, albeit rarely. Human-to-human transmission of H5N1 was reported in Thailand with the transmission of infection from an 11-year-old girl to her mother and aunt who developed febrile symptoms (Gottlieb 2005). Virus transmission from the environment to humans is limited but has occurred (Beigel et al. 2005; Zhou et al. 2018; Sandrock and Kelly 2007; Thomas and Noppenberger 2007). The persistence of influenza A virus in various environments depends on minor variations in humidity, temperature, salinity, pH, solar radiations, and air pollution. Seasonality is largely decided by humidity and temperature, with humid-rainy conditions favoring outbreaks in low latitudes, as recognized in tropical and subtropical regions, whereas cool-dry conditions enhance the survival and transmissibility of the virus in temperate conditions in high latitudes. In mid-latitudes, seasonal outbreaks result from changing humid-rainy and cool-dry conditions (Sooryanarain and Elankumaran 2015). Other important factors involved in the spread of AI across geographical regions are migration of wild birds, trade and movement of live birds and poultry products, and movement of human, animal, and fomites (van der Kolk 2019; Hautefeuille et al. 2020).

5.3 Role of Birds

The natural reservoirs of avian influenza virus are wild water birds mainly of the orders *Anseriformes* (geese, ducks, swans) and *Charadriiformes* (gulls, terns, shorebirds, and auks) (Pantin-Jackwood et al. 2016). As the virus is host-adapted, after entering into the host via feco-oral route, the virus starts replicating in the gastrointestinal and respiratory epithelium, inducing disease in some instances but mostly causing asymptomatic infections with shedding of virus in feces, resulting for high prevalence of virus among the wild aquatic birds (Short et al. 2015; Sandrock and Kelly 2007).

AI viruses are transmitted periodically from wild aquatic birds to domestic poultry which may produce subclinical infections due to lowly pathogenic strain (LPAIV); or due to accumulation of specific mutations, the LPAIV strain may transform into highly pathogenic strain (HPAIV) and produce severe systemic infections leading to inflammation and necrosis of skin, brain, and viscera of gallinaceous poultry. Although HPAI are not common in wild birds, there have been instances of HPAI infection in wild birds. It is suspected that wild birds may be the victims of spill back from infected poultry (Hassan et al. 2017). The first report of HPAI in wild birds can be traced back to 1961, where HPAI H5N3 was associated with deaths in common terns (Becker 1966). Since 2002, captive and wild birds are found infected with HPAI Eurasian-African H5N1 viruses causing illness and death. The lesions of HPAI H5N1 virus infection outbreak that occurred in 2002 in Hong Kong among the captive birds were similar and even more severe than those previously reported in chickens especially with severe pathological lesions in respiratory tract along with widespread neurological lesions. In 1999 and 2000, in Italy, Muscovy ducks (*Cairina moschata*) are found naturally infected with HPAIV H7N1 leading to mortality. Domestic ducks are less frequently infected with HPAIV before 2002. HPAIV H5N1 were also isolated from individual wild birds belonging to Order *Charadriiformes* (shorebirds and gulls), Order: *Columbiformes* (doves and pigeons), Order: *Ciconiiformes* (Egrets, storks and herons), Order: *Falconiformes* (eagles, falcons, kites, kestrel, goshawks, buzzard), Order: Gruiformes (swamp hen, coots, moorhen), Order: *Pelecaniformes* (pelicans, cormorants), Order: *Passeriformes* (sparrows, finches, crows, magpies, mynahs, starlings), Order: *Strigiformes* (owls), Order: *Phoenicopteriformes* (flamingos), and Order: *Podicipediformes* (grebes) (Pantin-Jackwood and Swayne 2009; Horman et al. 2018).

5.4 Disease

5.4.1 Pathogenesis

Pathogenesis of AI infection is well documented with a dysregulation in the molecules of the immune system of the host which is highly significant in host susceptibility and disease outcome. HPAI infection causes severe disease pathogenesis mainly due to "cytokine storm" or hypercytokinemia producing an inflammatory immune response, leading to cellular infiltration and diminished leucocytes (lymphopenia) at the site of infection which in turn facilitate the systemic dissemination of virus and death of the infected host (To et al. 2001). HPAI infections contribute to severe disease pathology when compared to LPAI infections. In vitro studies prove that infection with H5N1 (HPAI) virus resulted in upregulation or increased expression of proinflammatory cytokines, predominantly IL6 and IL8, in chicken lungs when compared to infection with H2N3 (LPAI) signifying that IL6 and IL8 cytokines would be the key regulators involved in severe disease pathology of HPAI infection (Pantin-Jackwood and Swayne 2009; Kuchipudi et al. 2014).

The disease severity in humans due to HPAI H5N1 infection may be due to receptor tropism of HA protein in virus to α 2,6-linked sialic acid or α2,3-linked sialic acid as mentioned earlier. However, the disease pathogenesis, attachment, and transmissibility pattern in humans differ from that of H5N1 in H7N9 infection with the attachment of the virus to the epithelial cells of the upper and lower respiratory tract. Though the H7N9 virus attaches to the upper respiratory tract, it does not facilitate aerosol transmission among humans or in animal models like ferrets. But the disease severity in humans due to H7N9 infection is mainly attributed to the attachment of the virus to lower respiratory tract. Experiments in ferret model also showed that mutated HPAI H5N1 viruses attaches to ciliated epithelium of upper respiratory tract but is not transmitted by aerosol route (van Riel et al. 2013; Richard et al. 2013; Hu et al. 2014; Chutinimitkul et al. 2010; Herfst et al. 2012).

5.4.2 Clinical Signs/Clinical Profile

LPAI infection in chickens shows mild clinical signs with excess mucus secretions, congested trachea, respiratory inflammation, and watery droppings as the infection is localized in the mucosa of GI tract, allowing the virus to replicate and excrete into the environment 2–3 days after infection with minimal gross lesions in the host species. The clinical severity due to LPAI infections varies even among avian hosts with ducks showing milder clinical signs, chicken showing moderate to severe signs than waterfowls (Wang et al. 2014; Horman et al. 2018). In contrast, HPAI infection in poultry cause acute, severe disease outbreak with increased mortality rates and clinical signs ranging from mild to severe systemic outcomes. Clinical signs include ruffled feather, dehydration, depression, nasal discharge, respiratory distress followed by the systemic spread of virus leading to hemorrhage and edema in various organs with continuous shedding of a high level of virus (Horman et al. 2018; Suzuki et al. 2009). Whereas the clinical signs in ducks infected with H5N1 virus may be as severe as in chickens with the dissemination of the virus to the nervous system and hemorrhages in body extremities. Wild ducks, in spite of shedding high levels of virus, exhibit mild clinical signs following infection with H5N1 virus than domestic ducks and other gallinaceous birds suggesting their role as reservoir hosts (Horman et al. 2018). Jeong et al. (2009) reported that experimental infection of the H5N1 virus in ducks displayed mild clinical signs without any complications. Another study by Yamamoto et al. (2016) reported that domestic ducks developed corneal opacity and the hemorrhages were less severe unlike those found in chickens following H5N1 infection.

In humans, the clinical symptoms vary with each subtype of avian influenza and usually range from slight conjunctivitis to severe encephalitis and pneumonia with multiple organ failure. In the year 2003, during H7N7 outbreak in the Netherlands, 92% patients presented with clinical symptoms of conjunctivitis, whereas in 1997, during HPAI outbreak in Hong Kong, and recently in Southeast Asia, the predominant symptoms were pneumonia, acute respiratory distress syndrome (ARDS), multiple organ failure, and death within 9–10 days. Symptoms of pulmonary hemorrhage, nausea, emesis, and diarrhea further complicated these cases. Laboratory picture revealed lymphopenia and thrombocytopenia; chest X-ray revealed lobar consolidation, interstitial infiltration, and air bronchograms (Sandrock and Kelly 2007; Short et al. 2015; Horman et al. 2018).

5.4.3 Pathology/Lesions

5.4.3.1 LPAI Birds

The gross pathological lesions vary depending upon the virus strain, host species affected, invading secondary pathogens, and period leading to death. Rhinitis, sinusitis, congested, and edematous trachea with occasional hemorrhage, tracheal plugs occluding the airway are the most frequently observed gross pathological lesions. When accompanied by secondary bacterial infections, nasal discharge, airsacculitis, fibrinopurulent-bronchopneumonia, and peritonitis may be observed. In hens, egg yolk peritonitis, swollen oviduct with luminal exudates, regressed ovaries, swollen kidneys, visceral gout, and few eggs may be misshapen, thin-shelled, and may lack pigment. In turkeys, swollen infraorbital sinuses and mild enteritis are seen, and rarely, pancreas may turn pale and mottled with random hemorrhages (Pantin-Jackwood and Swayne 2009).

5.4.3.2 HPAI Birds

As HPAIV infections are systemic, hemorrhages, edema, and necrosis of cardiovascular and nervous system and the integument are observed with no gross lesions in per-acute cases. In acute disease, edema of wattles, comb, neck, shank, intermandibular and periorbital areas, and feet with subcutaneous hemorrhages especially in featherless areas and egg yolk peritonitis are observed. Cyanotic comb, wattles, and snood due to ischemic necrosis with petechial-to-ecchymotic hemorrhages and foci of necrosis may be seen. Some virus strain may produce hyperemia and edema of eyelids, conjunctiva, and trachea. Internally, necrotic foci in visceral organs, hemorrhages on serosal and mucosal surfaces, especially on the epicardium, coronary fat, proventricular, and ventricular mucosa, and pectoral muscles are common, but the hemorrhages are less frequently observed in inner surface of sternum, Meckel's diverticulum, cecal tonsils, and pancreas which may be necrotic with red to brown mottling. Hemorrhagic and necrotic Peyer's patches of the small intestine, hemorrhagic and edematous lungs, and edematous brain are the unique and distinctive lesions seen in the classic AI viruses and recent Eurasian-African H5N1 HPAIV lineage. With most of the HPAIV infections, white necrotic foci in heart, occasionally congested, edematous, hemorrhagic lung with interstitial pneumonia, necrotic kidney with urate deposits are the gross lesions seen. Atrophy of cloacal bursa and thymus with or without hemorrhage, normal-sized or enlarged, pale spleen with necrotic foci may be seen in young birds (Pantin-Jackwood and Swayne 2009).

5.4.3.3 Human

Postmortem findings are consistent in humans which are mainly due to overwhelming inflammatory response. Pathological lesions include disseminated intravascular coagulation, diffuse alveolar damage, acute tubular atrophy and necrosis, and multiorgan damage. ARDS, respiratory insufficiency, and acute lung injury are the main features that coincide with the receptor tropism of AI virus to α-2,3sialic acids present in the cuboidal cells of lower respiratory tract resulting in a cytokine storm. Though virus isolations from brain, lungs, spleen, and intestine have been reported earlier, viral replication has been restricted to lungs (To et al. 2001; Cheung et al. 2002; Sandrock and Kelly 2007). Histological lesions characterized in ARDS are necrotic alveolar epithelium with diffuse alveolar damage, intraluminal hemorrhage, edema, fibrin deposits, alveolar macrophages and neutrophils, interstitial lymphocyte and plasma cell infiltration, type II hyperplasia, and fibrosis (Short et al. 2015).

5.5 Disease Management

5.5.1 Diagnosis

Diagnosis is a significant challenge in AI due to antigenic and genetic variation of virus, wide host range, and lack of specific clinical signs. Conventional diagnostic methods such as virus isolation followed by identification and characterization is the gold standard, but they are time-consuming making delay in availability of rapid results, therefore hindering the execution of control measures mainly stamping out policies. Though the recommended diagnostic method in manuals of the EU (CEC 2006a, b) and OIE is conventional technique, molecular techniques like real-time RT-PCR (RRT-PCR) and reverse transcriptase-polymerase chain reaction (RT-PCR) are now being employed in many laboratories as they provide rapid results and are sensitive. Handling of H5N1 requires biosafety level 3 (BSL3) facilities. Laboratories lacking BSL3 are recommended to detect the virulent HPAI strains using RT-qPCR for rapid diagnosis and characterization of the virus. The diagnostic techniques reported for avian influenza virus detection is presented in Table 5.2.

Any AI virus or influenza A virus (H5/H7 subtype) infection in poultry is considered notifiable when the intravenous pathogenicity index (IVPI) is greater than 1.2 or the mortality rate is at least 75%. Thus, notifiable avian influenza viruses

Table 5.2 Diagnostic methods involved in the detection and typing of avian influenza virus

Diagnostic methods	Procedure
Virus isolation	Chick embryo inoculation and HA test Samples from live birds include tracheal and cloacal swabs, feces, and pooled organ samples from dead birds Samples are inoculated into 9–11-day-old embryonated chicken egg via allantoic route and incubated at 37 °C for 4–7 days. After incubation, the infectious allantoic fluid is collected and subjected for HA test
Virus identification	Hemagglutination inhibition (HI) tests, immunodiffusion tests Identification of subtype using polyclonal chicken antisera against different influenza viruses
Assessment of pathogenicity	(i) Intravenous pathogenicity index (IVPI) in 6-week-old SPF chicken: It is the mean score of each bird over a 10 days observation. 0.1 mL of diluted infective allantoic fluid with HA titer >1/16 is injected intravenously into 10 nos. of 6-week-old SPF chickens and observed for up to 10 days. Each bird will be scored on examination: Score 0—Normal, score 1—Sick, score 2—Severely sick, score 3—Dead. Signs of sickness include depression, respiratory distress, cyanosis of skin and wattles, edematous head, and nervous signs. H10 nephropathogenic subtype of HPAI virus gives IVPI score of more than 1.2 with a high mortality rate (ii) Sequencing of the cleavage site (HA0) of HPAI H5 and H7 viruses to determine the presence or absence of multiple basic amino acids (Swayne and Alexander 1994)
Subtyping by reverse transcription PCR (RT-PCR)	RT-PCR is a powerful diagnostic method for rapid detection and subtyping of AI virus. Viral RNA is reverse transcribed into cDNA and amplified using Uni13 and Uni12 (conserved nucleotides at 5′ and 3′ terminus, respectively) (Desselberger et al. 1980) Similarly, Lee et al. (2001) and Hoffmann et al. (2001) have developed 157 primers for detecting HA subtypes 1–15 and 158 universal primers for full genome amplification of influenza A viruses, respectively To determine the pathotype of the virus in chickens, HA0 cleavage site was detected and sequenced using a universal primer developed by Gall et al. (2008) Tsukamoto et al. (2008), Alvarez et al. (2008), Qiu et al. (2009), and Fereidouni et al. (2009) have developed primer sets to amplify all HA and NA subtypes
Real-time RT-PCR (quantitative RT-PCR)	TaqMan probe-based or SYBR green-based real-time RT-PCR is used for detection of influenza A virus and subtyping Spackman et al. (2002) developed a modified real-time single-step RT-PCR for detection and identification of AIV subtype Real-time RT-PCR facilitated rapid detection and surveillance of influenza virus during an outbreak in Great Britain (Slomka et al. 2010)

(continued)

Table 5.2 (continued)

Diagnostic methods	Procedure
	According to genetic variations, primers were also updated for detection of influenza A viruses (van Elden et al. 2001; Yacoub et al. 2009; Xu et al. 2015; Takayama et al. 2015)
Loop-mediated isothermal amplification (LAMP)	Amplification of nucleic acid under isothermal conditions (Notomi et al. 2000). Reverse transcription LAMP (RT-LAMP) is a modified LAMP assay which is highly sensitive, specific, simple, and rapid assay for rapid detection and surveillance of RNA viruses including H5N1 HPAIV and also other avian influenza viruses. This technique involves the addition of reverse transcriptase enzyme besides DNA polymerase, stripping primers, and loop primers for reverse transcription (Imai et al. 2007; Chen et al. 2008; Postel et al. 2010; Yoshida et al. 2011; Luo et al. 2015)
Nucleic acid sequence-based amplification (NASBA)	NASBA, another rapid, sensitive isothermal amplification assay, amplifies the viral RNAs based on electrochemiluminescent detection to reduce cross-contamination. The disadvantage of the assay is highly costly kits and problems with the preparation of the master mix. NASBA assays require RNaseH, RTase, primer with T7 promoter, a reverse primer, and T7 RNA polymerase, and this assay had been successfully developed for detection H5 and H7 subtypes of AIV (Collins et al. 2002; Collins et al. 2003; Moore et al. 2004)
Immunochromatography	Immunochromatography assay is a rapid, simple, pen-side, antigen detection assay which uses monoclonal antibodies against HA or NP protein of influenza viruses (Bai et al. 2005, 2006; Tsuda et al. 2007; Manzoor et al. 2008; Wada et al. 2011; Sakurai et al. 2015) The detection limit of influenza A virus was found highly variable when poultry samples were employed in licensed commercial kits meant for human use (Woolcock and Cardona 2005). The disadvantage is that it is poorly sensitive and the kits are costly. This emphasizes that antigen detection in the field (screening assay) along with confirmation using laboratory-based amplification assays would be sensitive (Wada et al. 2011; Tsunetsugu-Yokota et al. 2014; Sakurai et al. 2015)
Antigen-capture ELISA	Antigen-capture ELISA, used for screening a large number of clinical samples, uses monoclonal antibody against the NP protein of the influenza A virus. Ag-ELISA was recently developed for detection of H5 and H7 viruses, and it was mainly developed to enhance the sensitivity of AGID test (Shafer et al. 1998; Zhou et al. 1998; He et al. 2007; Velumani et al. 2008; Ho et al. 2009; Lin et al. 2015)
Next-generation sequencing (NGS)	Next-generation sequencing (NGS) is a high-throughput sequencing technique which provides complete information of the viral genome including mutations (Barzon et al. 2011; Rutvisuttinunt et al. 2013; van den Hoecke et al. 2015; Zhao et al. 2015). Though it is an advantageous technique, it is

(continued)

Table 5.2 (continued)

Diagnostic methods	Procedure
	costlier than Sanger sequencing Tong et al. (2012, 2013) sequenced the whole genome of H17N10 and H18N11 viruses from bat samples using Illumina GAIIx platform

could be categorized into highly pathogenic notifiable avian influenza (HPNAI) and low pathogenicity notifiable avian influenza (LPNAI). Likewise, the new European Union Directive (CEC 2006a) extended the control measures to H5 and H7 subtype of viruses that satisfied neither vivo virulence nor molecular criteria. Now, the terms "notifiable avian influenza (NAI)," "high pathogenicity notifiable avian influenza," and "low pathogenicity notifiable avian influenza" have been dropped by the World Organisation for Animal Health (OIE 2021), in Terrestrial Animal Health Code 2018. Presently, OIE defines "avian influenza" as "an infection of poultry caused by any influenza A virus with high pathogenicity (HPAI), and by H5 and H7 subtypes with low pathogenicity (H5/H7 LPAI)."

A high pathogenicity influenza A virus (HPAI) is: "(i) any influenza A virus that is lethal for six, seven or eight of eight 4- to 8-week-old susceptible chickens within 10 days following intravenous inoculation with 0.2 mL of a 1/10 dilution of a bacteria-free, infective allantoic fluid or (ii) any influenza A virus that has an intravenous pathogenicity index (IVPI) greater than 1.2." For all H5 and H7 viruses of low pathogenicity in chickens, "the amino acid sequence of the connecting peptide of the haemagglutinin must be determined. If the sequence is similar to that observed for other HPAI isolates, the isolate being tested will be considered to be HPAI" (OIE 2021).

OIE uses the following classification system to identify influenza A viruses for which control measures and reporting is required: (1) All isolates that meet the above criteria are HPAI and notifiable. (2) Not highly pathogenic H5 and H7 isolates and do not have similar amino acid sequence as that of HPAI are notifiable and are designated as H5/H7 LPAI. (3) Non-H5/H7 influenza A (i.e., H1–4, H6, and H8–16) are not avian influenza and are not notifiable. (4) Influenza A viruses of high pathogenicity in wild and other birds other than poultry are notifiable (OIE 2021).

5.6 Prevention and Control

In spite of continuous surveillance, prevention, and control measures in domestic poultry, concern remains on the status of AI in wild and migratory birds that play a crucial role in dispersal of disease across the geographical areas. Some information available on host range and susceptibility are based on the samples collected from dead or sick birds. However, this information resulted from biased sampling does not provide insight into finding of the role played by the wild and migratory birds in the spread of AI virus. Recently, some surveillance programs explicitly designed to collect samples from apparently healthy wild birds have been carried out by a several nongovernmental organizations and international or national agencies. Active surveillance should be undertaken in healthy free-ranging wild birds especially in species known to have infected or epidemiological reservoirs of H5N1 and LPAI viruses, species that shares habitat with backyard and large poultry farms, species that are known to migrate or aggregate seasonally at roosting, breeding, and wintering sites (FAO 2007).

Prevention of AIV among the domestic poultry population is of major concern as they are responsible for influenza infection in humans due to handling or direct contact or interaction with infected poultry which facilitates the virus transmission to humans, reassorts events, and circulates among humans. Execution of biosecurity measures at farm level helps in prevention of AI virus because transmission of the virus may also occur by movement of contaminated vehicles, equipment, fomites, infected birds, and other infectious organic matter (Capua and Marangon 2006).

Control of AI in poultry birds is must for reasons such as to prevent the economic loss to the poultry industry, assured quality and safe food to consumers, and to prevent the pandemics. The basic control strategies for avian influenza include (i) biosecurity measures like the restricted movement of birds, proper cleaning, and disinfection, (ii) educating the farmers, (iii) surveillance and diagnosis, (iv) depopulating infected stock, and (v) vaccination.

Biosecurity which encompasses biocontainment and bio-exclusion represents to be an important and excellent means of prevention of AI. Strict biosecurity measures help in protecting the birds against AI and do not allow the virus to penetrate and perpetuate the virus in farm premises that are practically applied only in large-scale poultry industries which is not always possible especially in rural and semi-urban farms where the basic biosecurity measure like rearing of two or three species of birds together is not avoided that serve to be the constant source of virus, their spread, which in turn leads to endemic state (Capua and Marangon 2007).

Vaccination in combination with the control strategies like biosecurity and monitoring the existing plus evolution of new virus would support the eradication of the disease. In Mexico and Pakistan, during 1995, vaccination practices have decreased the susceptibility of the host to the disease (Swayne 2012). Implementation of vaccination against AI also reduced disease transmission, reduced virus shedding, and increased host resistance against field virus stating the contribution of vaccines and vaccination along with control measures in critical components successfully controlling of avian influenza (Capua et al. 2004; Van Der Goot 2005; Capua and Marangon 2006; Swayne 2012).

Emergency vaccination in combination with control measures mentioned above for AI helps in combating the spread and transmission and prevents culling of the susceptible healthy flock at risk for AI infection. The efficacy of this policy is based on flock density, biosecurity level, and virus strain in that area. In Italy, a work on DIVA strategy of emergency vaccination was carried out and was approved for use in the eradication of LPAI epidemics (H7N3 and H7N1) without preventive culling of healthy animals (Capua et al. 2004). In the year 2000, a heterologous vaccination against H7 virus was used as a "natural marker vaccine" in field level. Later, in Hong Kong, DIVA vaccination strategy was used against HPAI H5N1 outbreak and found successful in prevention and further spread of HPAI infection to the nearby farms (Ellis et al. 2004).

Prophylactic vaccination against H5 and H7 subtypes of viruses is completely a ground-breaking concept underlining the principle of generating protective immunity in the targeted population, increase resistance among susceptible birds when the virus is introduced into the flock, and to reduce shedding of virus, provided that same level of biosecurity measures is maintained (Capua and Marangon 2007). At the same time, it should also complement in maximizing the biosecurity measures when there is a high risk of exposure to virus exists. Eventually, prophylactic vaccination prevents the index case or reduces the secondary outbreak, thereby reducing the huge economic losses in the poultry dense areas.

References

Ai Y, Huang K, Xu D, Ren X, Qi HJ et al (2013) Case-control study of risk factors for human infection with influenza A(H7N9) virus in Jiangsu Province, China, 2013. Euro Surveill 18: 20510

Alexander DJ, Brown IH (2009) History of highly pathogenic avian influenza. Revue Scientifiqueet Technique 28:19–38. https://doi.org/10.20506/rst.issue.28.1.40

Alvarez AC, Brunck ME, Boyd V, Lai R, Virtue E, Chen W, Bletchly C, Heine HG, Barnard R (2008) A broad spectrum, one-step reverse-transcription PCR amplification of the neuraminidase gene from multiple subtypes of influenza A virus. Virol J 5:77

Areechokchai D, Jiraphongsa C, Laosiritaworn Y, Hanshaoworakul W, O'Reilly M (2006) Investigation of avian influenza (H5N1) outbreak in humans—Thailand, 2004. MMWR Morb Mortal Wkly Rep 55(Suppl 1):3–6

Bai GR, Sakoda Y, Mweene AS, Kishida N, Yamada T, Minakawa H, Kida H (2005) Evaluation of the ESPLINE INFLUENZA A and B-N Kit for the diagnosis of avian and swine influenza. Microbiol Immunol 49:1063–1067

Bai GR, Sakoda Y, Mweene AS, Fujii N, Minakawa H, Kida H (2006) Improvement of a rapid diagnosis kit to detect either influenza A or B virus infections. J Vet Med Sci 68:35–40

Balzli C, Lager K, Vincent A, Gauger P, Brockmeier S, Miller L et al (2016) Susceptibility of swine to H5 and H7 low pathogenic avian influenza viruses. Influenza Other Respi Viruses 10(4):346–352. https://doi.org/10.1111/irv.12386

Barzon L, Lavezzo E, Militello V, Toppo S, Palu G (2011) Applications of next-generation sequencing technologies to diagnostic virology. Int J Mol Sci 12:7861–7884

Becker WB (1966) The isolation and classification of Tern virus: influenza A-Tern South Africa–1961. J Hyg (Lond) 64(3):309–320. https://doi.org/10.1017/s0022172400040596

Beigel JH, Farrar J, Phil D et al (2005) Current concepts: avian influenza A (H5N1) infection in humans. N Engl J Med 353:1374–1385

Bonfanti L, Monne I, Tamba M et al (2014) Highly pathogenic H7N7 avian influenza in Italy. Vet Rec 174:382

Capua I, Alexander DJ (2007) Avian influenza infection in birds—a moving target. Influenza Other Respir Viruses 1:11–18. https://doi.org/10.1111/j.1750-2659.2006.00004.x

Capua I, Marangon S (2006) Control of avian influenza in poultry. Emerg Infect Dis 12:1319–1324

Capua I, Marangon S (2007) Control and prevention of avian influenza in an evolving scenario. Vaccine 25(30):5645–5652

Capua I, Cattoli G, Marangon S (2004) DIVA—a vaccination strategy enabling the detection of field exposure to avian influenza. Dev Biol (Basel) 119:229–233

CEC (2006a) Council Directive 2005/94/EC of 20 December 2005 on Community measures for the control of avian influenza and repealing Directive 92/40/EEC. Off J Eur Commission L10:16–65

CEC (2006b) Commission Decision 2006/437/EC approving a diagnostic manual for avian influenza as provided for in Council Directive 2005/94/EC. Off J Eur Commission L237:1–27

Chen HT, Zhang J, Sun DH, Ma LN, Liu XT, Cai XP, Liu YS (2008) Development of reverse transcription loop-mediated isothermal amplification for rapid detection of H9 avian influenza virus. J Virol Methods 151:200–203

Cheung TK, Poon LL (2007) Biology of influenza a virus. Ann N Y Acad Sci 1102:1–25

Cheung CY, Poon LL, Lau AS, Luk W, Lau YL, Shortridge KF, Gordon S, Guan Y, Peiris JS (2002) Induction of proinflammatory cytokines in human macrophages by influenza A (H5N1) viruses: a mechanism for the unusual severity of human disease? Lancet 360:1831–1837

Chutinimitkul S, Herfst S, Steel J, Lowen AC, Ye J, van Riel D et al (2010) Virulence associated substitution D222G in the hemagglutinin of 2009 pandemic influenza A(H1N1) virus affects receptor binding. J Virol 84(22):11802–11813

Collins RA, Ko LS, So KL, Ellis T, Lau LT, Yu AC (2002) Detection of highly pathogenic and low pathogenic avian influenza subtype H5 (Eurasian lineage) using NASBA. J Virol Methods 103: 213–225

Collins RA, Ko LS, Fung KY, Chan KY, Xing J, Lau LT, Yu AC (2003) Rapid and sensitive detection of avian influenza virus subtype H7 using NASBA. Biochem Biophys Res Commun 300:507–515

DEFRA (2007) Outbreak of highly pathogenic H5N1 avian influenza in Suffolk in January 2007. A report of the epidemiological findings by the National Emergency Epidemiology Group, DEFRA 5 April 2007. http://www.defra.gov.uk/foodfarm/farmanimal/diseases/atoz/ai/documents/epid_findings070405.pdf

DEFRA (2008) Highly pathogenic avian influenza, H7N7, Oxfordshire, June 2008. Situation at 12.30pm Wednesday 2nd July. http://www.defra.gov.uk/foodfarm/farmanimal/diseases/atoz/ai/documents/epireport-080711.pdf

Desselberger U, Racaniello VR, Zazra JJ, Palese P (1980) The 3′ and 5′ terminal sequences of influenza A, B, and C virus RNA segments are highly conserved and show partial inverted complementarity. Gene 8:315–328

Dhingra MS, Artois J, Dellicour S et al (2018) Geographical and historical patterns in the emergences of novel highly pathogenic avian influenza (HPAI) H5 and H7 viruses in poultry. Front Vet Sci 5:84. https://doi.org/10.3389/fvets.2018.00084

Dinh PN, Long HT, Tien NT, Hien NT, le Mai TQ, le Phong H et al (2006) Risk factors for human infection with avian influenza A H5N1, Vietnam, 2004. Emerg Infect Dis 12:1841–1847

Ellis TM, Leung CY, Chow MK, Bissett LA, Wong W, Guan Y et al (2004) Vaccination of chickens against H5N1 avian influenza in the face of an outbreak interrupts virus transmission. Avian Pathol 33(4):405–412

El-Shesheny R, Franks J, Marathe BM et al (2018) Genetic characterization and pathogenic potential of H10 avian influenza viruses isolated from live poultry markets in Bangladesh. Sci Rep 8(1):10693. Published 2018 Jul 16. https://doi.org/10.1038/s41598-018-29079-1

Eriksson P, Lindskog C, Engholm E et al (2018) Characterization of avian influenza virus attachment patterns to human and pig tissues. Sci Rep 8(1):12215. Published 2018 Aug 15. https://doi.org/10.1038/s41598-018-29578-1

FAO (2007) Wild birds and avian influenza: an introduction to applied field research and disease sampling techniques. In: Whitworth D, Newman SH, Mundkur T, Harris P (eds) FAO animal production and health manual, No. 5. Rome. www.fao.org/avianflu

Farooqui A, Liu W, Zeng T, Liu Y, Zhang L, Khan A et al (2016) Probable hospital cluster of H7N9 influenza infection. N Engl J Med 374(6):596–598. https://doi.org/10.1056/NEJMc1505359

Fereidouni SR, Starick E, Grund C, Globig A, Mettenleiter TC, Beer M, Harder T (2009) Rapid molecular subtyping by reverse transcription polymerase chain reaction of the neuraminidase gene of avian influenza A viruses. Vet Microbiol 135:253–260

Fouchier RA, Munster VJ (2009) Epidemiology of low pathogenic avian influenza viruses in wild birds. Rev Sci Tech 28(1):49–58

Gall A, Hoffmann B, Harder T, Grund C, Beer M (2008) Universal primer set for 320 amplification and sequencing of HA0 cleavage sites of all influenza A viruses. 321. J Clin Microbiol 46:2561–2567

Gonzales JL, Elbers ARW, Bouma A, Koch G, de Wit JJ et al (2010) Low-pathogenic notifiable avian influenza serosurveillance and the risk of infection in poultry—a critical review of the European Union active surveillance programme (2005–2007). Influenza Other Respir Viruses 4(2):91–99

Gottlieb S (2005) Research confirms human to human transmission of avian flu. BMJ 330(7485):211

Hassan MM, Hoque MA, Debnath NC, Yamage M, Klaassen M (2017) Are poultry or wild birds the main reservoirs for avian influenza in Bangladesh? EcoHealth 14(3):490–500. https://doi.org/10.1007/s10393-017-1257-6

Hautefeuille C, Dauphin G, Peyre M (2020) Knowledge and remaining gaps on the role of animal and human movements in the poultry production and trade networks in the global spread of avian influenza viruses—a scoping review. PLoS One 15(3):e0230567. Published 2020 Mar 20. https://doi.org/10.1371/journal.pone.0230567

He Q, Velumani S, Du Q, Lim CW, Ng FK, Donis R, Kwang J (2007) Detection of H5 avian influenza viruses by antigen-capture enzyme-linked immunosorbent assay using H5-specific monoclonal antibody. Clin Vaccine Immunol 14:617–623

Herfst S, Schrauwen EJ, Linster M, Chutinimitkul S, de Wit E, Munster VJ et al (2012) Airborne transmission of influenza A/H5N1 virus between ferrets. Science 336(6088):1534–1541

Ho HT, Qian HL, He F, Meng T, Szyporta M, Prabhu N, Prabakaran M, Chan KP, Kwang J (2009) Rapid detection of H5N1 subtype influenza viruses by antigen capture enzyme-linked immunosorbent assay using H5- and N1-specific monoclonal antibodies. Clin Vaccine Immunol 16: 726–732

Hoffmann E, Stech J, Guan Y, Webster RG, Perez DR (2001) Universal primer set for the full-length amplification of all influenza A viruses. Arch Virol 146:2275–2289. https://doi.org/10.1007/s007050170002

Hofle U, van de Bildt MW, Leijten LM, van Amerongen G, Verhagen JH, Fouchier RA, Osterhaus AD, Kuiken T (2012) Tissue tropism and pathology of natural influenza virus infection in black-headed gulls (Chroicocephalus ridibundus). Avian Pathol 41(6):547–553

Horman WSJ, Nguyen THO, Kedzierska K, Bean AGD, Layton DS (2018) The drivers of pathology in zoonotic avian influenza: the interplay between host and pathogen. Front Immunol 9:1812. https://doi.org/10.3389/fimmu.2018.01812

Hu J, Zhu Y, Zhao B, Li J, Liu L, Gu K et al (2014) Limited human-to-human transmission of avian influenza A(H7N9) virus, Shanghai, China, March to April 2013. Euro Surveill 19(25)

Imai M, Ninomiya A, Minekawa H, Notomi T, Ishizaki T, Van Tu P, Tien NT, Tashiro M, Odagiri T (2007) Rapid diagnosis of H5N1 avian influenza virus infection by newly developed influenza H5 hemagglutinin gene-specific loop-mediated isothermal amplification method. J Virol Methods 141:173–180

Imai M, Watanabe T, Kiso M, Nakajima N, Yamayoshi S, Iwatsuki-Horimoto K et al (2017) A highly pathogenic avian H7N9 influenza virus isolated from a human is lethal in some ferrets infected via respiratory droplets. Cell Host Microbe 22(5):615–26.e8. https://doi.org/10.1016/j.chom.2017.09.008

Jeong OM, Kim MC, Kim MJ, Kang HM, Kim HR, Kim YJ et al (2009) Experimental infection of chickens, ducks and quails with highly pathogenic H5N1 avian influenza virus. J Vet Sci 10(1):53–60. https://doi.org/10.4142/jvs.2009.10.1.53

Jones JC, Baranovich T, Zaraket H, Guan Y, Shu Y, Webby RJ et al (2013) Human H7N9 influenza A viruses replicate in swine respiratory tissue explants. J Virol 87(22):12496–12498. https://doi.org/10.1128/JVI.02499-13

Kuchipudi SV, Tellabati M, Sebastian S, Londt BZ, Jansen C, Vervelde L et al (2014) Highly pathogenic avian influenza virus infection in chickens but not ducks is associated with elevated host immune and pro-inflammatory responses. Vet Res 45:118. https://doi.org/10.1186/s13567-014-0118-3

Lai S, Qin Y, Cowling BJ, Ren X, Wardrop NA, Gilbert M et al (2016) Global epidemiology of avian influenza A H5N1 virus infection in humans, 1997–2015: a systematic review of individual case data. Lancet Infect Dis 16:e108–e118. https://doi.org/10.1016/S1473-3099(16)00153-5

Lam TT, Wang J, Shen Y, Zhou B, Duan L, Cheung CL et al (2013) The genesis and source of the H7N9 influenza viruses causing human infections in China. Nature 502(7470):241–244

Lee M-S, Chang P-C, Shien J-H, Cheng M-C, Shieh HK (2001) Identification and subtyping of avian influenza viruses by reverse transcription-PCR. J Virol Methods 97:13–22. https://doi.org/10.1016/S0166-0934(01)00301-9

Leung YH et al (2017) Epidemiology of human influenza A(H7N9) infection in Hong Kong. J Microbiol Immunol Infect 50:183–188. https://doi.org/10.1016/j.jmii.2015.06.004

Lin J, Wang R, Jiao P, Li Y, Li Y, Liao M, Yu Y, Wang M (2015) An 364 impedance immunosensor based on low-cost microelectrodes and specific 365 monoclonal antibodies for rapid detection of avian influenza virus H5N1 in 366 chicken swabs. Biosens Bioelectron 67: 546–552

Liu B, Havers F, Chen E, Yuan Z, Yuan H, Ou J et al (2014a) Risk factors for influenza A(H7N9) disease—China, 2013. Clin Infect Dis 59:787–794

Liu T, Bi Z, Wang X, Li Z, Ding S, Bi Z et al (2014b) One family cluster of avian influenza A (H7N9) virus infection in Shandong, China. BMC Infect Dis 14:98

Luo S, Xie Z, Xie L, Liu J, Xie Z, Deng X, Huang L, Huang J, Zeng T, Khan MI (2015) Reverse-transcription, loop-mediated isothermal amplification assay for the sensitive and rapid detection of H10 subtype avian influenza viruses. Virol J 12:145

Malik Peiris JS (2009) Avian influenza viruses in humans. Rev Sci Tech Off Int Epiz 28(1):161–174

Manzoor R, Sakoda Y, Sakabe S, Mochizuki T, Namba Y, Tsuda Y, Kida H (2008) Development of a pen-site test kit for the rapid diagnosis of H7 highly pathogenic avian influenza. J Vet Med Sci 70:557–562

Moore C, Hibbitts S, Owen N, Corden SA, Harrison G, Fox J, Gelder C, Westmoreland D (2004) Development and evaluation of a real-time nucleic acid sequence-based amplification assay for rapid detection of influenza A. J Med Virol 74:619–628

Naguib MM, Verhagen JH, Mostafa A, Wille M, Li R, Graaf A, Jarhult JD, Ellstrom P, Zohari S, Lundkvist A, Olsen B (2019) Global patterns of avian influenza A (H7): virus evolution and zoonotic threats. FEMS Microbiol Rev 19:1–14

Nicholls J, Chan M, Chan W, Wong H, Cheung C, Kwong D et al (2007) Tropism of avian influenza A (H5N1) in the upper and lower respiratory tract. Nat Med 13(2):147–149

Nielsen AA, Skovgard H, Stockmarr A, Handberg KJ, Jorgensen PH (2011) Persistence of low-pathogenic avian influenza H5N7 and H7N1 subtypes in house flies (Diptera: Muscidae). J Med Entomol 48(3):608–614

Notomi T, Okayama H, Masubuchi H, Yonekawa T, Watanabe K, Amino N, Hase T (2000) Loop-mediated isothermal amplification of DNA. Nucleic Acids Res 28:E63

OIE (2021). https://www.oie.int/fileadmin/Home/eng/Health_standards/tahm/3.03.04_AI.pdf

Olsen B, Munster VJ, Wallensten A, Waldenström J, Osterhaus AD, Fouchier RA (2006) Global patterns of influenza a virus in wild birds. Science 312(5772):384–388. https://doi.org/10.1126/science.1122438

Pantin-Jackwood MJ, Swayne D (2009) Pathogenesis and pathobiology of avian influenza virus infection in birds. Rev Sci Tech 28(1):113–136

Pantin-Jackwood MJ, Costa-Hurtado M, Shepherd E, DeJesus E, Smith D, Spackman E et al (2016) Pathogenicity and transmission of H5 and H7 highly pathogenic avian influenza viruses in mallards. J Virol 90:9967–9982. https://doi.org/10.1128/JVI.01165-16

Paul MC, Vergne T, Mulatti P, Tiensin T, Iglesias I (2019) Epidemiology of avian influenza viruses. Front Vet Sci 6:150. https://doi.org/10.3389/fvets.2019.00150

Postel A, Letzel T, Frischmann S, Grund C, Beer M, Harder T (2010) Evaluation of two commercial loop-mediated isothermal amplification assays for detection of avian influenza H5 and H7 hemagglutinin genes. J Vet Diagn Investig 22:61–66

Qiu BF, Liu WJ, Peng DX, Hu SL, Tang YH, Liu XF (2009) A reverse transcription-PCR for subtyping of the neuraminidase of avian influenza viruses. J Virol Methods 155:193–198

Ramos S, MacLachlan M, Melton A (2017) Impacts of the 2014–2015 highly pathogenic avian influenza outbreak on the poultry, U. S Sector. USDA ERS Outlook LDPM282-02

Richard M, Schrauwen EJ, de Graaf M, Bestebroer TM, Spronken MI, Van Boheemen S et al (2013) Limited airborne transmission of H7N9 influenza A virus between ferrets. Nature 501(7468):560–563

Rutvisuttinunt W, Chinnawirotpisan P, Simasathien S, Shrestha SK, Yoon IK, Klungthong C, Fernandez S (2013) Simultaneous and complete genome sequencing of influenza A and B with high coverage by Illumina MiSeq platform. J Virol Methods 193:394–404

Sakurai A, Takayama K, Nomura N, Kajiwara N, Okamatsu M, Yamamoto N, Tamura T, Yamada J, Hashimoto M, Sakoda Y et al (2015) Fluorescent immunochromatography for rapid and sensitive typing of seasonal influenza viruses. PLoS One 10:e0116715

Sandrock C, Kelly T (2007) Clinical review: update of avian influenza A infections in humans. Crit Care 11:209

Sato M, Maruyama J, Kondoh T et al (2019) Generation of bat-derived influenza viruses and their reassortants. Sci Rep 9(1):1158. Published 2019 Feb 4. https://doi.org/10.1038/s41598-018-37830-x

Scott AB, Singh M, Groves P, Hernandez-Jover M, Barnes B, Moloney KGB et al (2017) Comparisons of management practices and farm design on Australian commercial layer and meat chicken farms: cage, barn and free range. PLoS One 12(11):e0188505. https://doi.org/10.1371/journal.pone.0188505

Sendor AB, Weerasuriya D, Sapra A (2020) Avian influenza. In: StatPearls. StatPearls Publishing, Treasure Island (FL)

Shafer AL, Katz JB, Eernisse KA (1998) Development and validation of a 4competitive enzyme-linked immunosorbent assay for detection of type A influenza antibodies in avian sera. Avian Dis 42:28–34

Shinya K, Ebina M, Yamada S, Ono M, Kasai N, Kawaoka Y (2006) Avian flu: influenza virus receptors in the human airway. Nature 440(7083):435–436

Short KR, Richard M, Verhagen JH, van Riel D, Schrauwen EJA, van den Brand JMA, Mänz B, Bodewes R, Herfst S (2015) One health, multiple challenges: the inter-species transmission of influenza A virus. One Health 1:1–13

Sims LD, Domenech J, Benigno C, Kahn S, Kamata A, Lubroth J, Roeder P (2005) Origin and evolution of highly pathogenic H5N1 avian influenza in Asia. Vet Rec 6:159–164. https://doi.org/10.1136/vr.157.6.159

Slomka MJ, Irvine RM, Pavlidis T, Banks J, Brown IH (2010) Role of real-time RT-PCR platform technology in the diagnosis and management of notifiable avian influenza outbreaks: experiences in Great Britain. Avian Dis 54:591–596

Sooryanarain H, Elankumaran S (2015) Environmental role in influenza virus outbreaks. Annu Rev Anim Biosci 3(1):347–373

Spackman E, Senne DA, Myers TJ, Bulaga LL, Garber LP, Perdue ML, Lohman K, Daum LT, Suarez DL (2002) Development of a real-time reverse transcriptase PCR assay for type A influenza virus and the avian H5 and H7 hemagglutinin subtypes. J Clin Microbiol 40:3256–3260

Stallknecht DE, Brown JD (2007) Wild birds and the epidemiology of avian influenza. J Wildl Dis 43(Suppl):S15–S20

Sturm-Ramirez KM, Ellis T, Bousfield B, Bissett L, Dyrting K, Rehg JE, Poon L, Guan Y, Peiris M, Webster RG (2004) Re-emerging H5N1 influenza viruses in Hong Kong in 2002 are highly pathogenic to ducks. J Virol 78(9):4892–4901

Su S, Bi Y, Wong G, Gray GC, Gao GF, Li S (2015) Epidemiology, evolution, and recent outbreaks of avian influenza virus in China. J Virol 89:8671–8676

Suzuki K, Okada H, Itoh T, Tada T, Mase M, Nakamura K et al (2009) Association of increased pathogenicity of Asian H5N1 highly pathogenic avian influenza viruses in chickens with highly efficient viral replication accompanied by early destruction of innate immune responses. J Virol 83(15):7475–7486. https://doi.org/10.1128/JVI.01434-08

Swayne DE (2012) The role of vaccines and vaccination in high pathogenicity avian influenza control and eradication. Expert Rev Vaccines 11:877–880

Swayne DE, Alexander DJ (1994) Confirmation of nèphrotropism and nephropathogenicity of three low-pathogenic chicken-origin influenza viruses for chickens. Avian Pathol 23:345–352

Takayama I, Takahashi H, Nakauchi M, Nagata S, Tashiro M, Kageyama T (2015) Development of a diagnostic system for novel influenza A (H7N9) virus using a real-time RT-PCR assay in Japan. Jpn J Infect Dis 68:113–118

Thomas JK, Noppenberger J (2007) Avian influenza: a review. Am J Health Syst Pharm 64:149–165. https://doi.org/10.2146/ajhp060181

To KF, Chan PK, Chan KF, Lee WK, Lam WY, Wong KF, Tang NL, Tsang DN, Sung RY, Buckley TA et al (2001) Pathology of fatal human infection associated with avian influenza A H5N1 virus. J Med Virol 63:242–246

Tong S, Li Y, Rivailler P, Conrardy C, Castillo DA, Chen LM, Recuenco S, Ellison JA, Davis CT, York IA et al (2012) A distinct lineage of influenza A virus from bats. Proc Natl Acad Sci U S A 109:4269–4274

Tong S, Zhu X, Li Y, Shi M, Zhang J, Bourgeois M, Yang H, Chen X, Recuenco S, Gomez J et al (2013) New world bats harbor diverse influenza A viruses. PLoS Pathog 9:e1003657

Tsuda Y, Sakoda Y, Sakabe S, Mochizuki T, Namba Y, Kida H (2007) Development of an immunochromatographic kit for rapid diagnosis of H5 avian influenza virus infection. Microbiol Immunol 51:903–907

Tsukamoto K, Ashizawa H, Nakanishi K, Kaji N, Suzuki K, Okamatsu M, Yamaguchi S, Mase M (2008) Subtyping of avian influenza viruses H1 to H15 on the basis of hemagglutinin genes by PCR assay and molecular determination of pathogenic potential. J Clin Microbiol 46:3048–3055

Tsunetsugu-Yokota Y, Nishimura K, Misawa S, Kobayashi-Ishihara M, Takahashi H, Takayama I, Ohnishi K, Itamura S, Nguyen HL, Le MT et al (2014) Development of a sensitive novel diagnostic kit for the highly pathogenic avian influenza A (H5N1) virus. BMC Infect Dis 14:362

van den Hoecke S, Verhelst J, Vuylsteke M, Saelens X (2015) Analysis of the genetic diversity of influenza A viruses using next-generation DNA sequencing. BMC Genomics 16:79

van Der Goot JA, Koch G, De Jong MC, VanBoven M (2005) Quantification of the effect of vaccination on transmission of avian influenza (H7N7) in chickens. Proc Natl Acad Sci U S A 102:18141–18146

van der Kolk JH (2019) Role for migratory domestic poultry and/or wild birds in the global spread of avian influenza? Vet Q 39(1):161–167. https://doi.org/10.1080/01652176.2019.1697013

van Elden LJ, Nijhuis M, Schipper P, Schuurman R, van Loon AM (2001) Simultaneous detection of influenza viruses A and B using real-time quantitative PCR. J Clin Microbiol 39:196–200

van Riel D, Leijten LM, de Graaf M, Siegers JY, Short KR, Spronken MI et al (2013) Novel avian-origin influenza A (H7N9) virus attaches to epithelium in both upper and lower respiratory tract of humans. Am J Pathol 183(4):1137–1143

Velumani S, Du Q, Fenner BJ, Prabakaran M, Wee LC, Nuo LY, Kwang J (2008) Development of an antigen-capture ELISA for detection of H7 subtype avian influenza from experimentally infected chickens. J Virol Methods 147:219–225

Villanueva-Cabezas JP, Coppo MJC, Durr PA, McVernon J (2017) Vaccine efficacy against Indonesian Highly Pathogenic Avian Influenza H5N1: systematic review and meta-analysis. Vaccine 35(37):4859–4869. https://doi.org/10.1016/j.vaccine.2017.07.059

Wada A, Sakoda Y, Oyamada T, Kida H (2011) Development of a highly sensitive immunochromatographic detection kit for H5 influenza virus hemagglutinin using silver amplification. J Virol Methods 178:82–86

Wang J, Li CC, Diao YX, Sun XY, Hao DM, Liu X et al (2014) Different outcomes of infection of chickens and ducks with duck-origin H9N2 influenza A virus. Acta Virol 58:223–230. https://doi.org/10.4149/av_2014_03_223

Woolcock PR, Cardona CJ (2005) Commercial immunoassay kits for the detection of influenza virus type A: evaluation of their use with poultry. Avian Dis 49:477–481

Xu L, Bao L, Deng W et al (2014) Rapid adaptation of avian H7N9 virus in pigs. Virology 452–453:231–236. https://doi.org/10.1016/j.virol.2014.01.016

Xu X, Bao H, Ma Y, Sun J, Zhao Y, Wang Y, Shi J, Zeng X, Li Y, Wang X et al (2015) Simultaneous detection of novel H7N9 and other influenza A viruses in poultry by multiplex real-time RT-PCR. Virol J 12:69

Yacoub A, Kiss I, Zohari S, Hakhverdyan M, Czifra G, Mohamed N, Gyarmati P, Blomberg J, Belak S (2009) The rapid molecular subtyping and pathotyping of avian influenza viruses. J Virol Methods 156:157–161

Yamamoto Y, Nakamura K, Yamada M, Mase M (2016) Corneal opacity in domestic ducks experimentally infected with H5N1 highly pathogenic avian influenza virus. Vet Pathol 53(1):65–76. https://doi.org/10.1177/0300985815591077

Yoshida H, Sakoda Y, Endo M, Motoshima M, Yoshino F, Yamamoto N, Okamatsu M, Soejima T, Senba S, Kanda H et al (2011) Evaluation of the reverse transcription loop-mediated isothermal amplification (RT-LAMP) as a screening method for the detection of influenza viruses in the fecal materials of water birds. J Vet Med Sci 73:753–758

Yum J, Park EH, Ku KB, Kim JA, Oh SK, Kim HS et al (2014) Low infectivity of novel avian-origin H7N9 influenza virus in pigs. Arch Virol 159:2745–2749. https://doi.org/10.1007/s00705-014-2143-y

Zhao J, Ragupathy V, Liu J, Wang X, Vemula SV, El Mubarak HS, Ye Z, Landry ML, Hewlett I (2015) Nanomicroarray and multiplex next-generation sequencing for simultaneous identification and characterization of influenza viruses. Emerg Infect Dis 21:400–408

Zhou EM, Chan M, Heckert RA, Riva J, Cantin MF (1998) Evaluation of a competitive ELISA for detection of antibodies against avian influenza virus nucleoprotein. Avian Dis 42:517–522

Zhou L, Chen E, Bao C et al (2018) Clusters of human infection and human-to-human transmission of avian influenza A(H7N9) virus, 2013–2017. Emerg Infect Dis 24:397–400

Chapter 6
Newcastle Disease and Other Avian Paramyxoviruses

Abstract New castle disease (ND) (also "pseudofowl pest" and "avian pneumoencephalitis") is an OIE-listed disease caused by the virulent strains of *avian avulavirus 1* (formerly avian paramyxovirus-1) of genus *Avulavirus* (formerly *Paramyxovirus*) of the family *Paramyxoviridae*. It is an economically devastating viral disease of poultry industry which often led to trade restrictions and bans. Apart from domestic birds, natural or experimental NDV infection has been established in at least 250 bird species from 27 of the 50 Orders. In humans, especially in occupational risk groups, the virus causes a transient eye infection. Human NDV infections have typically occurred in result of direct contact with sick birds or infected carcasses. Ideal strategies to prevent and control ND in endemic countries include strict biosecurity measures, rigorous vaccination programs, active surveillance, culling, and proper disposal of infected carcasses. Efforts are being made to develop new NDV vaccine strategies that offer sterilizing immunity, which will prevent both clinical disease and shedding of the virus.

Keywords New castle disease · *Avulavirus* · Avian Paramyxoviruses · Occupational risk · Conjunctivitis · Poultry

6.1 Introduction

New castle disease (ND) (also "pseudofowl pest," "Ranikhet disease," and "avian pneumoencephalitis") is the very next disease to highly pathogenic avian influenza that creates fear among poultry farm owners and global regulatory agencies. Since its first authorized report in 1926 from Newcastle-upon-Tyne in England, thus far, ND has caused numerous epidemics in domestic poultry in six out of seven continents (Brown and Bevins 2017). It is generally believed that virulent NDV is either a regular cause of epizootics or enzootic in poultry all over Asia, Central or parts of South America, and Africa. In developed areas, like Western Europe, sporadic epizootics arise on a regular basis in spite of the extensive use of vaccination (Alexander 2009). The World Organisation for Animal Health (OIE) has included ND in list of notifiable diseases, which requires immediate notification (Bello et al.

Y. S. Malik et al., *Role of Birds in Transmitting Zoonotic Pathogens*, Livestock Diseases and Management, https://doi.org/10.1007/978-981-16-4554-9_6

2018). It is an economically devastating viral disease of poultry industry which often led to trade restrictions and bans (Brown and Bevins 2017). Newcastle disease is caused by virulent strains of avian *avulavirus 1* generally denoted as Newcastle disease virus (NDV). Worldwide, this virus affects both domestic and wild avian species with significantly high morbidity, mortality, and other production associated losses. Although NDV has zoonotic potential, it causes mild disease in humans, mostly characterized by conjunctivitis and in some cases a mild, self-limiting influenza-like illness (Dale and Brown 2013; Elmberg et al. 2017; Gupta et al. 2020). ND is endemic in majority of the developing countries, where agriculture is the primary activity and main source of income. Newer strains of NDV are constantly being isolated from all over the world (Kumar 2015; Absalón et al. 2019). Wide deployment of vaccines and rapid diagnostics, strict quarantine, and biosecurity measures have prevented many outbreaks. However, from 2008 to 2010 alone, 77 countries have confirmed outbreaks in domestic setup due to uneven vaccination and lack of biosecurity practices. And widespread live vaccine application throughout the world has also led to difficulty in assessing true geographical distribution (Alexander 2009; Gupta et al. 2020).

6.2 Epidemiology

6.2.1 Causative Agent

ND is caused by virulent viruses belonging to species *avian avulavirus 1* (formerly known as avian paramyxovirus 1 (APMV 1) and genus *Avulavirus* (formerly *Paramyxovirus*) of the family *Paramyxoviridae*. Avian avulaviruses are nonsegmented, single-stranded negative-sense RNA viruses with a ~15.2 kb genome. Although the length of the genome may vary, avian avulaviruses encode six structural and two nonstructural proteins. The structural proteins include nucleoprotein (NP), matrix protein (M), phosphoprotein (P), fusion protein (F), RNA polymerase (L), and a hemagglutinin-neuraminidase (HN) (Miller and Koch 2013). The NP, L, and P proteins together form a ribonucleoprotein (RNP) complex, which forms the nucleocapsid and also manages transcription and genome replication. There are three membrane proteins: the glycosylated F and attachment protein (HN), and the unglycosylated inner membrane M. The host range is determined by HN and F proteins by inducing attachment of virus and host cells fusion. A single avian paramyxovirus serotype 1 (APMV-1) comprises all the NDV strains. Based on the F gene sequence, the strains are divided into two major classes, I and II. Furthermore, class I has been subdivided into three (1a, 1b, 1c) subtypes, while class II strains are divided into at least 18 (I-XVII) genotypes. Class I are primarily low virulent strains (loNDV) typically found in wild birds, and class II strains exhibiting both virulence (vNDV) and low virulence are found in domestic poultry and wild birds (Zhu et al. 2014; Dimitrov et al. 2016; Li et al. 2020).

6.2.2 Hosts

Apart from domestic birds, natural or experimental NDV infection has been established in at least 250 bird species from 27 Orders (Kaleta and Baldauf 1988; Miller and Koch 2013). Probability of susceptibility may be present in all species of birds, but the disease outcome in terms of severity differs with different species. Isolations of vNDV have been recorded in almost all kinds of commercially reared birds ranging from broiler to pigeons and ostriches (Alexander 2009). Whereas trivial, self-limiting infections in poultry farm workers and researchers are sporadically documented, fortuitously the virus does not signify a potential or severe threat to public health, and person-to-person spread has not yet been described (Dale and Brown 2013). NDV has also been reported in cattle calf, pigs, sheep, small mammals, rabbits, nonhuman primates, and ferrets (Spickler and Rovid 2016; Yuan et al. 2012).

6.2.3 Transmission

NDV transmission is primarily through inhalation, ingestion, and direct mucous membrane contact, mainly the conjunctiva (Fig. 6.1). Availability of infectious virus from the infected bird determines the spread to another bird. Excretion of the virus from the infected bird is based on the organs in which virus multiplication occurs, and may vary with the virus pathotype. Birds with gastrointestinal symptoms presumably shed virus in feces, thereby contaminating feed and water, while viruses that are restricted to the respiratory system may be transmitted by respiratory secretions/aerosols. Infective virus particles entrapped in dried fecal material may

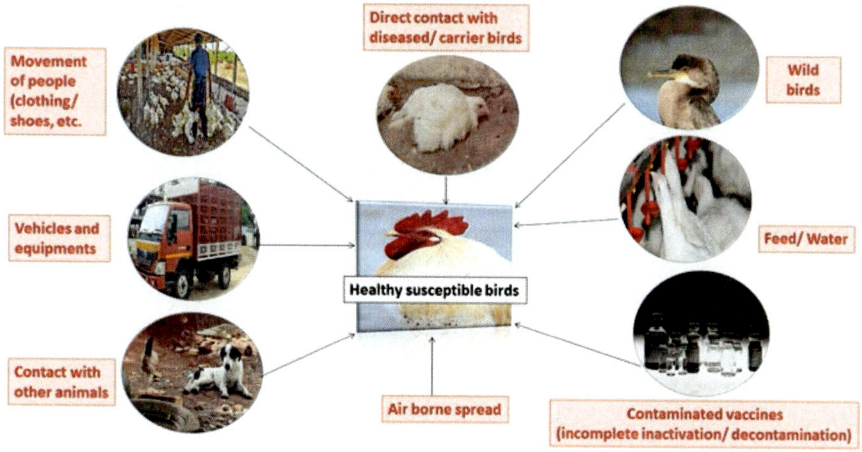

Fig. 6.1 Different source and mode of transmission of NDV in poultry

get inhaled also. Gallinaceous birds are considered to shed APMV-1 for 1–2 weeks; however, psittacine birds often excrete these viruses for a number of months, and occasionally for more than a year. Respiratory route of transmission may take a faster pace of spread compared to oral/fecal route (Alexander 2009; Spickler and Rovid 2016). Significance of virus getting transmitted vertically is not clear although virulent strain terminates egg-laying. Certain reports of isolation of vaccine strains from eggs laid by diseased birds are available (Pospisil et al. 1991). Another report described the isolation of vNDV strain from the cloacal swab, egg, and hatched progeny of birds with high antibody titers to NDV (Capua et al. 1993). In poultry setup, the secondary spread of the virus is attributed to the movement of live birds and poultry products, movement of people and equipment, contact with other animals, airborne spread, contaminated feed and water, mechanical transmission by flies, and contaminated vaccines (incomplete inactivation or decontamination) (Alexander 2009; Spickler and Rovid 2016). Figure 6.1 presents the different source and mode of transmission of NDV in poultry.

Human transmission is common among people who get frequent exposure of infected birds through their droppings, discharges from the eyes and beak. Occupationally, chances of exposure to the virus are higher in individuals who work in poultry farms, processing plants, laboratory workers (accidental splashing into the eye), veterinarians, vaccinators, etc. (Alexander 2009; Miskiewicz et al. 2018). Exposure to high doses of infection may cause clinical disease in humans. Pedersden et al. (1990) described that higher antibody titers to NDV in individuals who had recognized associations with poultry.

6.3 Role of Birds

Worldwide, ND is primarily an avian viral disease affecting domestic, peridomestic, and wild avifauna. NDV had been frequently isolated from wild birds, particularly migratory aquatic birds and other feral waterfowls. Many times, these viruses have been of mild virulence for chickens. However, vNDV has also been isolated from wild birds leading to the speculation that these species may play an important role in the spread of NDV (Snoeck et al. 2013). Nevertheless, virus spillovers from infected poultry are suggested to be responsible for isolations from wild birds (Miller and Koch 2013). Particularly, there are two identified exceptions where virulent type viruses have been endemic in wild birds. Columbiform birds are reported to be the reservoir for pigeon paramyxoviruses 1 (PPMV-1), a variant of vNDV. This variant is prevalent worldwide and extremely virulent for both domestic and wild pigeons (Ujvari et al. 2003). The other exceptional virus group are endemic in cormorant populations (genus *Phalacrocorax*) in North America, and periodic mortality episodes are recorded in this bird (Banerjee et al. 1994). Although adapted to pigeons and cormorants, these viruses have more potential to infect other species of birds and are hardly implicated with poultry infections (Ferreira et al. 2019). Incessant circulation and maintenance of lentogenic NDV between the feral and domesticated birds may facilitate the emergence of virulent NDV strains in poultry (Bello et al. 2018; Rahman et al. 2018). Human NDV infections have typically occurred in effect of direct contact with virus-infected birds or infected carcasses (Alexander 2009). To produce disease in humans, a large amount of virus exposure is necessary; that is why poultry farm workers and slaughterhouse workers are usual victims of infections (Brown and Bevins 2017). Phylogenetic analysis of NDV strains of wild and domestic origin supports the role of wild birds in the global redistribution of NDVs in poultry, and that wild birds play a significant role in the evolution and adaptation of these viruses (Rahman et al. 2018).

6.4 Disease

6.4.1 *Pathogenesis*

The variation in virulence among NDV is ascribed to amino acids at the cleavage spot on the F protein which arbitrates cell-cell, cell-virus interaction and fusion (Huang et al. 2004). In all the virulent NDV strains, 3 or more arginine or lysine residues occupy 113th position and phenylalanine occupies 117th position. Host cell proteases after recognizing particular amino acid motif at the cleavage site of F protein must cleave F_0 (precursor F protein) to F1 and F2 proteins to initiate infection via activation of hemolytic properties and cell fusion (Leighton and Heckert 2007; Brown and Bevins 2017). Proteases of infected cells can cleave viruses with a virulent cleavage site, thus allowing extensive replication of the virus and systemic infection. However, if the specific amino acid motif is absent in F protein cleavage site, trypsin and trypsin-like enzymes found in intestinal and respiratory passages can only mediate the cleavage which eventually leads to restricted host site replication. The HN protein also plays an important role in virulence and tissue tropism of NDV by stimulating the fusion activity of F protein, enabling penetration into the host cell and preventing self-agglutination by removal of sialic acid from virus progeny (Huang et al. 2004). Enveloped viruses like NDV have been known to enter the host cells by direct fusion mechanisms. The virus envelope fuses with the host cell plasma membrane or receptor-arbitrated endocytosis in which the host cell surface-specific receptor gets bonded with the virus and membrane fusion which results in nucleocapsid translocation into the host cell cytoplasm (Dortmans et al. 2011; Brown and Bevins 2017).

6.4.2 *Clinical Signs/Clinical Profile*

Based on clinical and pathologic profile, five distinct forms of ND are documented in birds (Marks et al. 2014). Velogenic viscerotropic ND (VVND) is the severest form characterized by high morbidity and mortality rates approaching 100%. The clinical signs noticed are conjunctivitis, ruffled feathers, dyspnea, nasal discharges, diarrhea, prostration, paralysis, and tremors (Falcon 2004). Velogenic neurotropic ND (VNND) is another form of the disease characterized by respiratory and neurological signs without the involvement of the gastrointestinal system. Affected birds typically manifest tremors, opisthotonus, twisting of head, and paralysis (Banerjee et al. 1994). Mesogenic ND (MND) is another form associated with respiratory and neurological signs with a very low mortality rate. In field conditions, an important indicator of this form is a drop in the egg production and slight to moderate respiratory ailments (Alexander and Senne 2008). Lentogenic ND (LND) and asymptomatic enteric ND (AEND) are the other trivial forms without any clinical disease. In LND, only the young chicks get a mild respiratory disease. In an experimental study, lentogenic Q4 and B1 strains failed to produce any apparent clinical signs in a 4-week-old chicken (Hamid et al. 1990). AEND is a complete asymptomatic and avirulent form with simple replication of the virus in the intestinal tract tissues of the chicken (Bello et al. 2018).

The most frequently reported clinical signs in human NDV infection have been eye infections, frequently comprising unilateral or bilateral reddening, conjunctivitis, edema of the eyelids, excessive lachrymation, and subconjunctival hemorrhage (Chang 1981). Although the eye infection is severe, it is usually for brief period lasting not more than 1 or 2 days, and the cornea is usually not affected. Infrequent generalized symptoms in humans are headache, chills, and fever, with or without conjunctivitis (Alexander 2009).

6.4.3 Pathology/Lesions

Postmortem findings in VVND forms in birds are ulcerative hemorrhages throughout the gastrointestinal tract, particularly at the junction of proventriculus and gizzard and in the cecal tonsils (Brown et al. 1999). Internal organs such as liver, spleen, and gut-associated lymphoid tissue show necrotic foci. Histological observations include microscopic evidence of hemorrhage and necrosis, such as those observed in the Peyer's patches and spleen. Except perivascular cuffing, neurological lesions have not been observed so far in the VVND form even among birds that died presenting neurological symptoms (Cattoli et al. 2011). Gross pathological lesions are usually not present among birds that died manifesting VNND signs. Nevertheless, during histology, perivascular cuffing and necrosis of Purkinje fibers are often encountered (Banerjee et al. 1994). In MND, gross lesions are also minimal, showing only mild splenomegaly and other secondary bacterial infection-related lesions. Histologically, perivascular cuffing and gliosis are observed with rare pancreatic necrosis (Brown et al. 1999). In LND, postmortem lesions may be completely absent or may show mild hemorrhages in the pulmonary and tracheal tissues. Important histological observations in the tracheal tissues include the proliferation of lymphoid follicles, loss of cilia, squamous cell metaplasia, and infiltration of lymphocytes (Hooper et al. 1999).

6.5 Disease Management

6.5.1 Diagnosis

Poultry respiratory disease-causing pathogens such as infectious bronchitis, avian influenza, and infectious laryngotracheitis viruses should be considered for differential diagnosis as clinical presentation of all these diseases are not so distinct. APMV-3 and -7 (other avian paramyxoviruses) may cross-react in regular serological diagnosis of NDV. Therefore, it is important to accurately identify the NDV strains rapidly to apply proper intervention strategies (Bello et al. 2018). Various diagnostic methods available to detect and characterize NDV are furnished in Table 6.1.

6.6 Prevention and Control

In the absence of treatment options, the most appropriate measures to control NDV are culling of diseased birds combined with aggressive vaccination program and strict biosecurity protocols. Worldwide, most of the epidemic outbreaks have been attributed to a group of causes comprising a lack of biosecurity, vaccine failure, poor vaccine coverage, antigenic variation, a short period of an immune response, immune suppression, and inhibition of live vaccines by maternal antibodies (Conan et al. 2012; Absalón et al. 2019). Thus far, several lentogenic NDV strains like LaSota, F, B, I2, V4 are widely used as live vaccines for the control of ND (OIE 2008). LaSota is extensively used in many countries due to its superior immunogenicity. Although not as immunogenic as LaSota, B1 strain-based live vaccines are known for its high attenuated nature with no postimmunization reactions in birds. The important feature of V4- and I2 strain-based vaccine is thermostability, as they can withstand higher temperatures in the deficiency of cold chain (Bensink and Spradbrow 1999). Other popular live vaccines comprise the Mukteswar and the Komarov mesogenic strains wherein they can be used as booster vaccines subsequent to priming with lentogenic strain-based live vaccines (Senne et al. 2004). All these live attenuated ND vaccines are good in stimulating both systemic and mucosal immune responses comparable to those of the natural infection, as they can replicate in chicken regardless of the administration site (Rauw et al. 2009). However, an excellent NDV vaccine is the one that not only prevents the disease but also decreases or eliminates virus shedding (Kapczynski et al. 2013). Regrettably, all the available vaccines of NDV can only prevent clinical disease but not the shedding of the virus (Roohani et al. 2015). Nevertheless, they remain the backbone of ND control for over six decades as they have good disease preventing capacity and are cost-effective. However, the quest for better alternatives still prevails and has led to the development of novel vaccine platforms with the advent of recombinant DNA technology. Among the newer vaccines, DNA vaccines (Zhao et al. 2014;

Table 6.1 Diagnostic methods available to detect and characterize NDV

Diagnostic methods	Techniques	References
Virus isolation	• Gold standard method for the definitive diagnosis of ND • Samples from live birds—oropharyngeal and cloacal swabs • Samples from moribund or dead birds—lungs, spleen, liver, kidney, cecal tonsils, and oronasal and cloacal swabs • Samples are inoculated into the allantoic cavity of 9–10-day-old chicken embryonated eggs (specific antibody free). After 4–7 days of incubation, hemagglutination test (HA) is deployed to detect the virus • As HA activity is possessed by other viruses like AIV, hemagglutination inhibition test (HI) with NDV-specific antisera or molecular tests is required to confirm the diagnosis • Isolation can also be accomplished in primary cell cultures permissive to NDV such as chicken embryo kidney, chicken embryo fibroblasts, avian myeloblasts, and DF-1. Cytopathic effect—syncytia formation, cell rounding, and cell death	Alexander (2000), Ravindra et al. (2009), OIE (2012)
Pathotyping	• As virulent NDV requires immediate OIE notification, pathotyping is necessary which can be done by in vivo pathogenicity assessment tests or molecular-based assays **NDV pathotyping tools** • Mean death time (MDT) test is done in 9–10-day-old embryonated chicken eggs (velogenic 40–60 h MDT; mesogenic 60–90 h MDT; lentogenic >90 h MDT) • Intracerebral pathogenicity index (ICPI) test is performed in 1-day-old SPF chicks (virulent strains 1.3–2; mesogenic strains 0.7–1.3; lentogenic strains 0.0–0.7) • Intravenous pathogenicity index (IVPI) performed in 4–6-week-old SPF chicken (virulent strains 0.5–3; mesogenic strains 0.0–0.5; lentogenic strains 0.0) [OIE notifiable virulent isolate—MDT	Alexander and Parsons (1986), OIE (2008), Samal et al. (2011)

(continued)

Table 6.1 (continued)

Diagnostic methods	Techniques	References
	value of 40–60 h/IVPI value of >0.5/ ICPI >1.3]	
Serological diagnosis	• The simplest serological test for NDV is HI. It determines the ability of NDV-specific antibodies to inhibit the RBCs agglutination by the NDV particles. A sudden increase in the titer might also be suggestive of exposure of birds to field NDV strain • ELISA is another robust test used for rapid diagnosis of ND • In HI, only the antibodies against HN protein are dctected, whereas ELISA platforms can detect antibodies against all the NDV proteins • Makkay et al. (1999) developed an indirect ELISA with recombinant NP protein as antigen and demonstrated the DIVA property	Makkay et al. (1999), Cross (2002), Berinstein et al. (2005), Tsunekuni et al. (2014)
Molecular assays	• Most common rapid molecular test to detect the NDV genome with higher sensitivity and accuracy is RT-PCR • RT PCRs are usually designed to detect NDV and identify the pathotype by targeting F gene followed by RFLP using *BglI* or analysis of the amino acid composition of the F cleavage site • RT-qPCR targeting fusion and matrix gene is also used often to screen and pathotype NDV from clinical samples • LAMP and RT-LAMP assays have been recently developed with greater sensitivity, rapidity, and cost-effectiveness • Genome sequencing-based studies have implications on understanding evolution and distribution of viruses, and implications on molecular diagnostics and vaccine strategies	Nanthakumar et al. (2000), Aldous et al. (2001), Wang et al. (2001), de Leeuw et al. (2005), Pham et al. (2005), Farkas et al. (2009), Miller et al. (2010), Kirunda et al. (2012), Bello et al. (2018), Tran et al. (2020)

Firouzamandi et al. 2016a; Firouzamandi et al. 2016b) and the virus-like particles (VLPs) (Pantua et al. 2006; Park et al. 2014; McGinnes et al. 2010) are acknowledged for their safety, but they are weakly immunogenic. Emerging recombinant viral-vectored NDV vaccines (Karaca et al. 1998; Sun et al. 2006; Esaki et al. 2013; Ewer et al. 2016) have also revealed promising protective ability, but their efficacy suffers from the activity of maternal antibodies against the vector. Till date, the most promising NDV vaccines are the genotype-matched live attenuated recombinant vaccine candidates made by reverse genetics as they explicitly target prevailing or dominant genotype in a specific region. Carcass disposal is equally important to control ND outbreaks as virulent NDV can remain viable in the infected bird tissues for weeks and turn into a source of environmental contamination. Another important biosecurity practice is the elimination of other avian species such as pigeons, ducks, etc. as they may become carriers of the virus (Afonso and Miller 2014; Absalón et al. 2019).

Ideally, strategies to prevent and control ND in endemic countries include strict biosecurity measures, rigorous vaccination programs, active surveillance, culling, and proper disposal of infected carcasses. Research should be direct toward developing a new NDV vaccine that offers sterilizing immunity, which will prevent both clinical disease and shedding of the virus.

References

Absalón AE, Cortés-Espinosa DV, Lucio E, Miller PJ, Afonso CL (2019) Epidemiology, control, and prevention of Newcastle disease in endemic regions: Latin America. Trop Anim Health Prod 51(5):1033–1048. https://doi.org/10.1007/s11250-019-01843-z

Afonso CL, Miller PJ (2014) Newcastle disease virus. In: Liu D (ed) Security sensitive microbes and toxins. CRC Press, Boca Raton, pp 689–698

Aldous EW, Collins MS, McGoldrick A, Alexander DJ (2001) Rapid pathotyping of Newcastle disease virus (NDV) using fluorogenic probes in a PCR assay. Vet Microbiol 80(3):201–212

Alexander DJ (2000) Newcastle disease and other avian paramyxoviruses. Int Off Epizootics 19(2):443–462

Alexander DJ (2009) Ecology and epidemiology of Newcastle disease. In: Capua I, Alexander DJ (eds) Avian influenza and Newcastle disease. Springer, Milan, pp 19–26

Alexander DJ, Parsons G (1986) Pathogenicity for chickens of avian paramyxovirus type i isolates obtained from pigeons in Great Britain during 1983–85. Avian Pathol 15(3):487–493

Alexander DJ, Senne DA (2008) Newcastle disease virus and other avian paramyxoviruses. In: Swayne DE, Glisson JR (eds) A laboratory manual for the isolation, identification and characterization of avian pathogens, Dufour-Zavala, 5th edn. American Association of Avian Pathologists, Jacksonville, pp 135–141

Banerjee M, Reed WM, Fitzgerald SD, Panigrahy B (1994) Neurotropic velogenic Newcastle-disease in cormorants in Michigan—pathology and virus characterization. Avian Dis 38:873–878. https://doi.org/10.2307/1592127

Bello MB, Yusoff K, Ideris A, Hair-Bejo M, Peeters BPH, Omar AR (2018) Diagnostic and vaccination approaches for Newcastle disease virus in poultry: the current and emerging perspectives. Biomed Res Int 2018:7278459

Bensink Z, Spradbrow P (1999) Newcastle disease virus strain I2—a prospective thermostable vaccine for use in developing countries. Vet Microbiol 68(1–2):131–139

Berinstein A, Vazquez-Rovere C, Asurmendi S et al (2005) Mucosal and systemic immunization elicited by Newcastle disease virus (NDV) transgenic plants as antigens. Vaccine 23(48–49):5583–5589

Brown VR, Bevins SN (2017) A review of virulent Newcastle disease viruses in the United States and the role of wild birds in viral persistence and spread. Vet Res 48:68

Brown C, King DJ, Seal BS (1999) Pathogenesis of Newcastle disease in chickens experimentally infected with viruses of different virulence. Vet Pathol 36(2):125–132

Capua I, Scacchia M, Toscani T, Caporale V (1993) Unexpected isolation of virulent Newcastle disease virus from commercial embryonated fowls' eggs. Zentralbl Veterinarmed B 40(9–10):609–612

Cattoli G, Susta L, Terregino C, Brown C (2011) Newcastle disease: a review of field recognition and current methods of laboratory detection. J Vet Diagn Investig 23(4):637–656

Chang PW (1981) Newcastle disease. In: Steele JH (ed) CRC handbook series in zoonoses. Section B: viral zoonoses, vol 2. CRC Press, Boca Raton, FL, pp 261–274

Conan A, Goutard FL, Sorn S, Vong S (2012) Biosecurity measures for backyard poultry in developing countries: a systematic review. BMC Vet Res 8:240

Cross G (2002) Hemagglutination inhibition assays. Semin Avian Exotic Pet Med 11(1):15–18

Dale E, Brown C (2013) Zoonotic diseases from poultry. Braz J Vet Pathol 6(2):76–82

de Leeuw OS, Koch G, Hartog L, Ravenshorst N, Peeters BPH (2005) Virulence of Newcastle disease virus is determined by the cleavage site of the fusion protein and by both the stem region and globular head of the haemagglutinin-neuraminidase protein. J Gen Virol 86(6):1759–1769

Dimitrov KM, Ramey AM, Qiu X, Bahl J, Afonso CL (2016) Temporal, geographic, and host distribution of avian paramyxovirus 1 (Newcastle disease virus). Infect Genet Evol 39:22–34. https://doi.org/10.1016/j.meegid.2016.01.008

Dortmans JCFM, Koch G, Rottier PJM, Peeters BPH (2011) Virulence of Newcastle disease virus: what is known so far? Vet Res 42:122

Elmberg J, Berg C, Lerner H, Waldenström J, Hessel R (2017) Potential disease transmission from wild geese and swans to livestock, poultry and humans: a review of the scientific literature from a one health perspective. Infect Ecol Epidemiol 7:1300450

Esaki M, Godoy A, Rosenberger JK et al (2013) Protection and antibody response caused by turkey herpesvirus vector Newcastle disease vaccine. Avian Dis 57(4):750–755

Ewer KJ, Lambe T, Rollier CS, Spencer AJ, Hill AVS, Dorrell L (2016) Viral vectors as vaccine platforms: from immunogenicity to impact. Curr Opin Immunol 41:47–54

Falcon MD (2004) Exotic Newcastle disease. Semin Avian Exotic Pet Med 13(2):79–85

Farkas T, Szekely E, Belak S, Kiss I (2009) Real-time PCR-based pathotyping of Newcastle disease virus by use of TaqMan minor groove binder probes. J Clin Microbiol 47(7):2114–2123

Ferreira HL, Taylor TL, Dimitrov KM, Sabra M, Afonso CL, Suarez DL (2019) Virulent Newcastle disease viruses from chicken origin are more pathogenic and transmissible to chickens than viruses normally maintained in wild birds. Vet Microbiol 235:25–34

Firouzamandi M, Moeini H, Hosseini D et al (2016a) Improved immunogenicity of Newcastle disease virus fusion protein genes. J Vet Sci 17(1):21–26

Firouzamandi M, Moeini H, Hosseini SD et al (2016b) Preparation, characterization, and in ovo vaccination of dextran-spermine nanoparticle DNA vaccine coexpressing the fusion and hemagglutinin genes against Newcastle disease. Int J Nanomed 11:259–267

Gupta A, Deka P, Kumar S (2020) Resiquimod inhibits Newcastle disease virus replication by modulating host cytokines: an understanding towards its possible therapeutics. Cytokine 125: 154811

Hamid H, Campbell RSF, Lamichhane C (1990) The pathology of infection of chickens with the lentogenic V4 strain of Newcastle disease virus. Avian Pathol 19(4):687–696

Hooper PT, Hansson E, Young JG, Russell GM, Della-Porta AJ (1999) Lesions in the upper respiratory tract in chickens experimentally infected with Newcastle disease viruses isolated in Australia. Aust Vet J 77(1):50–51

Huang Z, Panda A, Elankumaran S, Govindarajan D, Rockemann DD, Samal SK (2004) The hemagglutinin-neuraminidase protein of Newcastle disease virus determines tropism and virulence. J Virol 78:4176–4184

Kaleta EF, Baldauf C (1988) Newcastle disease in free-living and pet birds. In: Alexander DJ (ed) Newcastle disease. Kluwer Academic Publishers, Boston, pp 197–246

Kapczynski DR, Afonso CL, Miller PJ (2013) Immune responses of poultry to Newcastle disease virus. Dev Comp Immunol 41(3):447–453

Karaca K, Sharma JM, Winslow BJ et al (1998) Recombinant fowlpox viruses coexpressing chicken type I IFN and Newcastle disease virus HN and F genes: influence of IFN on protective efficacy and humoral responses of chickens following in ovo or post-hatch administration of recombinant viruses. Vaccine 16(16):1496–1503

Kirunda H, Thekisoe OMM, Kasaija PD et al (2012) Use of reverse transcriptase loop-mediated isothermal amplification assay for field detection of Newcastle disease virus using less invasive samples. Vet World 5(4):206–212

Kumar S (2015) Newcastle disease virus outbreaks in India: time to revisit the vaccine type and strategies. Vaccine 33(29):3268–3269

Leighton FA, Heckert RA (2007) Newcastle disease and related avian paramyxoviruses. In: Thomas NJ, Hunter DB, Atkinson CT (eds) Infectious diseases of wild birds. Blackwell Publishing, Oxford, pp 3–16

Li S, You G, Du J, Xia J, Wen Y, Huang X, Zhao Q, Han X, Yan Q, Wu R, Cao S, Huang Y (2020) A class I lentogenic Newcastle disease virus strain confers effective protection against the prevalent strains. Biologicals 63:74–80

Makkay AM, Krell PJ, Nagy E (1999) Antibody detection based differential ELISA for NDV-infected or vaccinated chickens versus NDV HN-subunit vaccinated chickens. Vet Microbiol 66(3):209–222

Marks FS, Rodenbusch CR, Okino CH (2014) Targeted survey of Newcastle disease virus in backyard poultry flocks located in wintering site for migratory birds from Southern Brazil. Prevent Vet Med 116(1–2):197–202

McGinnes LW, Pantua H, Laliberte JP, Gravel KA, Jain S, Morrison TG (2010) Assembly and biological and immunological properties of Newcastle disease virus-like particles. J Virol 84(9):4513–4523

Miller PJ, Koch G (2013) Newcastle disease, other avian paramyxoviruses and avian meta pneumovirus infections. In: Swayne DE, McDougal LR, Nolan LK, Suarez DL, Nair V (eds) Diseases of poultry, 13th edn. Wiley, New Jersey, pp 87–138

Miller PJ, Decanini EL, Afonso CL (2010) Newcastle disease: evolution of genotypes and the related diagnostic challenges. Infect Genet Evol 10(1):26–35

Miskiewicz A, Kowalczyk P, Oraibi SM, Cybulska K, Misiewicz A (2018) Bird feathers as potential sources of pathogenic microorganisms: a new look at old diseases. Anton Leeuw Int J G 111:1493–1507. https://doi.org/10.1007/s10482-018-1048-2

Nanthakumar T, Kataria RS, Tiwari AK, Butchaiah G, Kataria JM (2000) Pathotyping of Newcastle disease viruses by RT-PCR and restriction enzyme analysis. Vet Res Commun 24(4):275–286

OIE (2008) Newcastle disease. In: Manual of diagnostic tests and vaccines for terrestrial animals (mammals, birds and bees), vol 1, pp 576–589

OIE (2012) Newcastle disease (infection with newcastle disease virus). In: Manual of diagnostic tests and vaccines for terrestrial animals: (mammals, birds and bees), vol 1, pp 555–574

Pantua HD, McGinnes LW, Peeples ME, Morrison TG (2006) Requirements for the assembly and release of Newcastle disease virus-like particles. J Virol 80(22):11062–11073

Park JK, Lee DH, Yuk SS et al (2014) Virus-like particle vaccine confers protection against a lethal Newcastle disease virus challenge in chickens and allows a strategy of differentiating infected from vaccinated animals. Clin Vaccine Immunol 21(3):360–365

Pedersden KA, Sadasiv EC, Chang PW, Yates VJ (1990) Detection of antibody to avian viruses in human populations. Epidemiol Infect 104(3):519–525

Pham HM, Nakajima C, Ohashi K, Onuma M (2005) Loop-mediated isothermal amplification for rapid detection of Newcastle disease virus. J Clin Microbiol 43(4):1646–1650

Pospisil Z, Zendulkova D, Smid B (1991) Unexpected emergence of Newcastle disease virus in very young chicks. Acta Vet Brno 60:263–270

Rahman AU, Habib M, Shabbir MZ (2018) Adaptation of Newcastle Disease Virus (NDV) in feral birds and their potential role in interspecies transmission. Open Virol J 12:52–68. Published 2018 Aug 31. https://doi.org/10.2174/1874357901812010052

Rauw F, Gardin Y, Palya V (2009) Humoral, cell-mediated and mucosal immunity induced by oculo-nasal vaccination of oneday-old SPF and conventional layer chicks with two different live Newcastle disease vaccines. Vaccine 27(27):3631–3642

Ravindra PV, Tiwari AK, Ratta B, Chaturvedi U, Palia SK, Chauhan RS (2009) Newcastle disease virus-induced cytopathic effect in infected cells is caused by apoptosis. Virus Res 141(1):13–20

Roohani K, Tan SW, Yeap SK, Ideris A, Bejo MH, Omar AR (2015) Characterisation of genotype VII Newcastle disease virus (NDV) isolated from NDV vaccinated chickens, and the efficacy of LaSota and recombinant genotype VII vaccines against challenge with velogenic NDV. J Vet Sci 16(4):447–457

Samal S, Kumar S, Khattar SK, Samal SK (2011) A single amino acid change, Q114R, in the cleavage-site sequence of Newcastle disease virus fusion protein attenuates viral replication and pathogenicity. J Gen Virol 92(10):2333–2338

Senne DA, King DJ, Kapczynski DR (2004) Control of Newcastle disease by vaccination. Dev Biol 119:165–170

Snoeck CJ, Marinelli M, Charpentier E, Sausy A, Conzemius T, Losch S, Muller CP (2013) Characterization of Newcastle disease viruses in wild and domestic birds in Luxembourg from 2006 to 2008. Appl Environ Microbiol 79:639–645. https://doi.org/10.1128/AEM.02437-12

Spickler, Rovid A (2016) Newcastle disease. http://www.cfsph.iastate.edu/DiseaseInfo/factsheets.php

Sun HL, Wang YF, Miao DY, Zhang PJ, Zhi HD, Xu LL et al (2006) Construction and characterization of recombinant fowlpox virus co-expressing F and HN genes of Newcastle disease virus and gB gene of infectious larygnotracheitis virus. Chin J Biotechnol 22(6):931–938

Tran GTH, Sultan S, Osman N et al (2020) Molecular characterization of full genome sequences of Newcastle disease viruses circulating among vaccinated chickens in Egypt during 2011-2013. J Vet Med Sci 82(6). https://doi.org/10.1292/jvms.19-0623

Tsunekuni R, Hikono H, Saito T (2014) Evaluation of avian paramyxovirus serotypes 2 to 10 as vaccine vectors in chickens previously immunized against Newcastle disease virus. Vet Immun Immunopathol 160(3–4):184–191

Ujvari D, Wehmann E, Kaleta EF, Werner O, Savic V, Nagy E, Czifra G, Lomniczi B (2003) Phylogenetic analysis reveals extensive evolution of avian paramyxovirus type 1 strains of pigeons (Columba livia) and suggests multiple species transmission. Virus Res 96:63–73

Wang Z, Vreede FT, Mitchell JO, Viljoen GJ (2001) Rapid detection and differentiation of Newcastle disease virus isolates by a triple one-step RT-PCR. Onderstepoort J Vet Res 68(2):131–134

Yuan X, Wang Y, Yang J et al (2012) Genetic and biological characterizations of a Newcastle disease virus from swine in China. Virol J 9:129

Zhao K, Zhang Y, Zhang X et al (2014) Preparation and efficacy of Newcastle disease virus DNA vaccine encapsulated in chitosan nanoparticles. Int J Nanomed 9(1):389–402

Zhu J, Xu H, Liu H, Zhao Z, Hu S, Wang X, Liu X (2014) Surveillance of avirulent Newcastle disease viruses at live bird markets in Eastern China during 2008–2012 reveals a new sub-genotype of class I virus. Virol J 11(1):1–9. https://doi.org/10.1186/s12985-014-0211-2

Chapter 7
Usutu Virus

Abstract Usutu virus (USUV) is an emerging arbovirus that has been attracting interest from the scientific community because of its prevalence in larger areas of Europe. USUV belongs to the genus *Flavivirus* and is maintained through an enzootic cycle encompassing birds and mosquitoes. USUV has adversely affected the avian population, causing high rates of mortality in blackbirds and gray owls in France, Austria, the Netherlands, and Germany. Ornithophilic mosquito species of the genus *Culex* has been identified as the primary vector and is implicated in the transmission of the virus among birds, and to vulnerable mammals, specifically humans and horses which are considered as dead-end hosts, with short period and low levels of viremia. However, certain rare cases with the neurological disorder have been reported in humans, both in immunocompromised and healthy individuals. Compared to some other arboviruses, USUV has been relatively less studied, and therefore, multidisciplinary intervention strategies are required to understand further, detect, control, and prevent the spread of the virus in humans and birds, with an emphasis on routine surveillance including those at the reservoir–vector interface, and developing new therapeutic and prophylactic approaches.

Keywords Usutu virus · *Flavivirus* · *Culex* · Birds · Mosquitoes · Horses · Zoonotic

7.1 Introduction

Usutu virus (USUV), a member of the family *Flaviviridae*, is an emerging virus that has captured the interests of the scientific community because of its prevalence in Europe. USUV is an arbovirus of the genus *Flavivirus* that includes some of the most pathogenic arboviruses affecting humans, such as Japanese encephalitis virus (JEV), West Nile virus (WNV), yellow fever virus, Zika virus, and dengue (Gould and Solomon 2008; Cle et al. 2019). USUV belongs to the Japanese encephalitis serogroup and is phylogenetically related to WNV and JEV (Calisher and Gould 2003; Beck et al. 2013). The virus was termed after the Usutu River located in Swaziland, Southern Africa. In 1959, in South Africa, while researching the prevalence of viruses in arthropods, McIntosh first isolated USUV from *Culex neavei*

Y. S. Malik et al., *Role of Birds in Transmitting Zoonotic Pathogens*, Livestock Diseases and Management, https://doi.org/10.1007/978-981-16-4554-9_7

mosquitoes by intracerebral inoculation in new-born mice (McIntosh 1985). Thereafter, USUV was isolated from *Mansonia aurites*, a bird-biting mosquito, in Uganda (Williams et al. 1964).

USUV is a spherical enveloped virus of about 40–60 nm in diameter. The viral genome consists of a positive-sense single-stranded RNA that is ~11,064 nucleotides long, with an N7-methylguanosine-triphosphate cap at 5′ end and lacking a poly-A tail at the 3′ end (Bakonyi et al. 2004). The USUV genome contains a single open reading frame that encodes a polyprotein of 3434 amino acids which is eventually cleaved into eight nonstructural proteins (NS1/NS1′, NS2, NS2b, NS3, 2K, NS4a, NS4b, and NS5) and three structural proteins (premembrane prM, envelope E, and capsid C) (Calisher and Gould 2003). By phylogenetic analysis of NS5 sequences, USUV strains isolated across the world have been classified into at least eight lineages (five European and three African) (Cadar et al. 2017). The levels of genetic similarity appear to be influenced by the host and geographical origin of the isolate. Specific amino acid mutations associated with the host and geographical origin of the isolate have been observed, such as NS4B (M16I), C (A120V), prM (Y120N), and E (G195R) (Nikolay et al. 2013; Engel et al. 2016; Gaibani et al. 2013).

7.2 Hosts

Birds are the main reservoir of USUV. The virus has been identified in 93 different avian species corresponding to 35 families (Benzarti et al. 2019). House sparrows (*Passer domesticus*), blackbirds (*Turdus merula*), and gray owls (*Strix nebulosa*) are predominantly affected by USUV (Bakonyi et al. 2007; Weissenbock et al. 2003). In these avian species, the virus has been spotted in several organs, such as brain, heart, liver, and spleen, with the presence of necrotic lesions, suggesting severe pathogenesis (Becker et al. 2012; Chvala et al. 2004). USUV induced high mortality in blackbirds and gray owls in France (Lecollinet et al. 2016), Austria (Chvala et al. 2004), the Netherlands (Rijks et al. 2016), and Germany (Becker et al. 2012). Within 5 years of USUV entry into Germany, the country has lost 15% of its blackbird population (Cadar et al. 2017). On the other hand, the mortality rate of birds dropped abruptly after 2004 in Austria, although the number of birds infected with low viral titers grew (Bakonyi et al. 2007). More than 50% seropositivity was observed in birds of prey and owls. Herd immunity has been incriminated to play a role in these events (Bakonyi et al. 2007; Meister et al. 2008). Although there is no current information on USUV infection in the American robin (*Turdus migratorius*), a bird species which plays an imperative role in the spread of WNV in the United States (Bergsman et al. 2016), the virus is known to infect its European counterpart *Turdus merula*, and other Passeriformes birds which are vulnerable to WNV namely, *Sturnus vulgaris* and house sparrow (Nikolay 2015).

7.3 Pathology/Transmission

The experimental inoculation of USUV in chickens or geese resulted in mild pathogenicity, and large death episodes have been observed in only a few avian species (Chvala et al. 2005, 2006). Considering above, the variations in pathogenicity of USUV in different avian species remain to be properly elucidated. Coinfection with other pathogens, such as *Plasmodium* spp., may increase the severity of USUV infection in birds (Rouffaer et al. 2018; Roesch et al. 2019).

Other than humans and birds, USUV has also been detected in many mammalian species. Because of the similarities in the transmission cycles of USUV and WNV, horses are considered as a vulnerable host for USUV, and the virus has been reported in equids in Tunisia, Croatia, Serbia, Poland, and Spain (Roesch et al. 2019). USUV infection has been documented in wild ruminants (Garcia-Bocanegra et al. 2016), dogs (Montagnaro et al. 2019), wild boars (Escribano-Romero et al. 2015), and zoo mammals (Caballero-Gómeza et al. 2020). It has been suggested that bats (Cadar et al. 2014), rodents (Diagne et al. 2019), and tree squirrels (Romeo and Lecollinet 2018) may act as secondary reservoirs of USUV, as the virus did not exhibit severe pathogenicity in these species (Roesch et al. 2019). The ability of USUV to attain sufficient levels of viremia in these infected animals so as to initiate a new mosquito infection cycle remains to be elucidated.

Numerous species of mosquito have been associated with USUV infection in wild or captive avifauna (Cle et al. 2019). In general, these mosquitoes are ornithophilic species of the genus *Culex*. These vectors not only maintain the virus among avian species but are accountable for the spread to vulnerable mammals, especially humans and horses, which are dead-end hosts. USUV has been isolated from numerous mosquito species across the African continent, mainly in countries like Uganda, Senegal, and Kenya where entomological surveillance projects have been implemented, and in Southern and Central Europe including Italy and Austria (Camp et al. 2019; Cle et al. 2019). USUV has been more commonly detected in *Culex* species, and *Culex pipiens* has been identified as the main vector of USUV in Europe (Becker et al. 2012). Other mosquito species that may be infected with USUV include *Ae. minutus*, *Ae. japonicus*, *Ae. Albopictus*, *Anopheles maculipennis*, *Culiseta annulata*, *Coquilletidia aurites*, *Mansonia africana*, *Ochlerotatus detritus*, and *Ochlerotatus caspius* (Cle et al. 2019). Figure 7.1 schematically depicts the USUV transmission cycle.

Disorders of the CNS have been observed in birds infected with USUV disorders. Prominent clinical symptoms include prostration, weight loss, disorientation, and ataxia. Major macroscopic lesions are hepatomegaly and splenomegaly (Bakonyi et al. 2007; Cle et al. 2019). An inflammatory infiltrate consisting of histiocytic and lymphoid cells as well as necrotic lesions have been observed in kidneys, spleen, liver, heart, and brain of infected birds. Neuronophagia and glial nodules have also been found in the brain (Bakonyi et al. 2007; Cle et al. 2019). Because of its wide tropism and virulence in multiple organs and tissues, USUV is considered to be very pathogenic in captive and wild birds. In European countries, the spread of USUV has

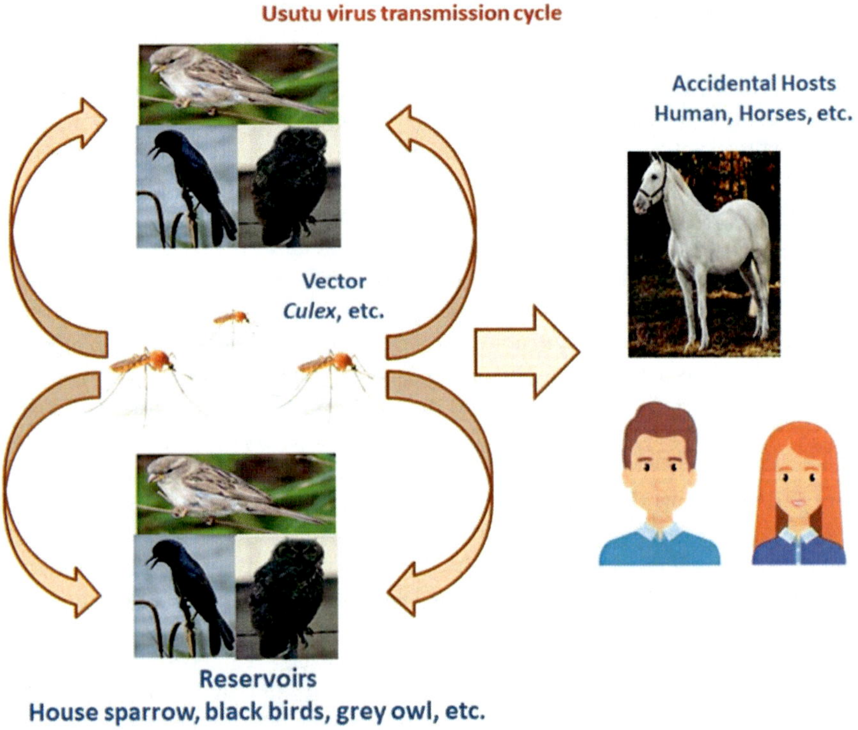

Fig. 7.1 Schematic representation of the USUV transmission cycle

resulted in extensive avian deaths, but the magnitude of USUV-associated mortality on the dynamics of the avian population is unclear. As with the case of WNV, the link between augmented bird mortality and speed of virus turnover in the natural reservoir that is a potential risk to incidental hosts like humans need to be studied (Chevalier et al. 2009; Cle et al. 2019).

In the 1980s and 2004, the first cases of USUV infection in humans were documented in the Central African Republic and Burkina Faso, respectively, and accompanied with mild symptoms such as rash and fever (Nikolay et al. 2011). A total of 28 acute USUV infections have been reported so far, with some cases showing severe complications such as facial paralysis and meningoencephalitis (Roesch et al. 2019). As evidenced by seroprevalence studies, USUV infections in humans may have been highly undervalued with many infected cases showing no symptoms. Based on (1) serological studies performed on samples from healthy individuals that included blood donors and forestry workers, and retrospective samples from 900 patients, one third of which were suspected for meningoencephalitis and encephalitis, and (2) failure to detect the virus in CSF of patients suffering from encephalitis, suggest that USUV may be significantly prevalent in humans, although rarely associated with neurological problems (Allering et al. 2012; Pierro et al. 2013; Maggi et al. 2015; Cvjetkovic et al. 2016; Grottola et al. 2017; Percivalle et al. 2017; Cordey et al. 2018). However, the tropism of the virus for the central nervous system in humans, especially in immunocompromised individuals, should be viewed with concern (Roesch et al. 2019).

The occurrence of USUV infection in humans has accentuated the need for the development of reliable and confirmatory diagnostic detection method. In this regard, a first USUV-specific and rapid real-time RT-PCR assay was developed, based on Hungarian and Austrian strain sequences by Cavrini et al. (2011). This assay was able to detect USUV in CSF and human blood with high sensitivity and specificity, but its detection was limited to a few European strains. To overcome this limitation, another real-time RT-PCR assay based on European and African strain sequence was developed (Nikolay et al. 2014). These techniques allow the detection of USUV in CSF and blood at the viremic stage. However, for diagnosis following the viremic stage, serological tests are important. The earliest IgG-capture ELISA assay specifically for the detection of USUV was developed by Gaibani et al. (2012). This assay could identify IgGs only specific for USUV in both German and Italian healthy blood donors (Allering et al. 2012; Gaibani et al. 2012). By fine-tuning a diagnostic algorithm and deploying the plaque reduction neutralization test (PRNT), the cross-reactivity between WNV and USUV was resolved (Martin et al. 2004; Gaibani et al. 2012). The PRNT is the existing standard to differentiate between intimately related flaviviruses (Gaibani et al. 2012). IgM-based assays specific to USUV must be developed as cross-reactivity for IgG is greater than for IgM in flaviviruses (Solomon 2004). No commercially licensed serological assay is available so far. The scope for development of novel USUV-related diagnosis is extensive as USUV infections have been documented across Europe and very few diagnostic methods are available for dependable diagnosis of USUV. Recent advances in next-generation sequencing (NGS) technologies have unbolted up new prospects for diagnosis and research of infectious diseases. Pathogen genomics can be explored to resolve decisive questions regarding the ecology of emerging viral diseases including origin and mode of transmission. In a recent study, using NGS, phylogenetic and molecular clock analysis and common USUV lineage were identified along with the evaluation of their most recent common ancestor (Oude Munnink et al. 2020).

Compared to some other arboviruses, USUV has been relatively less studied, and therefore, multidisciplinary intervention strategies are required to understand further, detect, control, and prevent the spread of the virus in humans and birds, with an emphasis on routine surveillance including those at the reservoir–vector interface, and developing new therapeutic and prophylactic approaches. All these efforts will certainly prepare us to face a possible large USUV outbreak in the coming years.

References

Allering L, Jöst H, Emmerich P, Günther S, Lattwein E, Schmidt M, Seifried E, Sambri V, Hourfar K, Schmidt-Chanasit J (2012) Detection of Usutu virus infection in a healthy blood donor from south-west Germany, 2012. Euro Surveill 17(50):20341

Bakonyi T et al (2004) Complete genome analysis and molecular characterization of Usutu virus that emerged in Austria in 2001: comparison with the South African Strain SAAR-1776 and other flaviviruses. Virology 328:301–310

Bakonyi T, Erdelyi K, Ursu K, Ferenczi E, Csorgo T, Lussy H, Chvala S, Bukovsky C, Meister T, Weissenbock H et al (2007) Emergence of Usutu virus in Hungary. J Clin Microbiol 45:3870–3874

Beck C et al (2013) Flaviviruses in Europe: complex circulation patterns and their consequences for the diagnosis and control of West Nile disease. Int J Environ Res Public Health 10:6049–6083

Becker N, Jost H, Ziegler U, Eiden M, Hoper D, Emmerich P, Fichet-Calvet E, Ehichioya DU, Czajka C, Gabriel M et al (2012) Epizootic emergence of Usutu virus in wild and captive birds in Germany. PLoS One 7:e32604

Benzarti E, Linden A, Desmecht D, Garigliany M (2019) Mosquito-borne epornitic flaviviruses: an update and review. J Gen Virol 100:119–132

Bergsman LD, Hyman JM, Manore CA (2016) A mathematical model for the spread of west nile virus in migratory and resident birds. Math Biosci Eng 13:401–424

Caballero-Gómeza J, Cano-Terriza D, Lecollinet S, Carbonell MD, Martínez-Valverde R et al (2020) Evidence of exposure to zoonotic flaviruses in zoo mammals in Spain and their potential role as sentinel species. Vet Microbiol 247:108763. https://doi.org/10.1016/j.vetmic.2020.108763

Cadar D, Becker N, Campos RM, Borstler J, Jost H, Schmidt-Chanasit J (2014) Usutu virus in bats, Germany, 2013. Emerg Infect Dis 20:1771–1773. https://doi.org/10.3201/eid2010.140909

Cadar D et al (2017) Widespread activity of multiple lineages of Usutu virus, Western Europe, 2016. Eur Secur 22:1–7

Calisher CH, Gould EA (2003) Taxonomy of the virus family Flaviviridae. Adv Virus Res 59:1–19

Camp JV, Kolodziejek J, Nowotny N (2019) Targeted surveillance reveals native and invasive mosquito species infected with Usutu virus. Parasit Vectors 12:46

Cavrini F, Della Pepa ME, Gaibani P, Pierro AM, Rossini G, Landini MP, Sambri V (2011) A rapid and specific real-time RT-PCR assay to identify Usutu virus in human plasma, serum, and cerebrospinal fluid. J Clin Virol 50:221–223

Chevalier V et al (2009) Predicting West Nile virus seroprevalence in wild birds in Senegal. Vector-Borne Zoonotic Dis 9:589–596

Chvala S, Kolodziejek J, Nowotny N, Weissenbock H (2004) Pathology and viral distribution in fatal Usutu virus infections of birds from the 2001 and 2002 outbreaks in Austria. J Comp Pathol 131:176–185

Chvala S, Bakonyi T, Hackl R, Hess M, Nowotny N, Weissenbock H (2005) Limited pathogenicity of Usutu virus for the domestic chicken (Gallus domesticus). Avian Pathol 34:392–395

Chvala S, Bakonyi T, Hackl R, Hess M, Nowotny N, Weissenbock H (2006) Limited pathogenicity of Usutu virus for the domestic goose (Anser anser f. domestica) following experimental inoculation. J Vet Med B Infect Dis Vet Public Health 53:171–175

Cle M, Beck C, Salinas S, Lecollinet S, Gutierrez S, Van de Perre P, Baldet T, Foulongne V, Simonin Y (2019) Usutu virus: a new threat? Epidemiol Infect 147:e232. https://doi.org/10.1017/S0950268819001213

Cordey S, Vieille G, Turin L, Kaiser L (2018) Usutu virus in cerebrospinal fluid: a 2-year survey in a Tertiary Care Hospital, Geneva, Switzerland. J Med Virol 90:609–611

Cvjetkovic IH, Petrovic T, Petric D, Cvjetkovic D, Kovacevic G, Radovanov J, Galovic AJ, Patic A, Nikolic N, Mikic SS (2016) Seroprevalence of mosquito-borne and tick-born microorganisms in human population of South Backa District. Arhiv Veterinarske Med 9:23–30

Diagne MM, Ndione MHD, Di Paola N, Fall G, Bedekelabou AP (2019) Usutu virus isolated from rodents in Senegal. Viruses 11:181. https://doi.org/10.3390/v11020181

Engel D et al (2016) Reconstruction of the evolutionary history and dispersal of Usutu virus, a neglected emerging arbovirus in Europe and Africa. mBio 7:1–12

Escribano-Romero E, Lupulovic D, Merino-Ramos T, Blazquez AB, Lazic G, Lazic S, Saiz JC, Petrovic T (2015) West Nile virus serosurveillance in pigs, wild boars, and roe deer in Serbia. Vet Microbiol 176:365–369. https://doi.org/10.1016/j.vetmic.2015.02.005

Gaibani P, Pierro A, Alicino R, Rossini G, Cavrini F, Landini MP, Sambri V (2012) Detection of Usutu virus-specific IgG in blood donors from northern Italy. Vector Borne Zoonotic Dis 12: 431–433

Gaibani P et al (2013) Comparative genomic and phylogenetic analysis of the first Usutu virus isolate from a human patient presenting with neurological symptoms. PLoS One 8(5). https://doi.org/10.1371/journal.pone.0064761

Garcia-Bocanegra I, Paniagua J, Gutierrez-Guzman AV, Lecollinet S, Boadella M, Arenas-Montes-A, Cano-Terriza D, Lowenski S, Gortazar C, Hofle U (2016) Spatio-temporal trends and risk factors affecting West Nile virus and related flavivirus exposure in Spanish wild ruminants. BMC Vet Res 12:249. https://doi.org/10.1186/s12917-016-0876-4

Gould E, Solomon T (2008) Pathogenic flaviviruses. Lancet 371:500–509

Grottola A, Marcacci M, Tagliazucchi S, Gennari W, Di AG, Orsini M, Monaco F, Marchegiano P, Marini V, Meacci M et al (2017) Usutu virus infections in humans: a retrospective analysis in the municipality of Modena, Italy. Clin Microbiol Infect 23:33–37

Lecollinet S, Blanchard Y, Manson C, Lowenski S, Laloy E, Quenault H, Touzain F, Lucas P, Eraud C, Bahuon C et al (2016) Dual emergence of Usutu virus in common blackbirds, Eastern France, 2015. Emerg Infect Dis 22:2225

Maggi F, Mazzetti P, Focosi D, Macera L, Scagnolari C, Manzin A, Antonelli G, Nelli LC (2015) Lack of Usutu virus RNA in cerebrospinal fluid of patients with encephalitis of unknown etiology, Tuscany, Italy. J Med Virol 87:913–916

Martin DA, Noga A, Kosoy O, Johnson AJ, Petersen LR, Lanciotti RS (2004) Evaluation of a diagnostic algorithm using immunoglobulin M enzyme-linked immunosorbent assay to differentiate human West Nile virus and St. Louis encephalitis virus infections during the 2002 West Nile virus epidemic in the United States. Clin Diagn Lab Immunol 11:1130–1133

McIntosh BM (1985) Usutu (SAAr 1776); nouvel arbovirus du groupe B. Int Cat Arboviruses 3: 1059–1060

Meister T, Lussy H, Bakonyi T, Sikutová S, Rudolf I, Vogl W, Winkler H, Frey H, Hubálek Z, Nowotny N, Weissenböck H (2008) Serological evidence of continuing high Usutu virus (*Flaviviridae*) activity and establishment of herd immunity in wild birds in Austria. Vet Microbiol 127(3–4):237–248

Montagnaro S, Piantedosi D, Ciarcia R, Loponte R, Veneziano V, Fusco G, Amoroso MG, Ferrara G, Damiano S, Iovane G et al (2019) Serological evidence of mosquito-borne flaviviruses circulation in hunting dogs in Campania Region, Italy. Vector Borne Zoonotic Dis 19:142–147. https://doi.org/10.1089/vbz.2018.2337

Nikolay B (2015) A review of West Nile and Usutu virus co-circulation in Europe: how much do transmission cycles overlap? Trans R Soc Trop Med Hyg 109:609–618

Nikolay B, Diallo M, Boye CS, Sall AA (2011) Usutu virus in Africa. Vector Borne Zoonotic Dis 11:1417–1423

Nikolay B et al (2013) Comparative full length genome sequence analysis of Usutu virus isolates from Africa. Virol J 10:1

Nikolay B, Weidmann M, Dupressoir A, Faye O, Boye CS, Diallo M, Sall AA (2014) Development of a Usutu virus specific real-time reverse transcription PCR assay based on sequenced strains from Africa and Europe. J Virol Methods 197:51–54

Oude Munnink BB, Münger E, Nieuwenhuijse DF et al (2020) Genomic monitoring to understand the emergence and spread of Usutu virus in the Netherlands, 2016–2018. Sci Rep 10:2798. https://doi.org/10.1038/s41598-020-59692-y

Percivalle E, Sassera D, Rovida F, Isernia P, Fabbi M, Baldanti F, Marone P (2017) Usutu virus antibodies in blood donors and healthy forestry workers in the Lombardy Region, Northern Italy. Vector Borne Zoonotic Dis 17:658–661

Pierro A, Gaibani P, Spadafora C, Ruggeri D, Randi V, Parenti S, Finarelli AC, Rossini G, Landini MP, Sambri V (2013) Detection of specific antibodies against West Nile and Usutu viruses in healthy blood donors in northern Italy, 2010-2011. Clin Microbiol Infect 19(10):E451–E453

Rijks JM, Kik ML, Slaterus R, Foppen R, Stroo A, Jzer JI, Stahl J, Grone A, Koopmans M, van der Jeugd HP et al (2016) Widespread Usutu virus outbreak in birds in the Netherlands, 2016. Euro Surveill 21:30391

Roesch F, Fajardo A, Moratorio G, Vignuzzi M (2019) Usutu virus: an arbovirus on the rise. Viruses 2019(11):640. https://doi.org/10.3390/v11070640

Romeo C, Lecollinet S (2018) Are tree squirrels involved in the circulation of flaviviruses in Italy? Transbound Emerg Dis 65:1372–1376. https://doi.org/10.1111/tbed.12874

Rouffaer LO, Steensels M, Verlinden M, Vervaeke M, Boonyarittichaikij R, Martel A, Lambrecht B (2018) Usutu virus epizootic and plasmodium coinfection in Eurasian blackbirds (Turdus merula) in Flanders, Belgium. J Wildl Dis 54:859–862

Solomon T (2004) Flavivirus encephalitis. N Engl J Med 351:370–378

Weissenbock H, Kolodziejek J, Fragner K, Kuhn R, Pfeer M, Nowotny N (2003) Usutu virus activity in Austria, 2001–2002. Microbes Infect 5:1132–1136

Williams MC et al (1964) The isolation of West Nile virus from man and of Usutu virus from the bird-biting mosquito Mansonia azurites (Theobald) in the Entebbe area of Uganda. Ann Trop Med Parasitol 58:367–374

Chapter 8
Neglected Bird-Associated Viral Zoonotic Infections

Abstract We here discuss some of the miscellaneous viral zoonoses involving birds either as reservoirs/natural/amplifying/carrier hosts. Western equine encephalomyelitis (WEE), Eastern equine encephalitis (EEE), Mayaro fever (MF), and sindbis fever (SF) are mosquito-transmitted viral zoonotic diseases caused by *Alphavirus* of *Togaviridae* family. St. Louis encephalitis (SLE) and Murray Valley encephalitis (MVE) are also mosquito-borne viral zoonoses caused by *Flavivirus* of *Flaviviridae* family. Kyasanur forest disease (KFD) and Crimean-Congo hemorrhagic fever (CCHF) are tick-borne viral zoonoses caused by *Flavivirus* (Family *Flaviviridae*) and *Orthonairovirus* (Family *Nairoviridae*), respectively. Epidemiology including transmission, geographic distribution, and role of birds, human disease spectrum, and prevention and control strategies for all the above diseases have been discussed in brief.

Keywords WEE · EEE · Mayaro fever · Sindbis fever · St. Louis encephalitis · Murray Valley encephalitis · Kyasanur forest disease · Crimean-Congo hemorrhagic fever

8.1 Western Equine Encephalomyelitis

Western equine encephalomyelitis (WEE) caused by WEE virus is a mosquito-transmitted arbovirus under the genus *Alphavirus* and family *Togaviridae* (Table 8.1). The name "WEE" is due to earlier major outbreaks in the western part of North America. In North America, the annual transmission cycle of WEE virus is well characterized with spring amplification, summer maintenance, fall decline, and winter quiescent. House sparrows (*Passer domesticus*) and *Culex tarsalis* mosquitoes are the principal enzootic host and vector, respectively (Barba et al. 2019; Bergren et al. 2020). During high enzootic periods, this virus can infect several mammals and start an independent *Aedes* spp.-mammal cycle. In South America, the ecology of WEE virus includes mammals and birds acting as amplification/ reservoir hosts and *Aedes* (*Ochlerotatus*) *albifasciatus* as a possible enzootic vector. Although several major outbreaks of equine and human WEE were documented in North

Y. S. Malik et al., *Role of Birds in Transmitting Zoonotic Pathogens*, Livestock Diseases and Management, https://doi.org/10.1007/978-981-16-4554-9_8

Table 8.1 Neglected viral zoonotic infections involving birds

Disease	Etiology (virus genus and family)	Distribution	Role of birds	Vector	Other reservoir/carrier/amplifier hosts	Major clinical signs in humans; case fatality rate (CFR)	Vaccines
Western equine encephalomyelitis	*Alphavirus, Togaviridae*	North and South America	House sparrows (*Passer domesticus*)—Principal enzootic host	*Culex tarsalis, Aedes (Ochlerotatus) albifasciatus*	Horses gets affected (dead-end hosts)	Mild pyrexia to encephalitis, followed by coma or death; 3–15% CFR	No human vaccines. Vaccines are available for equines
Eastern equine encephalomyelitis	*Alphavirus, Togaviridae*	North America	Tree-perching birds—Principal enzootic host	*Culiseta melanura*	Horses gets affected (dead-end or indicator hosts)	Neuroinvasive cases suffer from headache, fever, seizures, and altered mental status; 30% CFR in neuroinvasive EEE	No human vaccines. Vaccines are available for equines
Saint Louis encephalitis	*Flavivirus, Flaviviridae*	USA	Wild Passeriform (blue jays, robins, house sparrows, etc.) and Columbiform (pigeons) are the highly implicated enzootic and amplifying hosts	*Culex* spp.	Other vertebrates (horses and wild animals) are considered accidental or final hosts	Asymptomatic with rare presentations ranging from mild pyrexia to severe encephalitis; 5–15% CFR among encephalitic patients	No human or animal vaccines available
Murray Valley encephalitis	*Flavivirus, Flaviviridae*	Australia, New Guinea Island	Water birds (egrets and herons)—Principal enzootic hosts	*Culex annulirostris*	Rabbits and western gray kangaroos may act as an amplifier host	Asymptomatic with rare encephalitic presentations with fever, altered mental state, coma, and serious flaccid paralysis.	No human or animal vaccines available

						15–30% CFR among encephalitic patients	
Mayaro fever	*Alphavirus, Togaviridae*	South and Central Americas	Birds are one of the infected/susceptible hosts that may transmit infection to humans through mosquitoes	*Haemagogus* spp., *Aedes aegypti* (urban)	Other infected hosts are rodents, monkeys, sloths, and small mammals	Acute pyretic illness with arthritis/arthralgia and a maculopapular rash. The key symptom of chronic form of Mayaro fever is debilitating arthritis (50% cases)	No human or animal vaccines available
Sindbis fever	*Alphavirus, Togaviridae*	Eurasia, Africa, New Zealand, and Australia	Birds are the reservoir and amplifying hosts (fieldfare *Turdus pilaris*, redwing *Turdus iliacus*, and other passerines)	*Culex*, and occasionally *Aedes, Anopheles*, and *Culiseta* species	Not reported in other animals	Mild fever, headache, maculopapular rash, muscle pain, malaise, fatigue, and arthritis persisting for months to years	No human or animal vaccines available
Crimean-Congo hemorrhagic fever	*Orthonairovirus, Nairoviridae*	Asia, Africa, and Europe	Migratory birds—Transporter hosts transporting CCHFV-infected immature ticks	*Hyalomma* ticks	Small vertebrates such as rats, hedgehogs, and hares may act as amplifying hosts allowing immature ticks to feed on them	In severe cases—Hemorrhage, anemia, dehydration, myocardial infarction, lung edema, pleural effusion, kidney failure, capillary fragility, capillary toxicosis, and thrombocytopenia; >40% CFR	No human or animal vaccines available

(continued)

Table 8.1 (continued)

Disease	Etiology (virus genus and family)	Distribution	Role of birds	Vector	Other reservoir/carrier/amplifier hosts	Major clinical signs in humans; case fatality rate (CFR)	Vaccines
Kyasanur forest disease	*Flavivirus, Flaviviridae*	Western Ghats of South India	Ground-dwelling birds—Principal enzootic hosts along with other small mammals	Hard ticks (*Hemaphysalis spinigera*)	Red-faced bonnet monkey (*Macaca radiata*) and black-faced langur (*Presbytis entellus*)—Amplifying hosts	Biphasic clinical symptoms with initial acute flu-like phase accompanied by diarrhea, vomiting, bleeding, and severe muscle pain followed by afebrile stage during which most cases convalesce. In the second phase, meningoencephalitis affects approximately 10–20%; 3–10% CFR	Formalin-inactivated vaccine produced in chick embryo fibroblasts is currently in use for prophylaxis in the infected areas

America during the early-to-mid-twentieth century, there is a noticeable reduction in the virus activity in the late-twentieth century with the last case of human reported in 1998, and significantly reduced amounts of infected mosquito pools noticed in surveillance programs (Reisen and Wheeler 2016). WEEV causes encephalitis in horses and humans: both species are dead-end hosts and do not amplify sufficient viremia to infect mosquito vectors. However, few studies have reported that equids like burros and ponies develop viremia that can infect mosquito vectors (Reisen and Monath 1988; Bergren et al. 2020). Thus, there is an uncertainty in the role played by equids as amplification hosts. The disease spectrum in humans ranges from mild pyrexia to encephalitis, followed by coma or death. The severity is usually skewed toward elderly, infants, and young children with the case fatality range of 3–15%. The case fatality rate in equines ranges from 10 to 50%. No licensed vaccine is available for human use. However, formalin-inactivated vaccines are available for equines and are usually administered annually (Minke et al. 2004; Bergren et al. 2020).

8.2 Eastern Equine Encephalomyelitis

Eastern equine encephalitis (EEE) caused by EEE virus is a mosquito-transmitted arbovirus under the genus *Alphavirus* and family *Togaviridae* (Table 8.1). As the name suggests, EEE tends to occur more in the eastern part of North America. In North America, EEE is considered among the severe arboviral encephalitic diseases. *Culiseta melanura*, which breeds in hardwood swamps and freshwater, is the mosquito vector. Generally, EEE virus is sustained in a mosquito-bird-mosquito enzootic cycle, especially among tree-perching birds (Lindsey et al. 2018). Small mammals, amphibians, or reptiles may also be associated in the environmental circulation of virus (Morens et al. 2019). Spread to mammals is usually by bridging mosquito species such as *Aedes* or *Coquillettidia* that feed on both mammals and birds. Some studies provide evidence in support of the existence of alternative epizootic and enzootic cycles (Molaei et al. 2015; Bingham et al. 2016). Horses and humans are the dead-end hosts that generally do not develop adequate viremia to allow transmission of the virus to feeding mosquitoes. Barring 5% infected of infected humans developing meningitis or encephalitis, others show no apparent illness. Rare systemic infections are characterized by sudden fever, malaise, myalgia, chills, and arthralgia. Neuroinvasive human cases suffer from headache, fever, seizures, and altered mental status (Deresiewicz et al. 1997; Lindsey et al. 2018). The case fatality rate in EEE neuroinvasive disease is 30%, with 50% of survivors suffering from neurologic sequelae. Among other encephalitic alpha viruses, EEE virus causes most severe disease and mortality in equines. In fact, horses develop overt disease very early and thus serve as an indicator of epidemic start (Zacks and Paessler 2010). Although there is no licensed human vaccine, vaccines are existing for use in equines (Lindsey et al. 2018).

8.3 Saint Louis Encephalitis

St. Louis encephalitis (SLE) is a mosquito-borne zoonotic viral disease caused by SLE virus of genus *Flavivirus* under *Flaviviridae* family. This disease is endemic in the United States with some cases occurring in a wide area ranging from Argentina to Canada. In nature, the virus is maintained in an enzootic cycle between birds and mosquitoes (*Culex* sp.) with humans regarded as dead-end hosts. Among birds, wild Passeriform (blue jays, robins, house sparrows, etc.) and Columbiform (pigeons) are the highly implicated and amplifying hosts (Spinsanti et al. 2003; Curren et al. 2018). There are two distinct transmission patterns in western and eastern United States. In the eastern United States, SLEV transmission showed greater epidemic potential with *Culex quinquefasciatus, Culex pipiens,* and *Culex nigripalpus* as primary vectors. In contrast, in the western United States, the virus exhibited endemic transmission causing sporadic cases with *Culex tarsalis* as the primary vector (Davis et al. 2008; Curren et al. 2018). Other vertebrates (horses and wild animals) are considered accidental or final hosts (Rosa et al. 2013). In humans, the infection is usually asymptomatic with rare presentations ranging from mild pyrexia to severe encephalitis, with 5–15% case fatality rates among encephalitic patients. Severity increases with age, and people above 60 years old are more vulnerable to neuroinvasive disease. Currently, there are no licensed vaccines available to prevent SLE (Spinsanti et al. 2003; Curren et al. 2018; Diaz et al. 2018).

8.4 Murray Valley Encephalitis

Murray Valley encephalitis (MVE) is a mosquito-transmitted disease caused by MVE virus which is a member of the Japanese encephalitis serogroup of genus *Flavivirus* under *Flaviviridae* family. MVE is considered as a dreadful endemic in Australia and has been responsible for four large outbreaks of encephalitis on the east coast of Australia in the early twentieth century. Since then, MVEV has been circulating in enzootic foci at the top end of Northern Territory and the north of Western Australia in transmission cycles between *Culex annulirostris and* water birds. MVEV spread outside these foci is assumed to be due to flooding and rainfall that permitted infected water birds to formerly arid environments. Persistence in dehydration-resistant eggs of mosquitoes may also lead to outbreaks in previously arid areas, and the presence of enigmatic enzootic foci has also been suggested. Although MVEV has been reported in New Guinea Island, little is known about the epidemiology of the virus there (Smith et al. 2011; Selvey et al. 2014). MVEV infections are mostly asymptomatic, or cause a febrile illness with myalgia, headache, and occasional rash. Encephalitis occurs in 1:150 to 1:1000 cases presented with fever, altered mental state, coma, and serious flaccid paralysis. The outcome of encephalitis is death in 15–30% cases and 30–50% cases suffer from long-term neurological sequelae, whereas psychiatric illnesses and depression are common in all survivors. It also causes severe encephalitis in horses with insufficient viremia to infect mosquitoes. The main vertebrate hosts are supposed to be egrets and herons, mainly the night heron (*Nycticorax caledonicus*). Several domestic and wild animals have developed MVEV antibodies, but only rabbits and perhaps western gray kangaroos developed higher levels of viremia considered sufficient to upkeep local transmission cycles (Knox et al. 2012; Selvey et al. 2014; Mackenzie et al. 2017). Currently, no effective vaccination is available against MVEV, and prevention depends on mosquito control and sentinel surveillance using chickens (Knox et al. 2012).

8.5 Mayaro Fever

Mayaro fever caused by Mayaro virus (genus *Alphavirus*, family *Togaviridae*) is endemic to South America and Caribbean islands with occasional reports of outbreaks. Human cases have been primarily restricted to South and Central Americas, particularly in areas around the Amazon basin, with some cases in travelers returning from the Amazon basin emphasizing a possible global transmission (Mackay and Arden 2016). The virus is transmitted by mosquitoes, with *Haemagogus* as the principal vector, which transmits the virus from infected vertebrates (birds, rodents, monkeys, sloths, and small mammals) to susceptible individuals. The experimental finding of *Aedes aegypti*, being a probable vector of Mayaro virus, highlights the possibility of wider spread in urban areas (Esposito and Fonseca 2017). Izurieta (2018) proposed three dynamic transmission cycles: sylvatic cycle maintained by *Haemagogus* spp., which may further spread to urban setting through people visiting forests or living in peri-urban areas (intermediate cycle), and urban cycle mediated by *Aedes aegypti.* The clinical display of Mayaro virus is difficult to differentiate from other arboviruses such as chikungunya, dengue, and zika, possibly leading to underreporting (Lorenz et al. 2019). Human Mayaro viral infections are characterized by acute pyretic illness with arthritis/arthralgia and a maculopapular rash. Headache, retro-orbital pain, myalgia, diarrhea, and vomiting are other associated clinical manifestations (Acosta-Ampudia et al. 2018; Lorenz et al. 2019). The key symptom of Mayaro fever is debilitating arthritis, which can be established in the long-term recovery stage as a chronic form, occurring in 50% cases and persisting for a minimum of 1 year (Ferreira et al. 2020). As there is no effective antiviral therapy and licensed vaccines, control of mosquito populations and personal protection from mosquito bite are dependable preventive measures. Case management relies on symptomatic treatment with analgesics (Ganjian and Riviere-Cinnamond 2020).

8.6 Sindbis Fever

Sindbis fever is caused by sindbis virus (SINV), a member of the genus *Alphavirus* of the family *Togaviridae*. Sindbis fever is a bird-associated mosquito-borne zoonotic disease distributed in Eurasia, Africa, New Zealand, and Australia, forming five (SINV-I to SINV-V) genotypes, each of which appears to be restricted to a definite geographical region. All these regions have reported human cases; however, outbreaks have only been limited to northern Europe and South Africa and are associated with genotype SINV-I. SINV causes rash and polyarthritis, known as Pogosta disease in Finland and Ockelbo disease in Sweden. Birds are the reservoir and amplifying hosts of SINV and play a key role in spreading the virus to nonendemic areas. Mainly *Culex*, and occasionally *Aedes, Anopheles,* and *Culiseta* mosquitoes, have been associated in the transmission of the virus. Humans are dead-end hosts to receive accidental infection (Ziegler et al. 2019; Lundstrom et al. 2019). Local occurrence of sindbis fever requires the presence of SINV-I in the enzootic vector (*Culex torrentium*) and the amplifying bird species including fieldfare *Turdus pilaris*, redwing *Turdus iliacus*, and other passerines. Additionally, for tangential spread to humans from viremic birds, bridge vector like *A. cinereus* needs to be sufficiently abundant and infected (Hesson et al. 2015; Lundstrom et al. 2019). Clinical manifestations of sindbis fever in humans are generally mild with spontaneous recovery. Symptoms include mild fever, headache, maculopapular rash, muscle pain, malaise, fatigue, and arthritis persisting for months to years in several cases (Storm et al. 2013).

8.7 Crimean-Congo Hemorrhagic Fever

Crimean-Congo hemorrhagic fever (CCHF) is a tick-transmitted virulent viral zoonotic disease caused by CCHF virus belonging to the genus *Orthonairovirus* of the family *Nairoviridae.* Around 50 countries in Asia, Africa, and Europe are currently documented as endemic, or potentially endemic for CCHF, and the travelers visiting these areas have been infected with the severe hemorrhagic syndrome in some cases (Nasirian 2019). In nature, CCHF virus typically circulates in an enzootic cycle involving tick and vertebrates with *Hyalomma* ticks as a principal vector. A range of wild and domestic animals act as asymptomatic hosts in an endemic CCHF transmission cycle, playing a critical role in passing the virus from infected to a newer population of ticks (King et al. 2018; Nasirian 2019). Small vertebrates such as rats, hedgehogs, and hares may act as amplifying hosts allowing immature ticks to feed on them. Reptiles and birds (except ostriches) appear to be refractory to infection (Nasirian 2019). Emergence of CCHF in new areas and geographical expansion is mainly due to climate change, international animal trades, transportation of infected ticks by migratory birds, and human travel to habitats of migratory birds, or tourist places recognized as risky sites for transmission (Okely et al. 2019). CCHF outbreak may be created through migratory birds as they can transport CCHFV-infected immature ticks, and substantial importation of such ticks by migratory birds into the United Kingdom seems to occur (Jameson et al. 2012; Nasirian 2019). Apart from tick's bite, transmission of the virus to humans occurs through direct contact with the blood of an infected animal (slaughterhouse workers, farmers, veterinarians, etc.), nosocomial, and person-to-person transmission through body fluids like blood, saliva, and semen. Three cases of venereal transmission among spouses have also been observed (Pshenichnaya et al. 2016; Wahid et al. 2019). CCHF patients, during the prehemorrhagic period, experience chills, fever, abdominal pain, vomiting, loss of appetite, diarrhea, hearing loss, memory loss, poor vision, back pain, mood swings, aggression, violent behavior, confusion, and labored breathing. In severe cases of CCHF, common clinical observations are hemorrhage, anemia, dehydration, myocardial infarction, lung edema, pleural effusion, kidney failure, capillary fragility, capillary toxicosis, and thrombocytopenia. The case fatality rate of CCHF is >40% (Wahid et al. 2019). Although approved CCHF vaccine is not available for use, effective antiviral treatment (ribavirin) that inhibits virus replication is available. The synergistic effect of two FDA-approved molecules, i.e., ribavirin and chloroquine or chlorpromazine, against CCHF virus have been demonstrated recently (Ferraris et al. 2015; Tezer et al. 2016).

8.8 Kyasanur Forest Disease

Kyasanur forest disease (KFD) or monkey fever is caused by KFD virus (KFDV), a member of the genus *Flavivirus* and family *Flaviviridae*. KFD is geographically limited to the Western Ghats of South India; however, in recent decades, the zone has been expanding. The virus was first isolated in 1957 from a sick monkey from the Kyasanur Forest in Mysore (presently Karnataka) State of India. Since then, India reports 400–500 cases every year. The reservoir of the KFDV is hard ticks, particularly *Hemaphysalis spinigera*. Small mammals, such as shrews and rodents, or ground-dwelling birds and ticks maintains the natural enzootic cycle. The virus can cause epizootics with high mortality in monkeys, predominantly the red-faced bonnet monkey (*Macaca radiata*) and black-faced langur (*Presbytis entellus*) (Sharma et al. 2019). Following the death of the nonhuman primates, infected ticks droop down and form a hotspot for potential transmission to humans involved in clearing the sick or dead monkeys, or others entering the hotspot zone. Human transmission is possible by infected tick bite, direct contact with feces, or infected carcasses (Chakraborty et al. 2019). Common ancestry between KFDV in India and Alkhurma virus in Saudi Arabia was described in spite of their vast geographic separation, signifying an extensive movement of the virus, probably carried by migratory birds (Mehla et al. 2009; Madani et al. 2014; Murhekar et al. 2015). Viremic birds play a key role in the distant spread of viruses and may also transmit virus infected ticks over a long distance (Gould and Solomon 2008). KFDV-infected humans present biphasic clinical symptoms with initial acute flu-like phase accompanied by diarrhea, vomiting, bleeding, and severe muscle pain followed by afebrile stage during which most cases convalesce. In the second phase, meningoencephalitis affects approximately 10–20% cases 1–2 weeks consequent to the acute phase. The mortality rate of KFD is about 3–10% (Shah et al. 2018; Rajaiah 2019). Although specific antiviral treatment against KFDV is not established till date, a formalin-inactivated vaccine made in chick embryo fibroblasts is currently in use for prophylaxis in the infected areas (Mourya et al. 2019). Another control measure is the application of acaricides around 5-meter radius of the area where dead and infected monkeys are found (Rajaiah 2019).

References

Acosta-Ampudia Y, Monsalve DM, Rodríguez Y, Pacheco Y, Anaya JM, Ramírez-Santana C (2018) Mayaro: an emerging viral threat? Emerg Microbes Infect 7(1):1–11

Barba M, Fairbanks EL, Daly JM (2019) Equine viral encephalitis: prevalence, impact, and management strategies. Vet Med (Auckl) 10:99–110. https://doi.org/10.2147/VMRR.S168227. eCollection 2019

Bergren NA, Haller S, Rossi SL, Seymour RL, Huang J, Miller AL et al (2020) "Submergence" of Western equine encephalitis virus: evidence of positive selection argues against genetic drift and fitness reductions. PLoS Pathog 16(2):e1008102. https://doi.org/10.1371/journal.ppat.1008102

Bingham AM, Burkett-Cadena ND, Hassan HK, Unnasch TR (2016) Vector competence and capacity of Culex erraticus (Diptera: Culicidae) for eastern equine encephalitis virus in the southeastern United States. J Med Entomol 53:473–476

Chakraborty S, Andrade FCD, Ghosh S, Uelmen J, Ruiz MO (2019) Historical expansion of Kyasanur forest disease in India from 1957 to 2017: a retrospective analysis. GeoHealth 3(2):44–55

Curren EJ, Lindsey NP, Fischer M, Hills SL (2018) St. Louis encephalitis virus disease in the United States, 2003-2017. Am J Trop Med Hyg 99(4):1074–1079

Davis LE, Beckham JD, Tyler KL (2008) North American encephalitic arboviruses. Neurol Clin 26: 727–757

Deresiewicz RL, Thaler SJ, Hsu L, Zamani AA (1997) Clinical and neuroradiographic manifestations of eastern equine encephalitis. N Engl J Med 336:1867–1874

Diaz A, Coffey LL, Burkett-Cadena N, Day JF (2018) Reemergence of St. Louis encephalitis virus in the Americas. Emerg Infect Dis 24(12)

Esposito DLA, Fonseca BALD (2017) Will Mayaro virus be responsible for the next outbreak of an arthropod-borne virus in Brazil? Braz J Infect Dis 21(5):540–544

Ferraris O, Moroso M, Pernet O, Emonet S, Rembert AF, Paranhos-Baccalà G et al (2015) Evaluation of Crimean-Congo hemorrhagic fever virus in vitro inhibition by chloroquine and chlorpromazine, two FDA approved molecules. Antivir Res 118:75–81

Ferreira JM, Campos-Ferreira DS, Figueiredo EVMDS, Lima Filho JLD (2020) Mayaro fever: a brief review on the immune profile. Asian Pac J Trop Biomed 10(3):95–100

Ganjian N, Riviere-Cinnamond A (2020) Mayaro virus in Latin America and the Caribbean. Rev Panam Salud Publica 44:e14. https://doi.org/10.26633/RPSP.2020.14

Gould EA, Solomon T (2008) Pathogenic flaviviruses. Lancet 371:500–509

Hesson JC, Verner-Carlsson J, Larsson A, Ahmed R, Lundkvist Å, Lundstrom JO (2015) Culex torrentium mosquito role as major enzootic vector defined by rate of Sindbis virus infection, Sweden, 2009. Emerg Infect Dis 21:875–878

Izurieta R (2018) Mayaro virus: the jungle flu, vol 10. Dove Press, pp 9–17

Jameson LJ, Morgan PJ, Medlock JM, Watola G, Vaux AGC (2012) Importation of *Hyalomma marginatum*, vector of Crimean-Congo haemorrhagic fever virus, into the United Kingdom by migratory birds. Ticks Tick Dis 3:95–99

King AM, Lefkowitz EJ, Mushegian AR, Adams MJ, Dutilh BE, Gorbalenya AE, Harrach B, Harrison RL, Junglen S, Knowles NJ (2018) Changes to taxonomy and the international code of virus classification and nomenclature ratified by the international committee on taxonomy of viruses. Arch Virol 163:2601–2631

Knox J, Cowan RU, Doyle JS, Ligtermoet MK, Archer JS et al (2012) Murray Valley encephalitis: a review of clinical features, diagnosis and treatment. Med J Aust 196:322–326

Lindsey NP, Staples JE, Fischer M (2018) Eastern equine encephalitis virus in the United States, 2003-2016. Am J Trop Med Hyg 98:1472–1477

Lorenz C, Freitas Ribeiro A, Chiaravalloti-Neto F (2019) Mayaro virus distribution in South America. Acta Trop 198:105093

Lundstrom JO, Hesson JC, Schafer ML, Ostman O, Semmler T, Bekaert M et al (2019) Sindbis virus polyarthritis outbreak signalled by virus prevalence in the mosquito vectors. PLoS Negl Trop Dis 13(8):e0007702. https://doi.org/10.1371/journal.pntd.0007702

Mackay IM, Arden KE (2016) Mayaro virus: a forest virus primed for a trip to the city? Microbes Infect 18(12):724–734

Mackenzie JS, Lindsay MDA, Smith DW, Imrie A (2017) The ecology and epidemiology of Ross River and Murray Valley encephalitis viruses in Western Australia: examples of One Health in Action. Trans R Soc Trop Med Hyg 111:248–254. https://doi.org/10.1093/trstmh/trx045

Madani TA, Azhar EI, Abuelzein E-TME, Kao M, Al-Bar HMS, Farraj SA, Masri BE et al (2014) Complete genome sequencing and genetic characterization of Alkhumra hemorrhagic fever virus isolated from Najran, Saudi Arabia. Intervirology 57(5):300–310. https://doi.org/10.1159/000362334

Mehla R, Kumar SRP, Yadav PD, Barde PV, Yergolkar PN, Erickson BR, Carroll SA et al (2009) Recent ancestry of Kyasanur Forest disease virus. Emerg Infect Dis 15(9):1431–1437. https://doi.org/10.3201/eid1509.080759

Minke JM, Audonnet J-C, Fischer L (2004) Equine viral vaccines: the past, present and future. Vet Res 35(4):425–443. https://doi.org/10.1051/vetres:2004019. PMID: 15236675

Molaei G, Armstrong PM, Abadam CF, Akaratovic KI, Kiser JP, Andreadis TG (2015) Vector-host interactions of Culiseta melanura in a focus of eastern equine encephalitis virus activity in southeastern Virginia. PLoS One 10:e0136743

Morens DM, Folkers GK, Fauci AS (2019) Eastern equine encephalitis virus—another emergent arbovirus in the United States. N Engl J Med 381(21):1989–1992. https://doi.org/10.1056/NEJMp1914328

Mourya DT, Yadav PD, Ullas PT, Bhardwaj SD, Sahay RR, Chadha MS et al (2019) Emerging/re-emerging viral diseases & new viruses on the Indian horizon. Indian J Med Res 149:447–467

Murhekar MV, Kasabi GS, Mehendale SM, Mourya DT, Yadav PD, Tandale BV (2015) On the transmission pattern of Kyasanur Forest disease (KFD) in India. Infect Dis Pov 4:37. https://doi.org/10.1186/s40249-015-0066-9

Nasirian H (2019) Crimean-Congo hemorrhagic fever (CCHF) seroprevalence: a systematic review and meta-analysis. Acta Trop 196:102–120

Okely M, Anan R, Gad-Allah S, Samy AM (2019) Mapping the environmental suitability of etiological agent and tick vectors of Crimean-Congo hemorrhagic fever. Acta Trop 203: 105319. https://doi.org/10.1016/j.actatropica.2019.105319

Pshenichnaya NY, Sydenko IS, Klinovaya EP, Romanova EB, Zhuravlev AS (2016) Possible sexual transmission of Crimean-Congo hemorrhagic fever. Int J Infect Dis 45:109–111

Rajaiah P (2019) Kyasanur forest disease in India: innovative options for intervention. Hum Vaccin Immunother 15:2243–2248. https://doi.org/10.1080/21645515.2019.1602431

Reisen WK, Monath TP (1988) Western equine encephalomyelitis. In: Monath TP (ed) The arboviruses: epidemiology and ecology, vol V. CRC Press, Boca Raton, FL, pp 89–137

Reisen WK, Wheeler SS (2016) Surveys for antibodies against Mosquitoborne encephalitis viruses in California birds, 1996–2013. Vector-Borne Zoonotic Dis 16(4):264–282. https://doi.org/10.1089/vbz.2015.1888. PMID: 26974395

Rosa R, Costa EA, Marques RE et al (2013) Isolation of Saint Louis encephalitis virus from a horse with neurological disease in Brazil. PLoS Negl Trop Dis 7(11):e2537. Published 2013 Nov 21. https://doi.org/10.1371/journal.pntd.0002537

Selvey LA, Dailey L, Lindsay M, Armstrong P, Tobin S, Koehler AP et al (2014) The changing epidemiology of Murray Valley encephalitis in Australia: the 2011 outbreak and a review of the literature. PLoS Negl Trop Dis 8(1):e2656

Shah SZ, Jabbar B, Rahman Z, Nadeem S, Jabbar I, Azam S, Ahmed N, Nasir H, Rehman A (2018) Epidemiology, pathogenesis, and control of a tick-borne disease-Kyasanur Forest disease: current status and future directions. Front Cell Infect Microbiol 8. https://doi.org/10.3389/fcimb.2018.00149

Sharma SN, Kumawat R, Singh SK (2019) Kyasanur Forest disease: vector surveillance and its control. J Commun Dis 51(2):38–44

Smith DW, Speers DJ, Mackenzie JS (2011) The viruses of Australia and the risk to tourists. Travel Med Infect Dis 9:113–125

Spinsanti LI, Basquiera A, Bulacio S, Somale V, Kim SC, Re VE et al (2003) St. Louis encephalitis in Argentina: the first case reported in the last seventeen years. Emerg Infect Dis 9:271–273

Storm N, Weyer J, Markotter W, Kemp A, Leman PA, Dermaux-Msimang V et al (2013) Human cases of Sindbis fever in South Africa, 2006–2010. Epidemiol Infect 142:234–238

Tezer H, Ozkaya-Parlakay A, Gulhan B, Kanik-Yuksek S (2016) Ribavirin use in pediatric patients with Crimean-Congo hemorrhagic fever: is it really necessary? Braz J Infect Dis 20:222–223

Wahid B, Altaf S, Naeem N, Ilyas N, Idrees M (2019) Scoping review of Crimean-Congo hemorrhagic fever (CCHF) literature and implications of future research. J Coll Physicians Surg Pak 29(6):563–573

Zacks MA, Paessler S (2010) Encephalitic alphaviruses. Vet Microbiol 140(3–4):281–286. https://doi.org/10.1016/j.vetmic.2009.08.023

Ziegler U, Fischer D, Eiden M, Reuschel M, Rinder M, Muller K, Schwehn R, Schmidt V, Groschup MH, Keller M (2019) Sindbis virus—a wild bird associated zoonotic arbovirus circulates in Germany. Vet Microbiol

Part III
Bacterial Infections

Chapter 9
Clostridial Infections (Avian Botulism)

Abstract Avian botulism is a neuroparalytic disease caused by antigenically distinct toxins produced by *Clostridium botulinum*. The disease affects domestic fowls as well as wild birds for which botulism is most significant cause of mass mortality. The disease has been reported from many parts of the world. Birds usually contract the disease through consumption of contaminated feed, water, insects, and also from litter. Many epidemiological factors influence the occurrence of outbreak of botulism, and eutrophication of water bodies is a particular risk factor in wild birds. Domestic chicken and contaminated litter are important risk factors for spread, even to nonavian species. The disease is manifested as a flaccid paralytic illness resulting from blocking of the neurotransmitter, acetylcholine, by the botulinum toxin at synaptic junctions. Clinical signs include progressive paralysis affecting skeletal muscles of legs, wings, neck, and face in order. Postmortem lesions are usually inconspicuous and may not be informative enough. Therefore, diagnosis of the disease is aided by a host of laboratory tests including microbiological, immunological, and molecular tests. Despite success of antitoxin treatment for individual birds, application remains limited in farmed and wild birds. Prevention of the botulism principally involves general hygienic measures at farm level especially provisioning safe feed and prompt disposal of carcasses to prevent insect breeding. Botulism is an equally important disease in humans though involved toxin types are different for birds and humans. However, rare cases of human disease with avian types and vice versa have been documented.

Keywords Avian botulism · Clostridial diseases · Botulinum toxin · Wild birds · Human botulism · Domestic birds · Eutrophication

9.1 Introduction

Botulism, a severe paralytic disease, is an intoxication caused by ingestion of botulinum neurotoxins (BoNTs) produced by *Clostridium botulinum* (Anniballi et al. 2013; Pattison et al. 2008; Skarin et al. 2015). The intoxication affects many different hosts including wild and domestic birds including commercial chickens,

Y. S. Malik et al., *Role of Birds in Transmitting Zoonotic Pathogens*, Livestock Diseases and Management, https://doi.org/10.1007/978-981-16-4554-9_9

mammals, and humans worldwide (Pattison et al. 2008). Avian botulism is the most significant disease of the migratory birds, particularly the waterfowl and the shorebirds. The disease is known by many names such as limberneck, Western duck disease, duck disease, as well as the alkali poisoning (Pattison et al. 2008). Among wild and migratory birds, the disease is known to cause heavy mortality though severity of the disease varies according to host species involved and from place to place. Avian botulism (Type C) was responsible for over million deaths in some water birds in North America in a single year. High mortality rates are always expected in any avian botulism outbreaks as up to 50,000 deaths may be reported in only one outbreak. Other factors also affect the outcomes of each outbreak such as the type of circulating bacteria and the environmental factors.

9.2 Epidemiology

9.2.1 Agent

The cause of botulism is the toxins produced by the Gram-positive, spore-forming, and anaerobic bacteria *Clostridium botulinum*. The bacteria produce several types of toxins. There are eight well-known antigenically distinct toxigenic groups (A, B, C1, C2, D, E, F, and G). In birds, botulism is mostly caused by toxin type C, though types A, D, and E are involved. Human botulism, on the other hand, is mostly caused by A, B, E, and F (Pattison et al. 2008; Sato et al. 2016; Souillard et al. 2015). The botulinum toxin is highly potent with minimum lethal dose (MLD) as measured for guinea pigs is 12 × 10–5 mg/kg. Type C toxin is usually produced under certain conditions such as anaerobic environment; 35–37 °C are the optimum temperature for the toxin production (Sato et al. 2016).

9.2.2 Host Range and Distribution

Botulism affects a wide range of hosts including humans and birds. Wide range of wild birds are known to be susceptible to botulism, especially with type C toxin, and poultry may act as asymptomatic carrier for type D botulism (Badagliacca et al. 2018; Pattison et al. 2008; Sato et al. 2016; Silva et al. 2018). Susceptibilities of wild birds to *C. botulinum* also vary perhaps in relation to feeding habits, arrival season, and other factors (Anza et al. 2016). Almost all birds are susceptible to *C. botulinum* barring scavenger birds such as vultures. Outbreaks and cases of botulism have been reported in many species of birds including a recent mass mortality in migratory birds in Iran (Parchizadeh and Belant 2020).

Avian botulism, in fact, have been reported from many parts of the world often causing mass mortality in wild birds and farmed chickens (Lima et al. 2020; Parchizadeh and Belant 2020; Rasetti-Escargueil et al. 2019). In the Netherlands, botulism in waterfowls have been recorded since 1970 (Bongers and Tetenburg 1996). In Montana, United States, an investigation into mortality among water birds revealed involvement of type C botulism (Docherty et al. 2012). Intensive outbreaks of botulism in native chicken farms, pheasant farm, and in a community of spot-billed ducks were reported from South Korea (Jang et al. 2014). In an extensive South Korean outbreak of botulism among wild birds, 697 birds were killed in Namdong reservoir and Gong-gu region. Various types of birds were involved including spot-billed ducks, gray herons, Vega gulls, and sandpiper (Son et al. 2018). In the UK, an outbreak of botulism in water birds caused death of 990 gulls (June et al. 2016). Similarly, in another outbreak of botulism among farmed ducks and pheasants in the UK, approximately 800 duck and 800 pheasants died resulting in heavy economic loss (Otter et al. 2018). Outbreaks of botulism were also observed in farmed turkeys in Germany, where the mortality ranged from initial 12 to 50% (Popp et al. 2013).

Cases of botulism are also common among captive wild birds. In Brazil, an outbreak of type C botulism killed captive black-necked swans, Muscovy ducks, and whistling duck in a dam at a southern Brazilian zoo (Raymundo et al. 2012). Botulism may pose a threat to farmed laying hens too. In 2016, Sato and coworkers reported an outbreak of type C botulism in commercial layer chickens that occurred in two free-ranging farms raising chickens organically in Midwest United States (Sato et al. 2016). Similar outbreaks of botulism were also documented in Swedish laying hens affected by rather unusual mosaic toxin from type C/D *C. botulinum* isolates (Skarin et al. 2015). Intoxication of domestic chickens were also reported in a mixed species outbreak involving dogs, marmoset, and chickens (Silva et al. 2018). Laying hens of free-ranging farms were also known to be affected with mixed type C/D botulism in France (Souillard et al. 2017b). In Italy, an outbreak among free-living population of mallards and geese was documented with 86.8% mortality involving the chimeric type (C/D) *C. botulinum* (Badagliacca et al. 2018). In central European region of Poland, two outbreaks of botulism were reported which affected large number of water birds (5500 in one outbreak and 1600 in another) involving multiple species such as waterfowl, shorebirds, rallids, larids, mallards, and coots (Włodarczyk et al. 2014). In fact, globally, avian botulism is perhaps the most important disease of wild migratory birds (Anniballi et al. 2013).

9.2.3 Transmission and Role of Birds

Botulism spreads chiefly through feed, water, and litters in poultry sheds, though insects and maggots feeding on feces are also implicated (Pattison et al. 2008). Common feed ingredients such as rice hulls can also be potential sources for botulism in birds (Bano et al. 2013). The spread, however, is influenced by multiple factors, depending on the species involved and the settings in which the outbreaks occur. In an intensive longitudinal analysis of *C. botulinum* in commercial poultry farms from France, it was observed that darkling beetles, ventilation systems, soils around the houses, and the water were important source of infection among chickens (Souillard et al. 2014). The role of poultry birds in dissemination of botulism was demonstrated when an outbreak of botulism was traced back to poultry outbreak that had happened previously in a mixed farm (Souillard et al. 2015). Analysis of water birds for carriage and dissemination indicated that birds might excrete *C. botulinum* for up to 7 days with rails and ducks being more prolific excretor of the pathogen than gulls (Anza et al. 2016).

In an investigation to identify potential risk factors for botulism in wild water birds, it was observed that eutrophication of wetlands through discharge of inefficiently treated effluents led to planktonic build up which in turn attracted water birds which fell prey to botulism (Anza et al. 2014).

Domestic chickens and their litters play important role in dissemination of botulism (Rasetti-Escargueil et al. 2019). In France, outbreak of cattle botulism was attributed to flocks of chickens which were reared in nearby farm, and *C. botulinum* type D was demonstrated from the digestive tract of the affected cattle as well as from hens of the farm (Rasetti-Escargueil et al. 2019). Similarly, in the UK, poultry litters, farm dusts, and pasture contaminated with poultry manures were found to harbor a spread *C. botulinum* type C and type D to cattle (Relun et al. 2017; Souillard et al. 2017a). As chickens are somewhat resistant to type C and type D toxins, it is apparent that birds resistant to certain types of botulism may act as healthy carriers of botulism to other susceptible species (Rasetti-Escargueil et al. 2019).

9.3 Disease Features

9.3.1 Pathogenesis

The fundamental cause of botulism is botulinum neurotoxin (BoNT) elaborated by *C. botulinum*. The intoxication occurs through ingestion of food/feed containing preformed toxin. However, toxin-infection is also possible when material containing *C. botulinum* is ingested and the microorganism produces toxin inside the host. The binding of the *C. botulinum* neurotoxin to the cell membrane of the nerve cells is the most critical point in establishing paralysis. As soon as the neurotoxin reaches the presynaptic membrane, it binds irreversibly to the cholinergic nerve fibers and subsequently the paralysis occurs (Wobeser 1997). Botulinum toxin acts on nerve endings and blocks the neurotransmitter acetylcholine thus interfering with neural conduction resulting in a flaccid paralysis. The toxin is secreted by the bacteria during its vegetative growth and is activated by proteases of the host tissues. Following ingestion, the toxin is absorbed through intestinal epithelium and reaches circulation finally progressing to peripheral cholinergic nerve endings, where it

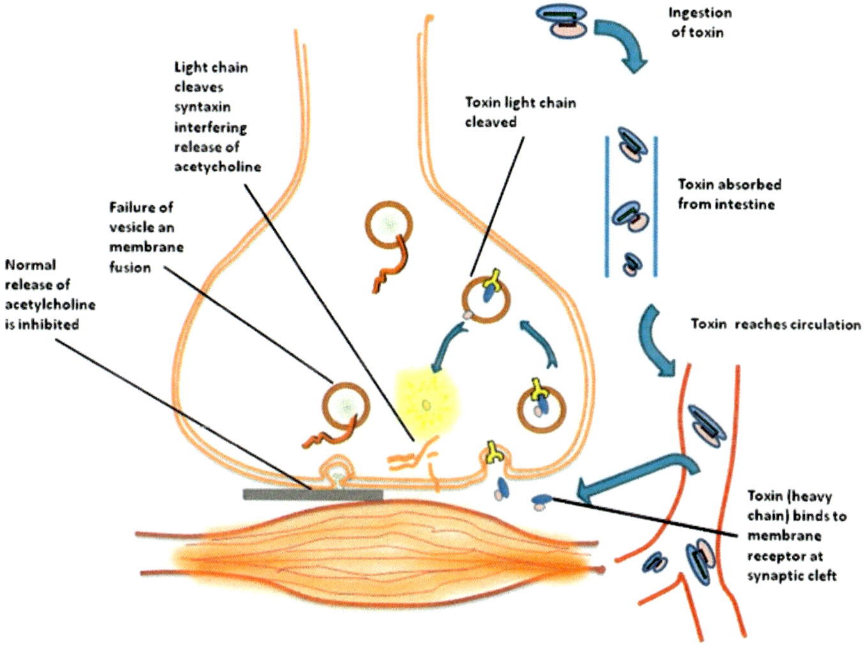

Fig. 9.1 Mechanism of flaccid paralysis caused by *Clostridium botulinum* toxin

binds to receptors, and enters the cytoplasm of the nerve cells to block the acetylcholine (Anniballi et al. 2013; Pattison et al. 2008). The signs of botulism are thus associated with failure of neural conduction (Fig. 9.1).

9.3.2 Clinical Signs and Pathology

Predominant clinical sign in botulism is flaccid type paralysis. As long as the toxins of the *C. botulinum* prevent the transmission of the signals in the peripheral nervous system mainly in the cholinergic nerves of both the automatic and the motor nerves, the expected outcome is the paralysis of the affected nerve. The affected birds usually develop a rapid onset of the flaccid paralysis of the skeletal muscles. In birds, this is manifested by progressive paralysis first affecting legs, followed by wings, neck, and face symptomized by bird's inability to walk or stand and uncoordinated flights with marked reluctance to move. Paralysis of neck muscles leads to drooping of head, and birds may be seen to bear head weight with beak support. Lateral or ventral recumbence is common. In broiler chickens, diarrhea may be seen (Anniballi et al. 2013; Pattison et al. 2008; Duff and Wildlife 2017).

Usually, postmortem findings are inconspicuous and nonspecific to the intoxication. Hyperemia and congestion of the internal organs may be seen in the carcasses of the dead birds. As the disease is often associated with ingestion of rotting carcasses and maggots, the gizzards may be examined for the same though such findings are uncommon (Anniballi et al. 2013; Pattison et al. 2008).

9.4 Disease Management

9.4.1 Diagnosis

Avian botulism is an important disease of birds that still presents diagnostic challenges. In case of birds, diagnosis of botulism is based presumptively on clinical signs. In addition, lack of pathological lesions in organs also point to possible occurrence of botulism in birds. Among many paralytic symptoms, paralysis of eyelids is considered a definitive sign indicating avian botulism (Pattison et al. 2008). Laboratory tests are helpful in establishing a diagnosis. Laboratory diagnosis of botulism centers around two approaches—detection of toxin and detection of the organism.

There are several tests available for the diagnosis of botulism. For detection of botulinum toxin, one of the most popular test is mouse lethality test (Dembek et al. 2007; Johnson et al. 2016). However, other tests including ELISA, chemiluminescent blot assay, immuno-PCR, electro-chemiluminescence, endopeptidase assay, lateral flow assay, and radioimmune assay have also been reported with varying successes (Lindström and Korkeala 2006). Among these assays, ELISA appears to be more commonly employed test due to ease of operation and speed of result delivery.

For detection of *C. botulinum* from suspected samples, classical bacteriological methods are useful. Culture and isolation of *C. botulinum* requires strict anaerobic conditions, and specialized laboratory is usually recommended. Common bacteriological media such as blood agar or egg yolk agar with or without selective agents are generally used for isolation and culture, though strict anaerobiosis is required at all stages. To overcome the limitation of traditional cultural methods, a number of DNA-based detection protocols have been reported (Lindström and Korkeala 2006). These assays are chiefly PCR based or are based on hybridization-specific probes or harness the power of both techniques. Sensitivities of these assays also vary considerably from as low as 12.5 fg to 0.3 ng for extracted DNA and 10^{-1} to 10^{3} cells per gram of sample (Lindström and Korkeala 2006; Palma et al. 2019; Rimawi 2019). Due to lethality and serious nature of botulism, the time-consuming nature of available detection techniques poses a challenge for utility of the results in treatment and control. However, recently a novel methodology for rapid isolation of *C. botulinum* strains involved in avian botulism have been described with innovative use of InstaGene matrix for isolation (Le Gratiet et al. 2020).

9.4.2 *Treatment and Control*

The mainstay of treatment for botulism is antitoxin therapy which need to be started as soon as botulism is suspected in birds (Anniballi et al. 2013). The mildly affected birds can be treated and protected from the other predators with a great success. Antitoxin therapy has also been reported to be successful in ostriches (Pattison et al. 2008). Successful treatment of wild birds (black-fronted Piping-guan, wild duck) and domestic chickens has been reported from Brazil with high success rates (Silva et al. 2017). Additional supportive therapy include oral rehydration, vitamins (A, D_3, and E), and provision of rest under shade (Pattison et al. 2008). Antibiotic therapy is usually not recommended though beta-lactam antibiotics have been used sometimes (Anniballi et al. 2013; Pattison et al. 2008). Though vaccinations are practiced for other animals, it is usually not practiced in birds.

Provision of safe feed free from *C. botulinum* spores and proper disposal of the carcasses of the dead birds are crucial steps for commercially farmed birds. For wild birds, avoidance of pollution and eutrophication of wetlands and water bodies where birds nest and rest may help in reducing the incidences. Early removal of carcasses is especially important as a study from North America showed that due to the rapid presence of the maggots in the dead birds as early as 3–5 days post death, early removal of the carcasses is one of the most important control measures against botulism though the cleanup process may be costly. Recently it has been shown that *C. botulinum* can survive lengthy periods in poultry manure, thus increasing the chances of environmental dissemination of spores and recurrence of the disease (Souillard et al. 2020). Therefore, proper decontamination and management of bird manure should be taken up for avoiding future disease outbreaks.

9.5 Public Health Concerns

Botulism is a deadly disease in avian and also in humans manifested by flaccid paralysis causing respiratory failure and death in fatal cases (Ambrožová 2019; Dembek et al. 2007; Palma et al. 2019; Rasetti-Escargueil et al. 2019; Wendt et al. 2017). Though rare cases of human disease with type C and D toxin and avian botulism due to types A and E have been documented, birds are not considered major risks for human botulism (Pattison et al. 2008). For humans, production and accumulation of botulinum neurotoxin in infected food is a significant risk. Toxin tends to accumulate during the period of food preservation prior to consumption when there is a favorable condition for bacterial growth. Botulism due to type C *C. botulinum*, though rare, does happen, especially in infants, and the toxin is produced in the intestinal tract leading to botulism. In humans, wound botulism is also reported where the wound is infected with toxigenic *C. botulinum* spores that eventually lead to poisoning. However, botulism is rarer in humans compared to birds. As such for safety of humans, birds from poultry flocks affected with botulism should not be allowed as human food for 2 weeks (Ambrožová 2019; Dembek et al. 2007; Lindström and Korkeala 2006; Palma et al. 2019; Rasetti-Escargueil et al. 2019; Rimawi 2019; Wendt et al. 2017).

References

Ambrožová H (2019) Botulism—a rare but still present, life-threatening disease. Epidemiol Mikrobiol Imunol 68:33–38

Anniballi F, Fiore A, Löfström C, Skarin H, Auricchio B, Woudstra C, Bano L, Segerman B, Koene M, Båverud V, Hansen T, Fach P, Åberg AT, Hedeland M, Engvall EO, De Medici D (2013) Management of animal botulism outbreaks: from clinical suspicion to practical countermeasures to prevent or minimize outbreaks. Biosecur Bioterror 11:S191–S199. https://doi.org/10.1089/bsp.2012.0089

Anza I, Vidal D, Laguna C, Díaz-Sánchez S, Sánchez S, Chicote Á, Florín M, Mateo R (2014) Eutrophication and bacterial pathogens as risk factors for avian botulism outbreaks in wetlands receiving effluents from urban wastewater treatment plants. Appl Environ Microbiol 80:4251–4259. https://doi.org/10.1128/aem.00949-14

Anza I, Vidal D, Feliu J, Crespo E, Mateo R (2016) Differences in the vulnerability of waterbird species to botulism outbreaks in Mediterranean wetlands: an assessment of ecological and physiological factors. Appl Environ Microbiol 82:3092–3099. https://doi.org/10.1128/aem.00119-16

Badagliacca P, Pomilio F, Auricchio B, Sperandii AF, Di Provvido A, Di Ventura M, Migliorati G, Caudullo M, Morelli D, Anniballi F (2018) Type C/D botulism in the waterfowl in an urban park in Italy. Anaerobe 54:72–74. https://doi.org/10.1016/j.anaerobe.2018.07.010

Bano L, Drigo I, Tonon E, Agnoletti F, Giovanardi D, Morandini E (2013) Rice hulls as a possible source of Clostridium botulinum type C spores for poultry. Vet Rec 173:427–428. https://doi.org/10.1136/vr.f6521

Bongers JH, Tetenburg GJ (1996) Botulism in waterfowl. Vet Q 18:156–157. https://doi.org/10.1080/01652176.1996.9694725

Dembek ZF, Smith LA, Rusnak JM (2007) Botulism: cause, effects, diagnosis, clinical and laboratory identification, and treatment modalities. Disaster Med Public Health Prep 1:122–134. https://doi.org/10.1097/DMP.0b013e318158c5fd

Docherty DE, Franson JC, Brannian RE, Long RR, Radi CA, Krueger D, Johnson RF (2012) Avian botulism and avian chlamydiosis in wild water birds, Benton Lake National Wildlife Refuge, Montana, USA. J Zoo Wildl Med 43:885–888. https://doi.org/10.1638/2011-0200r1.1

Duff P, Wildlife A, Group E (2017) Recognising clinical avian botulism in wild waterbirds. Vet Rec 181:15. https://doi.org/10.1136/vr.j3069

Jang I, Kang M-S, Kim H-R, Oh J, Lee J-I, Lee H-S, Kwon Y-K (2014) Occurrence of avian botulism in Korea during the period from June to September 2012. Avian Dis 58:666–669. https://doi.org/10.1637/10793-020414-case

Johnson AL, McAdams-Gallagher SC, Aceto H (2016) Accuracy of a mouse bioassay for the diagnosis of botulism in horses. J Vet Intern Med 30:1293–1299. https://doi.org/10.1111/jvim.13950

June D, Influenza A, Bird W (2016) Suspected avian botulism outbreaks in wild waterbirds during the summer. Vet Rec 179:144. https://doi.org/10.1136/vr.i4173

Le Gratiet T, Poezevara T, Rouxel S, Houard E, Mazuet C, Chemaly M, Le Maréchal C (2020) Development of an innovative and quick method for the isolation of Clostridium botulinum strains involved in avian botulism outbreaks. Toxins (Basel) 12. https://doi.org/10.3390/toxins12010042

Lima PC, Dutra IS, Araújo FAA, Lustosa R, Zeppelini CG, Franke CR (2020) First record of mass wild waterfowl mortality due to Clostridium botulinum in Brazilian semiarid. An Acad Bras Cienc 92:e20180370. https://doi.org/10.1590/0001-3765202020180370

Lindström M, Korkeala H (2006) Laboratory diagnostics of botulism. Clin Microbiol Rev 19:298–314. https://doi.org/10.1128/CMR.19.2.298-314.2006

Otter A, Payne J, Carr S (2018) Botulism in farmed ducks and pheasants. Vet Rec 183:452–453. https://doi.org/10.1136/vr.k4266

Palma NZ, da Cruz M, Fagundes V, Pires L (2019) Foodborne Botulism: Neglected Diagnosis Eur J case reports Intern Med 6:1122. https://doi.org/10.12890/2019_001122

Parchizadeh J, Belant JL (2020) Mass mortality of migratory birds in Iran. Science. https://doi.org/10.1126/science.abb4887

Pattison M, McMullin P, Bradbury J, Alexander D (eds) (2008) Poultry diseases, 6th edn. https://doi.org/10.1016/B978-0-7020-2862-5.50031-3

Popp C, Hauck R, Gad W, Hafez HM (2013) Type C botulism in a commercial turkey farm: a case report. Avian Dis Dig 7:e41–e42. https://doi.org/10.1637/10366-1027412-digest.1

Rasetti-Escargueil C, Lemichez E, Popoff MR (2019) Public health risk associated with botulism as foodborne zoonoses. Toxins (Basel) 12. https://doi.org/10.3390/toxins12010017

Raymundo DL, Von Hohendorf R, Boabaid FM, Both MC, Sonne L, Assis RA, Caldas RP, Driemeier D (2012) Outbreak of type C botulism in captive wild birds. J Zoo Wildl Med 43: 388–390. https://doi.org/10.1638/2010-0084.1

Relun A, Dorso L, Douart A, Chartier C, Guatteo R, Mazuet C, Popoff MR, Assié S (2017) A large outbreak of bovine botulism possibly linked to a massive contamination of grass silage by type D/C Clostridium botulinum spores on a farm with dairy and poultry operations. Epidemiol Infect 145:3477–3485. https://doi.org/10.1017/s0950268817002382

Rimawi BH (2019) Botulism in pregnancy—a clinical approach to diagnosis and management. J Matern Neonatal Med 32:3125–3132. https://doi.org/10.1080/14767058.2018.1457641

Sato Y, Wigle WL, Gallagher S, Johnson AL, Sweeney RW, Wakenell PS (2016) Outbreak of type C botulism in commercial layer chickens. Avian Dis 60:90–94. https://doi.org/10.1637/11293-100415-case.1

Silva ROS, Gómez SYM, Medeiros LB, Marques MVR, Silva ASG, Mureb EN, Oliveira Junior CA, Favoretto SM, Lobato FCF, Martins NRS (2017) Antitoxin therapy of natural avian botulism outbreaks occurred in Brazil. Anaerobe 48:115–117. https://doi.org/10.1016/j.anaerobe.2017.08.005

Silva ROS, Martins RA, Assis RA, Oliveira Junior CA, Lobato FCF (2018) Type C botulism in domestic chickens, dogs and black-pencilled marmoset (Callithrix penicillata) in Minas Gerais, Brazil. Anaerobe 51:47–49. https://doi.org/10.1016/j.anaerobe.2018.03.013

Skarin H, Lindgren Y, Jansson DS (2015) Investigations into an outbreak of botulism caused by Clostridium botulinum type C/D in laying hens. Avian Dis 59:335–340. https://doi.org/10.1637/10861-051214-case

Son K, Kim YK, Woo C, Wang S-J, Kim Y, Oem J-K, Jheong W, Jeong J (2018) Minimizing an outbreak of avian botulism (Clostridium botulinum type C) in Incheon, South Korea. J Vet Med Sci 80:553–556. https://doi.org/10.1292/jvms.17-0519

Souillard R, Woudstra C, Le Maréchal C, Dia M, Bayon-Auboyer MH, Chemaly M, Fach P, Le Bouquin S (2014) Investigation of Clostridium botulinum in commercial poultry farms in France between 2011 and 2013. Avian Pathol 43:458–464. https://doi.org/10.1080/03079457.2014.957644

Souillard R, Le Maréchal C, Hollebecque F, Rouxel S, Barbé A, Houard E, Léon D, Poëzévara T, Fach P, Woudstra C, Mahé F, Chemaly M, Le Bouquin S (2015) Occurrence of C. botulinum in healthy cattle and their environment following poultry botulism outbreaks in mixed farms. Vet Microbiol 180:142–145. https://doi.org/10.1016/j.vetmic.2015.07.032

Souillard R, Le Maréchal C, Ballan V, Mahé F, Chemaly M, Le Bouquin S (2017a) A bovine botulism outbreak associated with a suspected cross-contamination from a poultry farm. Vet Microbiol 208:212–216. https://doi.org/10.1016/j.vetmic.2017.07.022

Souillard R, Le Maréchal C, Ballan V, Rouxel S, Léon D, Balaine L, Poëzevara T, Houard E, Robineau B, Robinault C, Chemaly M, Le Bouquin S (2017b) Investigation of a type C/D botulism outbreak in free-range laying hens in France. Avian Pathol 46:195–201. https://doi.org/10.1080/03079457.2016.1240355

Souillard R, Le Marechal C, Balaine L, Rouxel S, Poezevara T, Ballan V, Chemaly M, Le Bouquin S (2020) Manure contamination with Clostridium botulinum after avian botulism outbreaks: management and potential risk of dissemination. Vet Rec. https://doi.org/10.1136/vr.105898

Wendt S, Eder I, Wölfel R, Braun P, Lippmann N, Rodloff A (2017) Botulism: diagnosis and therapy. Dtsch Med Wochenschr 142:1304–1312. https://doi.org/10.1055/s-0043-112232

Włodarczyk R, Minias P, Kukier E, Grenda T, Śmietanka K, Janiszewski T (2014) The first case of a major avian type C botulism outbreak in Poland. Avian Dis 58:488–490. https://doi.org/10.1637/10669-091913-case.1

Wobeser G (1997) Avian botulism—another perspective avian. J Wildl Dis 33:181–186

Chapter 10
Avian Campylobacteriosis

Abstract Avian campylobacteriosis is an important bacterial zoonotic infection of birds that affect both farmed and wild birds. The disease is caused by Gram-negative, microaerophilic bacteria of the genus *Campylobacter*—mostly *C. jejuni* and *C. coli*. Birds are considered natural and largest reservoir of *C. jejuni*. The infection is transmitted by contaminated food and water and also through direct contact with infected birds. Following oral infection, the organisms multiply in jejunum, ceca, and cloacae and invade epithelial layers of intestinal tract. Subsequent cellular damage is principally mediated by cytolethal distending toxin, a major virulence factor of *Campylobacter* spp. The disease is manifested by mucous diarrhea and drop in production in farmed birds. Wild birds are known to disseminate the disease far and wide through fecal excretion of the pathogen. Besides clinical signs, diagnosis of the avian campylobacteriosis may require laboratory detection methods ranging from microbiological techniques to immunological to modern molecular tools. Due to zoonotic nature of the organism, campylobacteriosis assumes great importance in humans and is one of the leading causes of foodborne illnesses in humans worldwide. Domestic chicken, and possibly wild birds, constitute important source of infections for humans. With rising antimicrobial resistance among *Campylobacter* spp. leading to difficult-to-treat infections in humans, avian campylobacteriosis poses significant public health challenge.

Keywords Avian campylobacteriosis · *Campylobacter* · *C. jejuni* · Foodborne · Zoonotic · Antimicrobial resistance

10.1 Introduction

Campylobacter spp. are one of the leading causes of foodborne gastroenteritis in humans worldwide affecting industrialized and developing nations (Fitzgerald 2015; Shane 2016). In the United States, the organism affects more than 1.3 million people each year (https://www.cdc.gov/foodsafety/diseases/campylobacter). The infection is mainly acquired through foodborne routes following consumption of undercooked poultry meat, meat products, dairy products, and occasionally through handling of

Y. S. Malik et al., *Role of Birds in Transmitting Zoonotic Pathogens*, Livestock Diseases and Management, https://doi.org/10.1007/978-981-16-4554-9_10

infected birds. Human infections are usually manifested by diarrhea, abdominal cramp, and fever, and occasionally long-term illnesses such as chronic arthritis and Guillain-Barré syndrome are also attributed to campylobacteriosis (Platts-Mills and Kosek 2014; Sheppard and Maiden 2015). Owing to the zoonotic nature of campylobacteriosis, domestic animals like pigs and poultry play important roles in spread of the pathogen, perhaps due to their relative abundance in the vicinity of human settlements. However, the role of nondomesticated and wild birds are also crucial as have been documented in a number of reports (Antilles et al. 2015; Fitzgerald 2015; Gorham and Lee 2016; Greig et al. 2015). In addition, the role of wild birds as a source of human infections are increasingly being documented (Cody et al. 2015; Sahin et al. 2015; Vogt et al. 2020).

10.2 Epidemiology

Epidemiology of *Campylobacter* spp. is complex due to the zoonotic nature of the infection with multiple vertebrate hosts, especially poultry and pigs playing crucial roles.

10.2.1 Etiology and Classification

Campylobacters are Gram-negative, curved rod-shaped, microaerophilic bacteria with polar flagellum (with exception of *Campylobacter gracilis* which is nonmotile and *Campylobacter showae* with multiple flagella) (Bolton 2015; Facciolà et al. 2017; Fitzgerald 2015; Hoepers et al. 2016). Though current listing of bacterial nomenclature records 39 species within the genus *Campylobacter* of family *Campylobacteriaceae* under Order *Camylobatriales* (Parte 2018), two species *C. jejuni* and *C. coli* are responsible for majority of infections (de Vries et al. 2017; Pattison et al. 2008), and these are most commonly isolated *Campylobacter* spp. from poultry birds and other avian species (Awad et al. 2018; Fitzgerald 2015; Hoepers et al. 2016; Skarp et al. 2016). However, less discussed species such as *C. concisus, C. upsaliensis, C. ureolyticus, C. hyointestinalis,* and *C. sputorum* are believed to be emerging (Facciolà et al. 2017). *C. coli* is usually associated with intestinal tract of the poultry and their meat products. *C. laridis* is designated as NATRC (nalidixic acid-resistant thermophilic campylobacter) and this species in mainly found in the free-living marine birds (Pattison et al. 2008; Sahin et al. 2015; Umaraw et al. 2017; Vogt et al. 2020).

10.2.2 Host Range and Reservoirs

The organism survives in domestic animals, pets, wild birds, and poultry. Birds are perhaps the natural and largest reservoir for the *Campylobacter jejuni* in the nature. In fact, *C. jejuni* is considered to be a commensal of the poultry gut. On the other hand, pigs are naturals hosts for *C. coli* (Fitzgerald 2015; Hermans et al. 2011; Nachamkin et al. 2008; Platts-Mills and Kosek 2014).

Intensive production of poultry birds such as broiler, breeders, as well as turkey may provide suitable environment favoring the presence of the *Campylobacter* spp. in the environment. The wild waterfowl and the free-living birds such as Passiriformes and Anseriformes and Columbiformes may act as a good reservoir for *C. jejuni* by contaminating the main water supplies by their fecal dropping (Sahin et al. 2015; Umaraw et al. 2017; Vogt et al. 2020; Antilles et al. 2015; Dipineto et al. 2014; Gargiulo et al. 2018; Greig et al. 2015; Hald et al. 2016; Keller and Shriver 2013; Whiley et al. 2013). Available evidences on geographic distribution of *Campylobacter* spp. in avian species indicate that the organism is prevalent worldwide including all continents except Antarctica (Abraham et al. 2020; Antilles et al. 2015; Cody et al. 2015; Dipineto et al. 2014; Gargiulo et al. 2018; Gorham and Lee 2016; Greig et al. 2015; Guirado et al. 2020; Hald et al. 2016; Hermans et al. 2011; Johnson et al. 2017; Jurado-Tarifa et al. 2016; Keller and Shriver 2013; Kwan et al. 2008; Llarena et al. 2015; Mudiyanselage et al. 2017; Platts-Mills and Kosek 2014; Ricke et al. 2018; Rutledge et al. 2013; Shane 2016; Steele et al. 2013; Tel et al. 2013; Tryjanowski et al. 2020; Vogt et al. 2020; Wei et al. 2019).

10.2.3 Transmission and Role of Birds

The infection is chiefly transmitted through contaminated feed and water or by direct contact. Campylobacterioses can spread rapidly within the flock following initial infection. In case of humans, the disease is principally foodborne and is acquired from ingestion of undercooked meat and meat products, unpasteurized milk and milk products, and through contaminated waters. Direct transmission to humans from animals is rare and may involve infected pets or occupational exposure at farms or slaughter houses (Bronowski et al. 2014; Facciolà et al. 2017; Hermans et al. 2011; Nachamkin et al. 2008; Pitkänen 2013; Platts-Mills and Kosek 2014; Skarp et al. 2016). Studies have shown that poultry meat constitute the greatest risk factor for campylobacter enteritis in humans (Skarp et al. 2016) (Fig. 10.1).

Transmission of *Campylobacter* spp. is facilitated by a number of survival mechanisms of the organism. These include various stress adaptation responses such as tolerance to desiccation and oxygen exposure, ability to thwart damage by reactive oxygen molecules, natural thermophilly, biofilm formation, and augmentation of viable but nonculturable (VBNC) state (Bolton 2015).

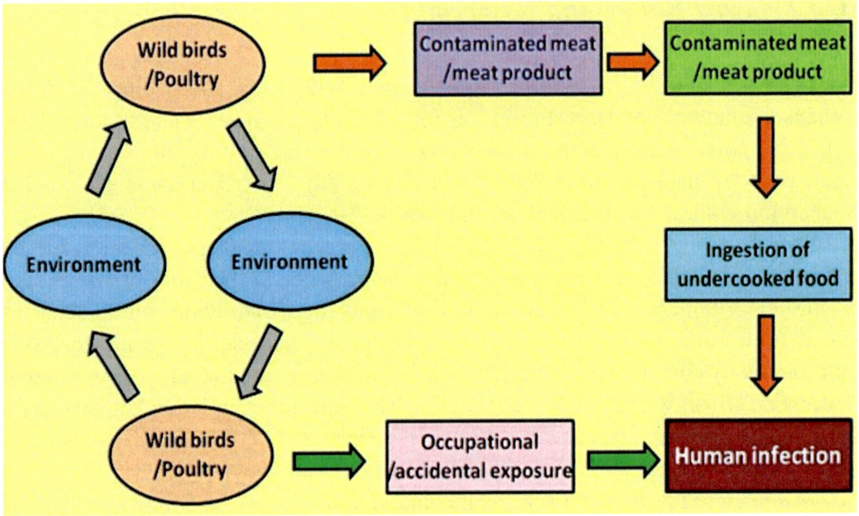

Fig. 10.1 Transmission routes of *Campylobacter* spp. from birds to humans

10.3 Disease Features

10.3.1 Pathogenesis and Pathological Features

Campylobacter infection is usually associated with oral infection; the bacteria is usually multiplicate at the last third of the jejunum, ceca, and cloacae. Process of disease causation by *Campylobacter* is intricate and not fully understood (Bolton 2015). The pathogenesis of *Campylobacter* infection starts with colonization of the intestinal mucosa followed by adherence. The motility of the cells does play a crucial role in colonization. Following adhesion, the Campylobacters invade the epithelial cells to cross the lamina propria to eventually reach the underlying connective tissue. The actual mechanism is however unclear, perhaps the bacterial cells follow both paracellular and transcellular routes (Bolton 2015; Facciolà et al. 2017; Hoepers et al. 2016). Cellular damage and death is mediated principally by cytolethal distending toxin (Cdt) through arrest of cell cycle (Facciolà et al. 2017). Studies on cell lines also revealed lethality of the toxins of *Campylobacter* on various cell line including Hela cell, CHO cells, as well as chicken embryos (Bolton 2015).

C. jejuni infection is usually associated with some pathological lesions in the intestine and the liver characterized by marked enlargement of the distal intestinal loops with accumulation of mucus and water in the lumen of the intestine. As a consequence of toxin production, hemorrhage may occur, which is commonly seen in some human cases infected with *C. jejuni*. Reddish or yellowish mottling of the liver parenchyma is usually associated with *C. jejuni* in chickens (Awad et al. 2018; Bolton 2015; Fitzgerald 2015; Pattison et al. 2008; Shane 2016).

10.3.2 Clinical Signs of Campylobacteriosis in Birds

The most important clinical signs of Campylobacteriosis in chicken is diarrhea with mucous tinged fecal matter that usually develop as early as 6 h after infection. These clinical manifestations are usually exaggerated in the presence of other immunosuppressive agents. Marked reduction in production is also reported in birds affected with campylobacterioses (Shane 2016; Umaraw et al. 2017). Many studies have documented *Campylobacter* spp. in birds. In Japan, an earlier study had reported that the isolation of *C. jejuni* from crows was much higher than other free-living birds such as sparrows and pheasants (Ito et al. 1988). In North America, one study that had been conducted to address the role of North America goose in transmission of several bacteria including *Campylobacter* species confirmed the potential role of North American goose in transmission of Campylobacteriosis over different places in North America due to the shedding of bacteria in their dropping (Feare et al. 1999).

Another expansive study conducted to investigate the role of wild birds as a potential reservoir of Campylobacteriosis, involving large numbers of wild bird species (1794 individual birds belonging to 107 species and 26 families), reported that 89 birds were positive for *C. jejuni*, 100 birds were positive for *C. lari*, as well as 17 birds were positive for *C. coli*, furthermore with dual infection of *C. jejuni* and *C. coli* in 10 birds (Waldenström et al. 2002). In Croatia, one study had been recently conducted to study the prevalence of *Campylobacter* sp. in free-living and wild birds. This study revealed that the prevalence of *C. jejuni* was 8.7% in the already tested birds (Vlahović et al. 2004). In Norway, another study had been recently conducted to figure out the role of feral pigeons, Mallard, and Graylag goose as a potential zoonotic reservoir. This study reported that six of the examined pigeons and one of the Mallards were positive for *C. jejuni* (Lillehaug et al. 2005). In northeastern Spain, 12.6% (40/318) of free-living waterfowls were reported to be positive for *Campylobacter* spp. with higher isolation rates for *C. coli* than *C. jejuni* (Antilles et al. 2015). *Campylobacter* spp. have also been isolated from 21% of common quails in Italy along with other enteric pathogens (Dipineto et al. 2014). Similarly, a study assessing carriage rates of *Campylobacter* spp. in Denmark involving 52 wild bird species in 12 farms indicated that almost 60% of the thrushes and one-fifth of sparrows harbored *Campylobacter* spp. (Hald et al. 2016).

However, a considerably lower prevalence was recorded in an investigation of wild birds in the mid-Atlantic region of United States, where only 9.2% carriage rate was observed (Keller and Shriver 2013). Higher carriage rate in the previous study of Hald et al. (2016) was perhaps due to vicinity of the farm animals which are also important reservoirs of *Campylobacter* spp. (Hald et al. 2016). Contrasting evidence, however, came from a molecular epidemiological study (Kwan et al. 2008) of *Campylobacter* spp. from farm animals, wildlife, and environment, which showed predominance of clonal complexes in particular niches, thus indicating epidemiological discontinuity as is commonly believed. Similar results were also reported from a study of *Campylobacter* spp. in barnacle geese which indicated clonality of

C. jejuni population in the birds dissimilar from commonly reported clonal groups *Campylobacter* in humans (Llarena et al. 2015). Again, genetic characterization and epidemiological analysis of isolates of *Campylobacter* spp. in South Korea highlighted frequent overlaps of clonal complexes (Sequence Types, ST) among human and wild bird isolates, indicating putative risk of infection to humans from the wild birds (Wei et al. 2019). Therefore, considering conflicting evidences from various studies and wide prevalence in the environment, it seems that the epidemiology of *Campylobacter* spp. needs further deeper study (Hermans et al. 2011). *Campylobacter* spp. were also isolated from chickens and commercial broiler birds, ducks, and occasionally from house crows almost from a large number of Southeast Asian countries including Brunei, Cambodia, Indonesia, Malaysia, Myanmar, Philippines, Thailand, Timor-Leste, and Vietnam (Mudiyanselage et al. 2017). In another study from Turkey, bald ibises (*Geronticus eremita*), a red list threatened species, revealed carriage rate of around 29% (Tel et al. 2013).

10.4 Disease Management

10.4.1 Diagnosis of Campylobacteriosis in Birds

Efficient and rapid diagnosis of infection by *Campylobacter* spp. in avian hosts is critical for treatment and control of disease at individual as well as at farm level. Moreover, optimal detection helps in proper surveillance and monitoring of *Campylobacter* infection which might pose a zoonotic health risk for humans.

While clinical signs can be of great help in raising a suspicion about *Campylobacter* infection in avian species including farmed poultry birds, the final diagnosis need to be confirmed by laboratory diagnosis. Laboratory diagnosis of *Campylobacter* spp. hinges upon traditional bacteriological and current molecular approaches including recent advances in biosensor technologies (Gharst et al. 2013; Hansson et al. 2018; Lin 2009; Pattison et al. 2008; Ricke et al. 2018; Vidic et al. 2017).

Classical microbiological detection of *Campylobacter* spp. involves enrichment of sample in suitable broth such as Bolton broth, followed by isolation by plating on selective media for which a number of commercial media are available (Bolton 2015; Gharst et al. 2013). Isolation of bacteria is usually followed by various biochemical tests in the laboratory, namely production of catalase and oxidase, reduction of nitrate and nitrite, expression of urease, H_2S production, and hydrolysis of hippurate and indoxyl acetate (Gharst et al. 2013; Pattison et al. 2008). However, these bacteriological tests are time-consuming and sometimes lead to confusing results.

To alleviate the bottlenecks, various other tests have been devised including molecular and immunological methods. Among the molecular methods, polymerase chain reaction (PCR) and real-time PCR techniques have been quite popular, and some commercial assays are also available (Ricke et al. 2018). On the other hand, immunological methods are mainly predominated by enzyme immunoassays, particularly ELISA, targeting various protein components of *Campylobacter* spp. (Josefsen et al. 2015; Pattison et al. 2008; Ricke et al. 2018). For typing of isolated *Campylobacter* spp., various typing methods have also been employed including pulsed-field gel electrophoresis, multilocus sequence typing, and whole genome sequencing, though these methods are mostly available to upgraded or reference laboratories (de Vries et al. 2017; Ghatak et al. 2017, 2020; Josefsen et al. 2015; Pattison et al. 2008; Ricke et al. 2018).

10.4.2 Treatment and Control

In humans, *Campylobacter* infection is usually a self-limiting diarrheal disease lasting for 3–5 days. Treatment with oral rehydration and correction of electrolyte imbalance is generally advocated. In severe cases, and in immunocompromised patients or children, administration of macrolide or fluoroquinolone antibiotics might be undertaken (Johnson et al. 2017). In farm animals and birds, except pets, treatment is usually not practiced.

Control of infection in birds, especially poultry, starts with good hygiene and sanitary management of farm premises and husbandry operations. Disinfection of farm equipment, especially hatcheries, is important. Chemical treatment of litters is also an effective tool to contain the spread of infection. Newer approaches involving use of probiotics in feed to reduce colonization, incorporation of fatty acids and bacteriocins, plant-derived substances, use of bacteriophage may also prove helpful, though requires further evaluations (Facciolà et al. 2017; Johnson et al. 2017; Meunier et al. 2016; Umaraw et al. 2017).

Apart from other approaches controlling *Campylobacters* in avian species, several vaccines or candidate vaccines have been employed for preemptive action against *Campylobacter* infection. A whole-cell vaccine is already in use for prevention, while a number of vaccine candidates including subunit vaccines are being intensely studied (Facciolà et al. 2017; Sahin et al. 2015; Shane 2016; Umaraw et al. 2017). Some novel vaccine approaches for control of campylobacterioses in poultry include innovative *in ovo* vaccination in broilers employing bacterin and subunit vaccine, a DNA prime/protein boost protocol for *C. jejuni* vaccination, and use of reverse vaccinology to identify potential novel targets for vaccination (Hansson et al. 2018; Lin 2009; Meunier et al. 2017, 2018; Sahin et al. 2015; Umaraw et al. 2017; Vandeputte et al. 2019).

10.5 Public Health Concerns

Campylobacter-induced enteritis is believed to be the most common of bacterial gastroenteritis worldwide and was estimated to cost approximately 7.5 million DALYs (disability-adjusted life years) (Platts-Mills and Kosek 2014). Poultry constitutes the largest reservoir and greatest risk factor for human infections (Hermans et al. 2011). While the primary route of infection in humans is oral, occupational nature of the disease has also been documented. While reviewing role of poultry meat as source of human campylobacterioses, Skarp and colleagues noted that occupational infection by *Campylobacter* spp. might vary (57–83%) among workers as was reported from the United States and Sweden (Skarp et al. 2016). Studies specifically identifying the risk of contracting *Campylobacter* and other enteric infections from wild birds are relatively scarce. However, Gorham and Lee (2016) identified that recreational exposure to water fecally contaminated by wild geese could present a potential health risk to humans (Gorham and Lee 2016). Similar risks from wild bird-associated *C. jejuni* have also been highlighted in the UK too (Cody et al. 2015). In addition to direct infection by the agent, wild bird-associated *Campylobacter* spp. have also been identified as potential disseminator of antibiotic resistance in the environment (Greig et al. 2015; Kaakoush et al. 2015; Whiley et al. 2013).

References

Abraham S, Sahibzada S, Hewson K, Laird T, Abraham R, Pavic A, Truswell A, Lee T, O'Dea M, Jordan D (2020) Emergence of fluoroquinolone-resistant Campylobacter jejuni and Campylobacter coli among Australian chickens in the absence of fluoroquinolone use. Appl Environ Microbiol 86. https://doi.org/10.1128/AEM.02765-19

Antilles N, Sanglas A, Cerdà-Cuéllar M (2015) Free-living waterfowl as a source of zoonotic bacteria in a dense wild bird population area in northeastern Spain. Transbound Emerg Dis 62: 516–521. https://doi.org/10.1111/tbed.12169

Awad WA, Hess C, Hess M (2018) Re-thinking the chicken–Campylobacter jejuni interaction: a review. Avian Pathol 47:352–363. https://doi.org/10.1080/03079457.2018.1475724

Bolton DJ (2015) Campylobacter virulence and survival factors. Food Microbiol 48:99–108. https://doi.org/10.1016/j.fm.2014.11.017

Bronowski C, James CE, Winstanley C (2014) Role of environmental survival in transmission of Campylobacter jejuni. FEMS Microbiol Lett 356:8–19. https://doi.org/10.1111/1574-6968.12488

Cody AJ, McCarthy ND, Bray JE, Wimalarathna HML, Colles FM, Jansen van Rensburg MJ, Dingle KE, Waldenström J, Maiden MCJ (2015) Wild bird-associated Campylobacter jejuni isolates are a consistent source of human disease, in Oxfordshire, United Kingdom. Environ Microbiol Rep 7:782–788. https://doi.org/10.1111/1758-2229.12314

de Vries SP, Gupta S, Baig A, Wright E, Wedley A, Jensen AN, Lora LL, Humphrey S, Skovgård H, Macleod K, Pont E, Wolanska DP, L'Heureux J, Mobegi FM, Smith DGE, Everest P, Zomer A, Williams N, Wigley P, Humphrey T, Maskell DJ, Grant AJ (2017) Genome-wide fitness analyses of the foodborne pathogen Campylobacter jejuni in in vitro and in vivo models. Sci Rep 7:1251. https://doi.org/10.1038/s41598-017-01133-4

Dipineto L, Russo TP, Gargiulo A, Borrelli L, De Luca Bossa LM, Santaniello A, Buonocore P, Menna LF, Fioretti A (2014) Prevalence of enteropathogenic bacteria in common quail (Coturnix coturnix). Avian Pathol 43:498–500. https://doi.org/10.1080/03079457.2014.966055

Facciolà A, Riso R, Avventuroso E, Visalli G, Delia SA, Laganà P (2017) Campylobacter: from microbiology to prevention. J Prev Med Hyg 58:E79–E92

Feare CJ, Saunders MF, Blasco R, Bishop JD (1999) Canada goose droppings as a potential source of pathogenic bacteria. J R Soc Promot Heal 119:146–155

Fitzgerald C (2015) Campylobacter. Clin Lab Med 35:289–298. https://doi.org/10.1016/j.cll.2015.03.001

Gargiulo A, Fioretti A, Russo TP, Varriale L, Rampa L, Paone S, Bossa LMDL, Raia P, Dipineto L (2018) Occurrence of enteropathogenic bacteria in birds of prey in Italy. Lett Appl Microbiol 66: 202–206. https://doi.org/10.1111/lam.12836

Gharst G, Oyarzabal OA, Hussain SK (2013) Review of current methodologies to isolate and identify Campylobacter spp. from foods. J Microbiol Methods 95:84–92. https://doi.org/10.1016/j.mimet.2013.07.014

Ghatak S, He Y, Reed S, Strobaugh TJ, Irwin P (2017) Whole genome sequencing and analysis of Campylobacter coli YH502 from retail chicken reveals a plasmid-borne type VI secretion system. Genomics Data 11:128–131. https://doi.org/10.1016/j.gdata.2017.02.005

Ghatak S, He Y, Reed S, Irwin P (2020) Comparative genomic analysis of a multidrug-resistant Campylobacter jejuni strain YH002 isolated from retail beef liver. Foodborne Pathog Dis. https://doi.org/10.1089/fpd.2019.2770

Gorham TJ, Lee J (2016) Pathogen loading from Canada geese faeces in freshwater: potential risks to human health through recreational water exposure. Zoonoses Public Health 63:177–190. https://doi.org/10.1111/zph.12227

Greig J, Rajić A, Young I, Mascarenhas M, Waddell L, Lejeune J (2015) A scoping review of the role of wildlife in the transmission of bacterial pathogens and antimicrobial resistance to the food chain. Zoonoses Public Health 62:269–284. https://doi.org/10.1111/zph.12147

Guirado P, Paytubi S, Miró E, Iglesias-Torrens Y, Navarro F, Cerdà-Cuéllar M, Attolini CS-O, Balsalobre C, Madrid C (2020) Differential distribution of the wlaN and cgtB genes, associated with Guillain-Barré syndrome, in Campylobacter jejuni isolates from humans, broiler chickens, and wild birds. Microorganisms 8. https://doi.org/10.3390/microorganisms8030325

Hald B, Skov MN, Nielsen EM, Rahbek C, Madsen JJ, Wainø M, Chriél M, Nordentoft S, Baggesen DL, Madsen M (2016) Campylobacter jejuni and Campylobacter coli in wild birds on Danish livestock farms. Acta Vet Scand 58:1–10. https://doi.org/10.1186/s13028-016-0192-9

Hansson I, Sandberg M, Habib I, Lowman R, Engvall EO (2018) Knowledge gaps in control of Campylobacter for prevention of campylobacteriosis. Transbound Emerg Dis 65(Suppl 1):30–48. https://doi.org/10.1111/tbed.12870

Hermans D, Pasmans F, Messens W, Martel A, Van Immerseel F, Rasschaert G, Heyndrickx M, Van Deun K, Haesebrouck F (2011) Poultry as a host for the zoonotic pathogen Campylobacter jejuni. Vector-Borne Zoonotic Dis 12:89–98. https://doi.org/10.1089/vbz.2011.0676

Hoepers PG, Medina G, Rossi DA, Fernandez H (2016) Campylobacter spp. and related organisms in poultry, Campylobacter spp. and related organisms in poultry. Springer. https://doi.org/10.1007/978-3-319-29907-5_1

Ito K, Kubokura Y, Kaneko K, Totake Y, Ogawa M (1988) Occurrence of *Campylobacter jejuni* in free-living wild birds from Japan. J Wildl Dis 24:467–470. https://doi.org/10.7589/0090-3558-24.3.467

Johnson TJ, Shank JM, Johnson JG (2017) Current and potential treatments for reducing Campylobacter colonization in animal hosts and disease in humans. Front Microbiol 8:1–14. https://doi.org/10.3389/fmicb.2017.00487

Josefsen MH, Bhunia AK, Engvall EO, Fachmann MSR, Hoorfar J (2015) Monitoring Campylobacter in the poultry production chain—from culture to genes and beyond. J Microbiol Methods 112:118–125. https://doi.org/10.1016/j.mimet.2015.03.007

Jurado-Tarifa E, Torralbo A, Borge C, Cerdà-Cuéllar M, Ayats T, Carbonero A, García-Bocanegra I (2016) Genetic diversity and antimicrobial resistance of Campylobacter and Salmonella strains isolated from decoys and raptors. Comp Immunol Microbiol Infect Dis 48:14–21. https://doi.org/10.1016/j.cimid.2016.07.003

Kaakoush NO, Castaño-Rodríguez N, Mitchell HM, Man SM (2015) Global epidemiology of campylobacter infection. Clin Microbiol Rev 28:687–720. https://doi.org/10.1128/CMR.00006-15

Keller JI, Shriver WG (2013) Prevalence of three Campylobacter species, C. Jejuni, C. Coli, and C. Lari, using multilocus sequence typing in wild birds of the mid-Atlantic region, USA. J Wildl Dis 50:31–41. https://doi.org/10.7589/2013-06-136

Kwan PSL, Barrigas M, Bolton FJ, French NP, Gowland P, Kemp R, Leatherbarrow H, Upton M, Fox AJ (2008) Molecular epidemiology of Campylobacter jejuni populations in dairy cattle, wildlife, and the environment in a farmland area. Appl Environ Microbiol 74:5130–5138. https://doi.org/10.1128/AEM.02198-07

Lillehaug A, Monceyron Jonassen C, Bergsjø B, Hofshagen M, Tharaldsen J, Nesse LL, Handeland K (2005) Screening of feral pigeon (*Colomba livia*), mallard (*Anas platyrhynchos*) and graylag goose (*Anser anser*) populations for *Campylobacter* spp., *Salmonella* spp., avian infuenza virus and avian paramyxovirus. Acta Vet Scand 46:193–202

Lin J (2009) Novel approaches for Campylobacter control in poultry. Foodborne Pathog Dis 6:755–765. https://doi.org/10.1089/fpd.2008.0247

Llarena AK, Skarp-de Haan CPA, Rossi M, Hänninen ML (2015) Characterization of the Campylobacter jejuni population in the barnacle geese reservoir. Zoonoses Public Health 62:209–221. https://doi.org/10.1111/zph.12141

Meunier M, Guyard-Nicodème M, Dory D, Chemaly M (2016) Control strategies against Campylobacter at the poultry production level: biosecurity measures, feed additives and vaccination. J Appl Microbiol 120:1139–1173. https://doi.org/10.1111/jam.12986

Meunier M, Guyard-Nicodème M, Vigouroux E, Poezevara T, Beven V, Quesne S, Bigault L, Amelot M, Dory D, Chemaly M (2017) Promising new vaccine candidates against Campylobacter in broilers. PLoS One 12:e0188472. https://doi.org/10.1371/journal.pone.0188472

Meunier M, Guyard-Nicodème M, Vigouroux E, Poezevara T, Béven V, Quesne S, Amelot M, Parra A, Chemaly M, Dory D (2018) A DNA prime/protein boost vaccine protocol developed against Campylobacter jejuni for poultry. Vaccine 36:2119–2125. https://doi.org/10.1016/j.vaccine.2018.03.004

Mudiyanselage J, Jayarukshi K, Premarathne K, Satharasinghe DA, Tang J, Huat Y, Basri DF, Rukayadi Y, Nakaguchi Y, Nishibuchi M, Radu S (2017) Impact of human Campylobacter infections in Southeast Asia: the contribution of the poultry sector. Crit Rev Food Sci Nutr 57: 3971–3986. https://doi.org/10.1080/10408398.2016.1266297

Nachamkin I, Szymanski CM, Blaser MJ (eds) (2008) Campylobacter, 3rd edn. American Society of Microbiology, Washington, DC. https://doi.org/10.1086/600392

Parte AC (2018) LPSN—list of prokaryotic names with standing in nomenclature (Bacterio.net), 20 years on. Int J Syst Evol Microbiol 68:1825–1829. https://doi.org/10.1099/ijsem.0.002786

Pattison M, McMullin P, Bradbury J, Alexander D (eds) (2008) Poultry diseases, 6th edn. https://doi.org/10.1016/B978-0-7020-2862-5.50031-3

Pitkänen T (2013) Review of Campylobacter spp. in drinking and environmental waters. J Microbiol Methods 95:39–47. https://doi.org/10.1016/j.mimet.2013.06.008

Platts-Mills JA, Kosek M (2014) Update on the burden of Campylobacter in developing countries. Curr Opin Infect Dis 27:444–450. https://doi.org/10.1097/QCO.0000000000000091

Ricke SC, Feye KM, Chaney WE, Shi Z, Pavlidis H, Yang Y (2018) Developments in rapid detection methods for the detection of foodborne Campylobacter in the United States. Front Microbiol 9:3280. https://doi.org/10.3389/fmicb.2018.03280

Rutledge ME, Siletzky RM, Gu W, Degernes LA, Moorman CE, DePerno CS, Kathariou S (2013) Characterization of Campylobacter from resident Canada geese in an urban environment. J Wildl Dis 49:1–9. https://doi.org/10.7589/2011-10-287

Sahin O, Kassem II, Shen Z, Lin J, Rajashekara G, Zhang Q (2015) Campylobacter in poultry: ecology and potential interventions. Avian Dis 59:185–200. https://doi.org/10.1637/11072-032315-Review

Shane SM (2016) Campylobacter infection of commercial poultry. Rev Sci Tech l'OIE 19:376–395. https://doi.org/10.20506/rst.19.2.1224

Sheppard SK, Maiden MCJ (2015) The evolution of campylobacter jejuni and Campylobacter coli. Cold Spring Harb Perspect Biol 7:1–13. https://doi.org/10.1101/cshperspect.a018119

Skarp CPA, Hänninen ML, Rautelin HIK (2016) Campylobacteriosis: the role of poultry meat. Clin Microbiol Infect 22:103–109. https://doi.org/10.1016/j.cmi.2015.11.019

Steele CM, Brown RN, Botzler RG (2013) Prevalences of zoonotic bacteria among seabirds in rehabilitation centers along the Pacific Coast of California and Washington, USA. J Wildl Dis 41:735–744. https://doi.org/10.7589/0090-3558-41.4.735

Tel OY, Bozkaya F, Keskin O (2013) Salmonella, Campylobacter, and Chlamydophila in bald ibis (Geronticus eremita) feces in turkey. J Zoo Wildl Med 44:21–26. https://doi.org/10.1638/1042-7260-44.1.21

Tryjanowski P, Nowakowski JJ, Indykiewicz P, Andrzejewska M, Śpica D, Sandecki R, Mitrus C, Goławski A, Dulisz B, Dziarska J, Janiszewski T, Minias P, Świtek S, Tobolka M, Włodarczyk R, Szczepańska B, Klawe JJ (2020) Campylobacter in wintering great tits Parus major in Poland. Environ Sci Pollut Res Int 27:7570–7577. https://doi.org/10.1007/s11356-019-07502-y

Umaraw P, Prajapati A, Verma AK, Pathak V, Singh VP (2017) Control of campylobacter in poultry industry from farm to poultry processing unit: a review. Crit Rev Food Sci Nutr 57:659–665. https://doi.org/10.1080/10408398.2014.935847

Vandeputte J, Martel A, Van Rysselberghe N, Antonissen G, Verlinden M, De Zutter L, Heyndrickx M, Haesebrouck F, Pasmans F, Garmyn A (2019) In ovo vaccination of broilers against Campylobacter jejuni using a bacterin and subunit vaccine. Poult Sci 98:5999–6004. https://doi.org/10.3382/ps/pez402

Vidic J, Manzano M, Chang C-M, Jaffrezic-Renault N (2017) Advanced biosensors for detection of pathogens related to livestock and poultry. Vet Res 48:11. https://doi.org/10.1186/s13567-017-0418-5

Vlahović K et al (2004) Campylobacter, Salmonella and Chlamydia in free-living birds of Croatia. Eur J Wildl Res 50:127–132. https://doi.org/10.1007/s10344-004-0052-1

Vogt NA, Stevens CPG, Pearl DL, Taboada EN, Jardine CM (2020) Generalizability and comparability of prevalence estimates in the wild bird literature: methodological and epidemiological considerations. Anim Health Res Rev 21(1):89–95. https://doi.org/10.1017/S1466252320000043

Waldenström J, Broman T, Carlsson I, Hasselquist D, Achterberg RP, Wagenaar JA et al (2002) Prevalence of *Campylobacter jejuni*, *Campylobacter lari*, and *Campylobacter coli* in different ecological guilds and taxa of migrating birds. Appl Environ Microbiol 68:5911–5917. https://doi.org/10.1128/AEM.68.12.5911-5917.2002

Wei B, Kang M, Jang HK (2019) Genetic characterization and epidemiological implications of Campylobacter isolates from wild birds in South Korea. Transbound Emerg Dis 66:56–65. https://doi.org/10.1111/tbed.12931

Whiley H, van den Akker B, Giglio S, Bentham R (2013) The role of environmental reservoirs in human campylobacteriosis. Int J Environ Res Public Health 10:5886–5907. https://doi.org/10.3390/ijerph10115886

Chapter 11
Avian Chlamydiosis (Psittacosis, Ornithosis)

Abstract Avian chlamydiosis, also known as Psittacosis or Ornithosis, is a zoonotic disease of birds caused by members of the genus *Chlamydia*. The organisms causing chlamydiosis are obligate intracellular pathogens with a biphasic developmental cycle characterized by the infectious form known as elementary body and the metabolically active form known as reticulate body. The disease affects wide variety of avian hosts, and outbreaks among psittacine birds are commonly reported from various parts of the world. Nasal discharge and fecal droppings from infected birds, including the asymptomatic ones, serve as main source of infection for other birds and humans. The organisms enter the host usually through nasal route and then affect multiple organs. With progression of disease, the affected birds exhibit depression, anorexia, nasal discharge, conjunctivitis, respiratory signs, and diarrhea. While clinical signs lead to a presumptive diagnosis, laboratory techniques, namely serological and molecular techniques are also employed for confirmation. Treatment of the disease in avian hosts is dependent on antimicrobial chemotherapy mostly by tetracyclines and fluoroquinolones. Avian chlamydiosis is an important zoonotic disease, and many cases or outbreaks have been reported especially in occupationally exposed groups of people and thus constitute a putative public health hazard.

Keywords Avian chlamydiosis · Psittacosis · Ornithosis · *Chlamydia* · Elementary body · Reticulate body

11.1 Introduction

Avian chlamydiosis is an infection of birds caused by members of the genus *Chlamydia.* The disease is important zoonoses for humans and is often transmitted from birds. Due to its association with psittacine birds, the disease is also known as psittacosis or ornithosis or parrot fever. Though other chlamydial diseases such as trachoma were known to humans for ages, the origin of psittacosis is less well documented. It is believed that the disease perhaps originated from psittacine pets in Latin America though counter narrative of Australian origin has also been documented. Earliest evidence of zoonotic psittacosis was reported in 1615 yet,

Y. S. Malik et al., *Role of Birds in Transmitting Zoonotic Pathogens*, Livestock Diseases and Management, https://doi.org/10.1007/978-981-16-4554-9_11

the family name Chlamydiaceae was proposed almost 350 years later in 1966 (Borel et al. 2018; Chu and Durrani 2019; OIE 2018). In birds the disease is usually characterized by nonspecific clinical signs such as lethargy, exertion, and nasal and ocular discharges. In human, avian chlamydiosis produces flu-like symptoms. The disease is highly contagious in nature, and widespread outbreaks of psittacosis were reported worldwide affecting many countries presumably due to unrestricted trade of psittacine birds in the 1930s (Hogerwerf et al. 2020). These widespread outbreaks in the past found their mention and psittacosis pandemic in the literature (Hogerwerf et al. 2020).

11.2 Epidemiology

11.2.1 Etiology and Classification

Chlamydiae are unique obligate intracellular organisms that are Gram negative with a biphasic developmental cycle. The classification of the organisms underwent several changes over the last few decades. Currently the organisms causing ornithosis or psittacosis belong to the family Chlamydiaceae with a single genus *Chlamydia* under which there are 11 recognized species and 3 candidate species (Borel et al. 2018). Recognized and defined species of *Chlamydia* include *C. abortus*, *C. avium*, *C. caviae*, *C. felis*, *C. gallinacea*, *C. muridarum*, *C. pecorum*, *C. pneumoniae*, *C. psittaci*, *C. suis*, and *C. trachomatis* (Borel et al. 2018). While it is traditionally believed that avian chlamydiosis is caused by *C. psittaci*, recent reports indicated that *C. gallinacea* and *C. avium* are also widely responsible (Borel et al. 2018; Li et al. 2017; Sachse and Laroucau 2015).

11.2.2 Morphology and Life Cycle

The organism runs through a unique life cycle with two distinct developmental forms – EB, elementary body, and RB, reticulate body (Fig. 11.1). While the former is the main infectious form, the reticulate body is the metabolically active multiplying form observed in host cells. EB being the infectious form attaches to the tissue and enters the host cells. It is approximately 0.2 μm in size and is characterized by the presence of condensed nucleoid material found at the periphery of the bodies. The reticulate body (RB), which measures 0.8 μm approximately, on the other hand undergoes binary division and maturation to give rise into new generations of EBs (Beeckman and Vanrompay 2009; Borel et al. 2018; Chu et al. 2020; OIE 2018; Rohde et al. 2018; Sachse and Laroucau 2015; Szymańska-Czerwińska and Niemczuk 2016).

Chlamydia life cycle includes two distinct stages. The EB is characterized by its resistance to extracellular environment, while the RB has the ability to multiply

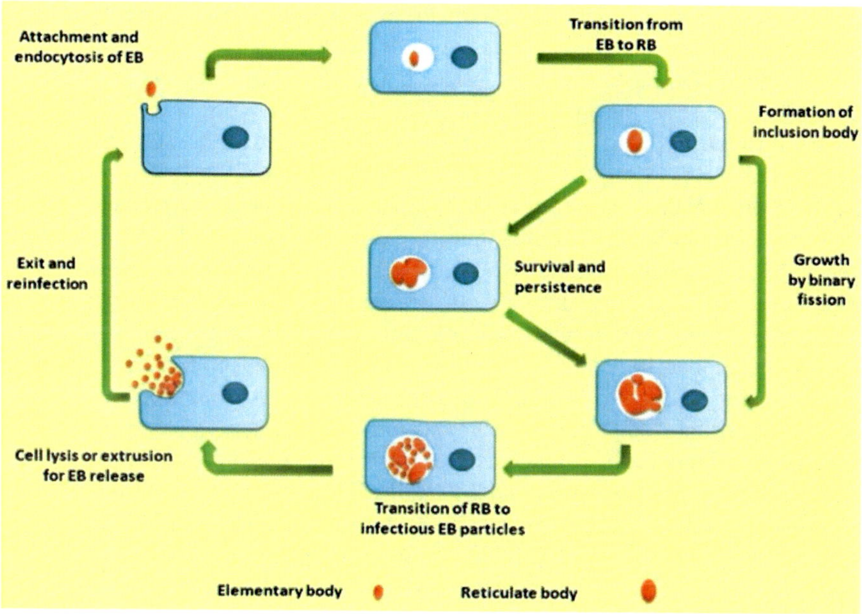

Fig. 11.1 Replication cycle of *Chlamydia* spp. inside hosts

inside the cell but unable to multiply outside the infected host cells. The infectious cycle usually starts when the EBs of *C. psittaci* attach to the epithelial lining of the target host tissue. They move along to the base of the microvilli where they are transported to inside of the plasma membrane to form endocytic vesicles. These vesicles are resistant to the action of lysosomal enzymes of the host cells. These vesicles carrying the EBs travel to the nuclear regions when the EBs undergo changes to morph into RBs. These RBs undergo multiplication by binary fission, giving rise to the characteristic "hour glass" appearance followed by formation of inclusion bodies which is characteristic features of chlamydiosis infection. Exocytosis of those inclusion bodies occurs, and, in some cases, persistent infections may develop due to the presence of the remaining EBs inside the cells (Vanrompay et al. 2007). Due to intracellular lifestyle and dependence on host metabolites, in vitro detection of *Chlamydiae* requires animal cell culture (Beeckman and Vanrompay 2009; Borel et al. 2018; Chu et al. 2020; OIE 2018; Rohde et al. 2018; Sachse and Laroucau 2015; Szymańska-Czerwińska and Niemczuk 2016).

11.2.3 Host Range and Distribution

Psittacosis has been reported in about 465 different avian species. The severity of the infection varies with age and species of the birds affected. Among poultry birds, turkeys are more susceptible followed by ducks and geese and chicken in which the disease is relatively rare. Young birds are more prone to clinical disease than mature birds. Among wild birds seashores, geese, gulls, as well as penguins appear to be more susceptible than the pheasants and quails (Borel et al. 2018; Chu et al. 2020; de Freitas Raso et al. 2006; Hogerwerf et al. 2020; OIE 2018; Pattison et al. 2008; Santos et al. 2014).

Avian chlamydiosis has been reported from all over the world including continents of North and South Americas, Europe, Australia, Africa, and Asia. In a recent review of evidence of zoonotic sources, Hogerwerf et al. noted that the disease is widespread (Hogerwerf et al. 2020). In Brazil, an outbreak of chlamydiosis was recorded in captive blue-fronted Amazon parrots that were recovered from illegal trading (de Freitas Raso et al. 2009). A similar outbreak of *C. psittaci* among psittacine birds rescued from illegal trade was also recorded with deaths of 36 psittacines (Ecco et al. 2009). Another study among captive Amazon parrots belonging to breeding centers in Brazil documented active infection of psittacosis with seroprevalence varying between 60 and 100% (de Freitas et al. 2014). Occurrence of *C. psittaci* infection in blue-fronted Amazon parrots and Hyacinth macaws was documented previously too (de Freitas Raso et al. 2006). Screening of Eurasian collared doves in northern Italy revealed a prevalence of 61% in urban and suburban areas (Donati et al. 2014). In a study on backyard chicken farms in Italy, incidence of *C. psittaci* and *C. gallinacea* infection was observed to be 15% (Donati et al. 2018). In Arizona, United States, serological evidence of *C. psittaci* infection was recorded among feral Rosy-faced lovebirds and other sympatric birds, for example, rock doves (Dusek et al. 2018). In Switzerland, a study of feral pigeon fecal droppings from six public sites indicated presence of *C. psittaci* as detected through PCR for *ompA* gene of the organism (Geigenfeind et al. 2012). Using a rapid real-time PCR for detection of chlamydial DNA, a survey of 369 free living birds in Poland revealed a 7.3% prevalence of chlamydiosis, of which majority was attributable to *C. psittaci* (Krawiec et al. 2015). Infection in psittacine birds can be asymptomatic without any clinical signs as was documented in Polish psittacine birds (Piasecki et al. 2012). However, review of data from 11 European countries indicated higher seroprevalence (19–96%) among feral pigeons (Magnino et al. 2009). Though reports of psittacosis are relatively scanty, in Iran, occurrence of *C. psittaci* infection was reported in companion and wild birds (Madani and Peighambari 2013). In addition, recent reports of psittacosis have surfaced from Faroe Islands in Europe, from domestic pigeons in Turkey, and also from Australian parrots, and other captive and wild birds in Australian zoo (Altıntaş et al. 2020; Amery-Gale et al. 2020; Stokes et al. 2020; Wang et al. 2020).

With emerging evidence of *C. gallinacea* as a dominant pathogen for avian chlamydiosis, many researchers documented incidence of this less studied pathogen. In the United States, the organism was first documented in 2017 from backyard poultry in Alabama (Li et al. 2017). In an extensive study in China, *C. gallinacea* was identified as endemic chlamydial species in chickens, while *C. psittaci* was predominant in pigeons (Guo et al. 2016).

11.2.4 Transmission

Psittacosis is characterized by induction of pneumo-enteritic clinical syndrome in the affected host. This means both respiratory and enteric manifestations are more common in association with *C. psittaci* infection. Infected birds, which may be asymptomatic, are principal source of infections (Borel et al. 2018; OIE 2018; Pattison et al. 2008). The organisms are usually shed in both the nasal discharge and fecal matter of the infected birds. Activation of bacterial shedding can by induced by several stress factors up on the infected bird such as nutritional deficiencies, transportation, overcrowding, extreme cold weather, and egg laying. The severity of the developed disease mainly depends on several factors such as the virulence of the infecting bacteria, the infectious bacterial doses, as well as the immune status of the bird. The main source of infection seems to be through inhalation or ingestion of infected material from the infected birds as *C. psittaci* is usually excreted in both the respiratory exudates and the fecal matters in large amounts. Infected birds may remain shedding the bacteria up to 60 days post infection. If the domestic birds sharing the water supply or even the aquatic environment with some of the infected wild birds, they can get infected through the contaminated water source by the shed bacteria. *C. psittaci* may be transmitted from the parents to the newly hatched birds inside the nest which may have a severe outcome on the young birds. Arthropod vectors such as flies, lice, and mites may play a role in disease transmissions well (Agunos et al. 2016; Borel et al. 2018; Chu and Durrani 2019; Pattison et al. 2008).

11.3 Disease Characteristics

11.3.1 Pathogenesis

The organisms primarily enter hosts through inhalation and reaches lungs, air sacs, and pericardium. Organisms multiply in these sites and eventually spread to other organs, for example, liver, spleen, and kidney via circulation. These secondary sites also serve as proliferation sites for the pathogen with huge numbers of RBs and EBs generated in the process. It has been suggested that mammalian strains of *C. psittaci* are less pathogenic for birds than avian strains. Though considerable variation in virulence among strains are recognized, it is commonly accepted that strains isolated from psittacine birds are more virulent than those originating from non-psittacine hosts (Pattison et al. 2008; Rohde et al. 2018; van Buuren et al. 1994).

11.3.2 Clinical Signs and Pathology

Incubation period of psittacosis in birds vary. In poultry birds including turkey the incubation period ranges between 5 and 14 days and may extend to 60 days. Psittacosis in affected birds is manifested by depression, anorexia, nasal exudation which may be purulent, conjunctivitis, respiratory distress characterized by rales, and diarrhea with grayish to green coloration. Nervous signs resembling uncoordinated gait and ataxia may be observed in ducks and pigeons. At times, there may not be any prominent clinical signs especially in case of psittacine birds and may remain asymptomatic. In such cases the birds may show marked retardation growth, loss of body weight, and reduction in egg production. In chickens, the disease is usually less severe (OIE 2018; Panzetta et al. 2018; Pattison et al. 2008; Szymańska-Czerwińska and Niemczuk 2016).

Pathological features of the disease depend on many factors including the organs affected, host, severity of the infection, and others. Usually there is serositis with yellowish fibrinous exudates, air sacculitis, pneumonia, splenomegaly, hepatomegaly, tracheitis, sinusitis, and pericarditis. Histopathological examination may reveal focal necrosis of liver and inclusion bodies (Pattison et al. 2008; van Buuren et al. 1994).

11.4 Disease Management

11.4.1 Diagnosis of Avian Chlamydiosis

Diagnosis of chlamydial infection of avian presents many challenges mostly due to unique characteristics of the agent which cannot be cultured in artificial medium unlike bacterial pathogens. However, over the decades there has been steady progress in the development of diagnostic techniques for this enigmatic pathogen. Moreover, due to nonspecific clinical signs exhibited by the hosts, the disease diagnosis needs to be confirmed by laboratory diagnosis. For the purpose a wide array of laboratory tests (Table 11.1) has been developed with varying degree of successes and their inherent advantages and disadvantages.

Despite availability of a plethora of tests for avian mycoplasmosis, researchers continued to develop newer and more accurate detection systems. These methods included development of real-time PCR techniques, modification, and refinement of PCR for surveillance studies, isothermal reaction for rapid detection, direction detection of chlamydial DNA from feces of caged birds, and others (De Puysseleyr et al. 2014; Elder and Brown 1999; Hewinson et al. 1997; Jelocnik et al. 2017; Madani and Peighambari 2013; McElnea and Cross 1999; Sareyyupoglu et al. 2007).

Table 11.1 Common tests used of laboratory diagnosis of avian chlamydiosis (Source: OIE 2018)

Approach	Category	Method	Test	Putative application
Agent/sub-component detection	Molecular	Nucleic acid detection	Polymerase chain reaction	Clinical case confirmation
			Real-time polymerase chain reaction	Clinical case confirmation, surveillance
			DNA microarray	Clinical confirmation
	Direct visualization	–	Modified Gimenez staining	Clinical confirmation
	Isolation in cell culture	–	–	Clinical confirmation
	Isolation in embryonated eggs	–	–	Clinical confirmation
	Protein antigen	–	Immunohistochemical staining	Clinical confirmation
	–	–	Enzyme-linked immunosorbent assay	Declaration of population freedom, surveillance
Serological methods	–	–	Modified direct complement fixation test	–
	–	–	Recombinant major outer membrane protein ELISA	–
Others	–	–	Agar gel immunodiffusion test	–
	–	–	Latex agglutination test	–
	–	–	Microfluorescence test	–

11.4.2 Treatment and Control

Principal approach to treat chlamydial infection is antibiotic chemotherapy. Choice of the drugs is usually limited to antibiotics of tetracycline group including doxycycline and chlortetracycline. Fluoroquinolone drugs are other options of antibiotic treatment. Both classes of drugs are known to be effective in birds and humans (Beeckman and Vanrompay 2009; Deschuyffeleer et al. 2012; Jawetz 1969; Pattison et al. 2008).

Hygienic management of flocks and aviaries crucial for effective control. Segregation of infected and healthy birds is important as the infection tends to spread through inhalation. In humans, occupational exposure may be prevented by awareness, education, prompt treatment of affected birds and humans, and most importantly protective gears for occupationally exposed persons. For birds, live and inactivated vaccines are available, though their practical use is only limited (Beeckman and Vanrompay 2009; Deschuyffeleer et al. 2012; Pattison et al. 2008; Phillips et al. 2019).

11.5 Public Health Concerns

Psittacosis is an important zoonotic infection in humans manifested by flu-like symptoms. The disease is usually transmitted from birds following close contact or inhalation of dusts laden with chlamydial EBs that were excreted with feces (Agunos et al. 2016; Bastidas et al. 2013; Beeckman and Vanrompay 2009; Rohde et al. 2018; Szymańska-Czerwińska and Niemczuk 2016). In humans, several outbreaks are usually associated with the introduction of some psittacine birds to a new human community or to new pet stores. Several instances of transmission of *C. psittaci* have been recorded. The disease has been reported from most of the European countries, Australia, the United States, Argentina, Chile, and Japan (Beeckman and Vanrompay 2009). A familial outbreak of psittacosis in Turkey was detected in mother and son who contracted the infection from pet parrots kept at home (Çiftçi et al. 2008). Zoonotic transmission of *C. psittaci* was documented also in Belgium where pigeon fanciers got infected from affected homing pigeons (Dickx et al. 2010). A similar outbreak of psittacosis was recorded in a Belgian parrot breeding facility where the manager, the veterinarian, and her assistant were infected thus highlighting the occupational zoonotic nature of the disease (Harkinezhad et al. 2007). In another extensive study of human psittacosis associated with pet birds in Belgium, chlamydial infections were documented in about 15% of the personnel associated with bird breeding facilities (Vanrompay et al. 2007). A review of literature indicated that psittacosis is also an important respiratory disease that occurs often in humans following direct or indirect exposure to infected birds of their droppings (Gorman et al. 2009). In fact, the increase in numbers of pigeons in various urban areas of Europe and subsequent urban environmental with pigeon droppings is considered to be a major risk for acquiring psittacosis in humans (Magnino et al. 2009). While reports are rare from African continent, a recent report from Egypt highlighted the risk of psittacosis in bird handlers (Tolba et al. 2019).

References

Agunos A, Pierson FW, Lungu B, Dunn PA, Diseases SA, Agunos A, Pierson AFFW, Lungu BB, Dunn CPA (2016) Review of non-foodborne zoonotic and potentially zoonotic poultry diseases. Avian Dis 60:553–575. https://doi.org/10.1637/11413-032416-Review.1

Altıntaş Ö, Ünal N, Karagöz A, Cantekin Z (2020) Investigation and genotyping of *Chlamydia psittaci ompA* gene in pigeon (*Columbia domestica*) feces. Mikrobiyol Bul 54:144–153. https://doi.org/10.5578/mb.68757

Amery-Gale J, Legione AR, Marenda MS, Owens J, Eden PA, Konsak-Ilievski BM, Whiteley PL, Dobson EC, Browne EA, Slocombe RF, Devlin JM (2020) Surveillance for *Chlamydia* spp. with multilocus sequence typing analysis in wild and captive birds in Victoria. Aust J Wildl Dis 56:16–26

Bastidas RJ, Elwell CA, Engel JN, Valdivia RH (2013) Chlamydial intracellular survival strategies. Cold Spring Harb Perspect Med 3:1–20. https://doi.org/10.1101/cshperspect.a010256

Beeckman DSA, Vanrompay DCG (2009) Zoonotic Chlamydophila psittaci infections from a clinical perspective. Clin Microbiol Infect 15:11–17

Borel N, Polkinghorne A, Pospischil A (2018) A review on chlamydial diseases in animals: still a challenge for pathologists? Vet Pathol 55:374–390. https://doi.org/10.1177/0300985817751218

Chu J, Durrani MI (2019) Psittacosis [WWW Document]. https://www.ncbi.nlm.nih.gov/books/NBK538305/. Accessed 17 Apr 2019

Chu J, Yarrarapu SNS, Durrani MI (2020) Psittacosis. Treasure Island (FL)

Çiftçi B, Güler ZM, Aydoğdu M, Konur Ö, Erdoğan Y (2008) Familial outbreak of psittacosis as the first *Chlamydia psittaci* infection reported from Turkey. Tuberk Toraks 56:215–220

de Freitas Raso T, Seixas GHF, Guedes NMR, Pinto AA (2006) Chlamydophila psittaci in free-living Blue-fronted Amazon parrots (*Amazona aestiva*) and Hyacinth macaws (*Anodorhynchus hyacinthinus*) in the Pantanal of Mato Grosso do Sul, Brazil. Vet Microbiol 117:235–241. https://doi.org/10.1016/j.vetmic.2006.06.025

de Freitas Raso T, Godoy SN, Milanelo L, de Souza CAI, Matuschima ER, Araújo JP, Pinto AA (2009) An outbreak of Chlamydiosis in captive Blue-fronted Amazon parrots (*Amazona aestiva*) in Brazil. J Zoo Wildl Med 35:94–96. https://doi.org/10.1638/02-090

de Freitas T, Sc M, Augusto A, Pintod VM (2014) Evidence of *Chlamydophila psittaci* infection in captive Amazon parrots in Brazil. J Zoo Wildl Med 33:118–121. https://doi.org/10.1638/1042-7260(2002)033[0118:eocpii]2.0.co;2

De Puysseleyr K, De Puysseleyr L, Geldhof J, Cox E, Vanrompay D (2014) Development and validation of a real-time PCR for Chlamydia suis diagnosis in swine and humans. PLoS One 9: 1–7. https://doi.org/10.1371/journal.pone.0096704

Deschuyffeleer TPG, Tyberghien LFV, Dickx VLC, Geens T, Saelen JMMM, Vanrompay DCG, Braeckman LACM (2012) Risk assessment and management of Chlamydia psittaci in poultry processing plants. Ann Occup Hyg 56:340–349. https://doi.org/10.1093/annhyg/mer102

Dickx V, Beeckman DSA, Dossche L, Tavernier P, Vanrompay D (2010) Chlamydophila psittaci in homing and feral pigeons and zoonotic transmission. J Med Microbiol 59:1348–1353. https://doi.org/10.1099/jmm.0.023499-0

Donati M, Laroucau K, Delogu M, Vorimore F, Aaziz R, Cremonini E, Biondi R, Cotti C, Baldelli R, Di Francesco A (2014) *Chlamydia psittaci* in Eurasian collared doves (*Streptopelia decaocto*) in Italy. J Wildl Dis 51:214–217. https://doi.org/10.7589/2014-01-010

Donati M, Laroucau K, Guerrini A, Balboni A, Salvatore D, Catelli E, Lupini C, Levi A, Di Francesco A (2018) Chlamydiosis in backyard chickens (*Gallus gallus*) in Italy. Vector-Borne Zoonotic Dis 18:222–225. https://doi.org/10.1089/vbz.2017.2211

Dusek RJ, Justice-Allen A, Bodenstein B, Knowles S, Grear DA, Adams L, Levy C, Yaglom HD, Shearn-Bochsler VI, Ciembor PG, Gregory CR, Pesti D, Ritchie BW (2018) *Chlamydia psittaci* in feral rosy-faced lovebirds (*Agapornis roseicollis*) and other backyard birds in Maricopa County, Arizona, USA. J Wildl Dis 54:248–260. https://doi.org/10.7589/2017-06-145

Ecco R, Preis IS, Martins NRS, Vilela DR, Shivaprasad HL (2009) An outbreak of chlamydiosis in captive psittacines. Braz J Vet Pathol 2:85–90

Elder J, Brown C (1999) Review of techniques for the diagnosis of *Chlamydia psittaci* infection in psittacine birds. J Vet Diagnostic Investig 11:539–541. https://doi.org/10.1177/104063879901100611

Geigenfeind I, Vanrompay D, Haag-Wackernagel D (2012) Prevalence of Chlamydia psittaci in the feral pigeon population of Basel, Switzerland. J Med Microbiol 61:261–265. https://doi.org/10.1099/jmm.0.034025-0

Gorman J, Cook A, Ferguson C, van Buynder P, Fenwick S, Weinstein P (2009) Pet birds and risks of respiratory disease in Australia: a review. Aust N Z J Public Health 33:167–172. https://doi.org/10.1111/j.1753-6405.2009.00365.x

Guo W, Li J, Kaltenboeck B, Gong J, Fan W, Wang C (2016) Chlamydia gallinacea, not *C. psittaci*, is the endemic chlamydial species in chicken (Gallus gallus). Sci Rep 6:1–10. https://doi.org/10.1038/srep19638

Harkinezhad T, Verminnen K, Van Droogenbroeck C, Vanrompay D (2007) Chlamydophila psittaci genotype E/B transmission from African grey parrots to humans. J Med Microbiol 56: 1097–1100. https://doi.org/10.1099/jmm.0.47157-0

Hewinson RG, Griffiths PC, Bevan BJ, Kirwan SES, Field ME, Woodward MJ, Dawson M (1997) Detection of *Chlamydia psittaci* DNA in avian clinical samples by polymerase chain reaction. Vet Microbiol 54:155–166. https://doi.org/10.1016/S0378-1135(96)01268-0

Hogerwerf L, Roof I, de Jong MJK, Dijkstra F, van der Hoek W (2020) Animal sources for zoonotic transmission of psittacosis: a systematic review. BMC Infect Dis 20:192. https://doi.org/10.1186/s12879-020-4918-y

Jawetz E (1969) Chemotherapy of chlamydial infections. Adv Pharmacol Chemother 7:253–282

Jelocnik M, Islam MM, Madden D, Jenkins C, Branley J, Carver S, Polkinghorne A (2017) Development and evaluation of rapid novel isothermal amplification assays for important veterinary pathogens: *Chlamydia psittaci* and *Chlamydia pecorum*. PeerJ 5:e3799. https://doi.org/10.7717/peerj.3799

Krawiec M, Piasecki T, Wieliczko A (2015) Prevalence of *Chlamydia psittaci* and other *Chlamydia* species in wild birds in Poland. Vector-Borne Zoonotic Dis 15:652–655. https://doi.org/10.1089/vbz.2015.1814

Li L, Luther M, MacKlin K, Pugh D, Li J, Zhang J, Roberts J, Kaltenboeck B, Wang C (2017) *Chlamydia gallinacea*: a widespread emerging *Chlamydia* agent with zoonotic potential in backyard poultry. Epidemiol Infect 145:2701–2703. https://doi.org/10.1017/S0950268817001650

Madani SA, Peighambari SM (2013) PCR-based diagnosis, molecular characterization and detection of atypical strains of avian *Chlamydia psittaci* in companion and wild birds. Avian Pathol 42:38–44. https://doi.org/10.1080/03079457.2012.757288

Magnino S, Haag-Wackernagel D, Geigenfeind I, Helmecke S, Dovč A, Prukner-Radovčić E, Residbegović E, Ilieski V, Laroucau K, Donati M, Martinov S, Kaleta EF (2009) Chlamydial infections in feral pigeons in Europe: review of data and focus on public health implications. Vet Microbiol 135:54–67. https://doi.org/10.1016/j.vetmic.2008.09.045

McElnea CL, Cross GM (1999) Methods of detection of *Chlamydia psittaci* in domesticated and wild birds. Aust Vet J 77:516–521. https://doi.org/10.1111/j.1751-0813.1999.tb12123.x

OIE (2018) Avian chlamydiosis. In: OIE terrestrial manual. OIE, pp 1–13

Panzetta ME, Valdivia RH, Saka HA (2018) Chlamydia persistence: a survival strategy to evade antimicrobial effects in-vitro and in-vivo. Front Microbiol 9:1–11. https://doi.org/10.3389/fmicb.2018.03101

Pattison M, McMullin P, Bradbury J, Alexander D (eds) (2008) Poultry diseases, 6th edn. https://doi.org/10.1016/B978-0-7020-2862-5.50031-3

Phillips S, Quigley BL, Timms P (2019) Seventy years of Chlamydia vaccine research—limitations of the past and directions for the future. Front Microbiol 10:1–18. https://doi.org/10.3389/fmicb.2019.00070

Piasecki T, Chrzastek K, Wieliczko A (2012) Detection and identification of Chlamydophila psittaci in asymptomatic parrots in Poland. BMC Vet Res 8. https://doi.org/10.1186/1746-6148-8-233

Rohde G, Straube E, Essig A, Reinhold P, Sachse K (2018) Chlamydial zoonoses. Dtsch Aerzteblatt Online:174–181. https://doi.org/10.3238/arztebl.2010.0174

Sachse K, Laroucau K (2015) Two more species of *Chlamydia*-does it make a difference? Pathog Dis 73:1–3. https://doi.org/10.1093/femspd/ftu008

Santos F, Leal DC, Raso TF, Souza BMPS, Cunha RM, Martinez VHR, Barrouin-Melo SM, Franke CR (2014) Risk factors associated with *Chlamydia psittaci* infection in psittacine birds. J Med Microbiol 63:458–463. https://doi.org/10.1099/jmm.0.060632-0

Sareyyupoglu B, Cantekin Z, Bas B (2007) Chlamydophila psittaci DNA detection in the faeces of cage birds. Zoonoses Public Health 54:237–242. https://doi.org/10.1111/j.1863-2378.2007.01060.x

Stokes HS, Martens JM, Jelocnik M, Walder K, Segal Y, Berg ML, Bennett ATD (2020) Chlamydial diversity and predictors of infection in a wild Australian parrot, the Crimson Rosella (*Platycercus elegans*). Transbound Emerg Dis. https://doi.org/10.1111/tbed.13703

Szymańska-Czerwińska M, Niemczuk K (2016) Avian chlamydiosis zoonotic disease. Vector-Borne Zoonotic Dis 16:1–3. https://doi.org/10.1089/vbz.2015.1839

Tolba HMN, Abou Elez RMM, Elsohaby I (2019) Risk factors associated with Chlamydia psittaci infections in psittacine birds and bird handlers. J Appl Microbiol 126:402–410. https://doi.org/10.1111/jam.14136

van Buuren CE, Dorrestein GM, van Dijk JE (1994) *Chlamydia psittaci* infections in birds: a review on the pathogenesis and histopathological features. Vet Q 16:38–41. https://doi.org/10.1080/01652176.1994.9694414

Vanrompay D, Harkinezhad T, Van De Walle M, Beeckman D, Van Droogenbroeck C, Verminnen K, Leten R, Martel A, Cauwerts K (2007) Chlamydophila psittaci transmission from pet birds to humans. Emerg Infect Dis 13:1108–1110. https://doi.org/10.3201/eid1307.070074

Wang H, Jensen J-K, Olsson A, Vorimore F, Aaziz R, Guy L, Ellström P, Laroucau K, Herrmann B (2020) *Chlamydia psittaci* in fulmars on the Faroe Islands: a causative link to South American psittacines eight decades after a severe epidemic. Microbes Infect. https://doi.org/10.1016/j.micinf.2020.02.007

Chapter 12
Avian Colibacillosis (*Escherichia coli*)

Abstract Avian colibacillosis caused by *Escherichia coli* is an economically important disease of birds especially the commercially farmed species. The disease is reported worldwide from many species of wild birds also. The disease is mainly transmitted through contaminated feed and water which are polluted by organisms shed via fecal droppings of the birds. Depending on the host condition and immune status, the disease assumes generalized and localized forms. Affected birds are generally underweight, with shabby appearance and lower productivity. Septicemic form of the disease results in moribund condition of the host, while intestinal form of the disease is manifested by watery diarrhea and concomitant fluid loss. Localized forms of the disease affect joints, bones, and other organs. Diagnosis of the disease is based on the clinical signs and history and is often complemented by laboratory tests. Besides bacteriological techniques for laboratory detection, many other methods including immunological and molecular techniques are available. Antimicrobial therapy is the mainstay of the treatment though increasingly frequent drug resistance in avian pathogenic *E. coli* pose clinical challenge. Since many strains of *E. coli* are important human pathogens, the disease is of public health importance both from possible zoonotic angle and also due to food safety concerns. Due to asymptomatic nature, wild and peri-domestic birds may serve as important source for human infections.

Keywords Colibacillosis · *Escherichia* · *E. coli* · Septicemia · Bone and joint infection · Zoonotic · Wild birds

12.1 Introduction

Colibacillosis is an important bacterial disease affecting poultry and other birds and is caused by *Escherichia coli*. Though *E. coli* are normal inhabitants of gastrointestinal tracts, the disease is caused by pathogenic strains, often referred as APEC (avian pathogenic *Escherichia coli*). Infections caused by APEC may be systemic or localized and is usually a result of compromised immune status in birds. The organisms might affect almost any organ or system and is of considerable economic

Y. S. Malik et al., *Role of Birds in Transmitting Zoonotic Pathogens*, Livestock Diseases and Management, https://doi.org/10.1007/978-981-16-4554-9_12

significance especially in intensive farming practices. Though poultry and other birds are not considered important source for human infections with Shiga toxin producing *E. coli*, concerns have been raised for enhanced vigil and possible human infection with Ex-PEC (extra-intestinal pathogenic *Escherichia coli*) (Guabiraba and Schouler 2015; Nolan et al. 2013). The primary difference between the mammalian and avian colibacillosis is that the mammalian infection is usually an enteric disease, while the avian infections generally occur secondary to immunosuppression by another pathogen followed by localized or systemic affection by *E. coli* (Collingwood et al. 2014; Guabiraba and Schouler 2015; Nolan et al. 2013; Pattison et al. 2008). Due to multiorgan involvement, colibacillosis is manifested in various forms including yolk sac infection, egg peritonitis, colisepticemia, coligranuloma (Hjärre's disease), and others (Guabiraba and Schouler 2015; Nolan et al. 2013; Pattison et al. 2008; Projahn et al. 2018).

Globally the disease is of considerable economic significance for commercial poultry industry with significant losses in production and performance. However, economic losses due to colibacillosis in wild and peri-domestic birds are difficult to assess, and ready data are not available.

12.2 Epidemiology

12.2.1 Etiology and Classification

The causative agent of colibacillosis is *E. coli* which is a member of the family Enterobacteriaceae and is Gram negative, non-sporing, motile bacilli with more than 200 recognized serotypes. The genus is named after Theodor Escherich who first described the bacteria from infant feces (Nolan et al. 2013). *E. coli* is the type of species in the genus *Escherichia* which is closely related to genus *Shigella*. In addition to *E. coli*, two other members of the genus *E. albertii* and *E. fergusonii* have also been implicated in diseases in avian hosts. Both these latter species of *Escherichia* have been demonstrated to cause intestinal diseases in poultry birds and harbor a number of well-known virulence factors (Adesina et al. 2019; Bhatt et al. 2019; Gaastra et al. 2014; Nolan et al. 2013; Ooka et al. 2019; Simmons et al. 2016). However, their role as agents of colibacillosis needs to be further established through scientific investigations, and *E. coli* remains overwhelmingly major and recognized causative agent of colibacillosis.

12.2.2 Hosts and Distribution

Colibacillosis had been reported in all poultry and avian species, including chickens, turkeys, ducks, quails, pheasants, pigeons, guinea fowls, waterfowls, ostriches, emu, peacocks, gulls, starlings, crows, doves, lapwings, wild turkeys, hawk, songbirds, sparrows, swans, thrushes, and others (Collingwood et al. 2014; Espinosa et al. 2018; Guabiraba and Schouler 2015; Nolan et al. 2013). Young birds usually express greater susceptibility to the disease compared to the older birds (Nolan et al. 2013; Pattison et al. 2008).

E. coli are ubiquitous in distribution. They are normal inhabitants of intestinal tracts of mammals including humans and poultry birds. While the commensal flora may be harmless for healthy hosts, almost all avian species are susceptible to colibacillosis (Guabiraba and Schouler 2015; Nolan et al. 2013).

12.2.3 Transmission

Colibacillosis is mostly transmitted through contaminated water and feed. *E. coli* being a normal inhabitant of intestinal tracts, environmental contamination with fecal matter is considered to play an important role in the spread of disease. In this context, the free-living birds play an important role in disease transmission. Moreover, mechanical means of transmission by house flies and insects are also important. The darkling beetles play an effective role in transmission of *E. coli* among poultry farms either through the larval or adult stage. Birds get infected through ingestion of the larval or the adult form or even contamination of poultry food with the beetles fecal matters (Nolan et al. 2013; Pattison et al. 2008).

Environmental survival in manure may also aid in sustained transmission of the pathogen as one study conducted to address the role of the soil or manure in the survival of the *E. coli* revealed that manure may have a preservative effect on the *E. coli* O:157 which can move with the manure for up to 2 months under the top manure layer if irrigation or rainfall had occurred (Collingwood et al. 2014; Kim et al. 2020; Nolan et al. 2013; Pattison et al. 2008).

12.3 Pathogenic Characteristics

12.3.1 Antigenic Structure of E. coli

Characterization of *E. coli* antigens allow for serotyping of the organism which is an important tool for bacterial typing for epidemiological purposes. Antigens of *E. coli* are divided into three types: O antigen (Ohne—comprising the LPS component of the cell wall), H antigen (Hauch, consisting of the flagellar protein), and K antigen, determined by the capsular proteins and the Pilus antigen (F) (Gyles 2007; Nolan et al. 2013). The current system of *E. coli* serotyping had been reported by Ewing (1986) which is the globally well recognized. Serotyping of *E. coli* remains an important tool in study of avian colibacillosis and APEC as certain serotypes (O1, O2, O18, O35, O36, O78, O111) are known to be predominantly associated with disease in avian species (Nolan et al. 2013).

12.3.2 Virulence Associated Factors

Most pathogenic strains of *E. coli* carry virulence factors to facilitate their pathogenesis. These virulence factors may be in the form of toxin production, attachment or adhesion molecules, iron uptake molecules, substances related to antiphagocytic activity or serum resistance, and others (Table 12.1) (Guabiraba and Schouler 2015; Gyles 2007; Nolan et al. 2013). In addition to the endotoxins produced by the bacteria, the cell wall structure of the pathogenic bacteria may also secrete some toxins. It has been reported that the pathogenic strains of avian *E. coli* produce less toxins than that of the mammalian ones; this may be due to the lack of the toxin production machinery in the avian serotypes or bias due to the lack of the appropriate test evaluating those toxins. According to the toxin produced, disease syndrome, and effects on cell culture, *E. coli* are classified into various virulence groups: (a) enteroaggregative (EAEC), (b) enterohemorrhagic (EHEC), (c) enteroinvasive (EIEC), (d) enteropathogenic (EPEC), and (e) enterotoxigenic (ETEC). In case of both the enteropathogenic (EPEC) and enteroinvasive (EIEC) which have the ability to invade the enterocytes, no toxins are produced by the bacteria and they induce their action through the adhesion to the intestinal mucosa. The other category of the virulence factors is the adhesins which may be either fimbrial or non-fimbrial. Although the exact functions of the fimbriae are not completely clear, some studies showed that the F1 is greatly expressed during the initial colonization stage of the bacteria to the tracheal epithelium, while P fimbriae gets expressed when the bacteria enter lower respiratory tract or other parts of the body. However, bacteria expressed fimbria F1 tend to undergo phagocytosis relatively quickly. The other category of *E. coli* are attaching and effacing ones as they have a non-fimbrial adhesion molecule called intimin by which the bacteria attach to the enterocytes and lead to attaching and effacing lesion, so they called it AEEC (Gao et al. 2020; Guabiraba and Schouler 2015; Hu et al. 2020; Kim et al. 2020; Liu et al. 2018; Mu et al. 2020; Nielsen et al. 2020; Nolan et al. 2013; Song et al. 2020; Thomrongsuwannakij et al. 2020).

APEC strains are also known to produce a number of toxins though the degree of expression is somewhat lesser than that of mammalian strains. These toxins include endotoxin, cytolethal distending toxin, cytotoxic necrotizing factor, various hemolysins, and vacuolating autotransporter toxin (Guabiraba and Schouler 2015; Liu et al. 2018; Nielsen et al. 2020; Nolan et al. 2013; Song et al. 2020).

Other virulence factors that also play role in APEC pathogenesis include various iron uptake mechanisms, protectins, invasins, and ability to form and survive in a biofilm (Guabiraba and Schouler 2015; Gyles 2007; Nolan et al. 2013). Emergence of antimicrobial resistance in APEC strains presents a rising concern all over the world. A number of reports from many parts of the world have proven that antimicrobial resistance in APEC strains has the potential to indirectly or directly affect public health (Awad et al. 2016; Dhaouadi et al. 2020; Kim et al. 2020; Kuznetsova et al. 2020; Nhung et al. 2017; Päivärinta et al. 2020; Sgariglia et al. 2019; Subedi et al. 2018; Thomrongsuwannakij et al. 2020; Vounba et al. 2019; Yoon et al. 2020; Yu et al. 2020).

Table 12.1 Major virulence factors of various types of pathogenic *E. coli*[a, b]

Functional category	DAEC	EAEC	EHEC	EIEC	EPEC	ETEC	NMEC	UPEC
Adherence	*Afa/Dr family*	*AAFs, dispersin*	*ECP, Efa-1/LifA, intimin, Paa, ToxB*	–	*BFP, intimin, lymphostatin/LifA, Paa*	*Adhesive fimbriae, EtpA*	*FdeC, S fimbriae*	*Dr adhesins, F1C fimbriae, P fimbriae, S fimbriae, type 1 fimbriae*
Toxin	–	*EAST1, Pet, Pic, ShET1*	*Hemolysin, Stx*	*ShET1, ShET2, Shiga toxin*	*CDT, EAST1*	*LT, ST*	*CNF-1*	*Alpha hemolysin, CNF-1*
Iron uptake	–	–	*Chu*	*Aerobactin, Shu*		–	–	*Aerobactin, Chu, enterobactin, IroN*
Protease	–	–	*EspP, StcE*	*IcsP (SopA), Pic, SigA*	*EspC*	–	–	*Pic, Sat, Tsh*
Regulation	–	–	*Ler*		*Ler, Per*	–	–	–
Secretion system	–	–	*TTSS*	*T2SS, T6SS, TTSS*	*TTSS*	–	–	–
Type III translocated protein	–	–	*Cif, EspA, EspB, EspD, EspF, EspG, EspH, Map, NleA/EspI, NleC, NleD, Tir*	–	*Cif, EspA, EspB, EspD, EspF, EspG, EspH, Map, NleA/EspI, NleC, NleD, Tir*	–	–	–
Pathogenicity island	–	–	*LEE*	*SHI-1 (she), SHI-2, SHI-3, SRL*	*EspC island, LEE*	–	–	*PAI I, PAI II, PAI III*
Actin-based motility	–	–	–	*IcsA (VirG)*	–	–	–	–
Endotoxin	–	–	–	*LPS*	–	–	–	–
Invasion	–	–	–	–	–	–	*AslA, Ibes, K1*	–

(continued)

Table 12.1 (continued)

Functional category	DAEC	EAEC	EHEC	EIEC	EPEC	ETEC	NMEC	UPEC
							capsule, OmpA, TraJ	
Immune evasion	–	–	–	–	–	–		*TcpC*

[a]*DAEC* diffusely aggregative *E. coli*, *EAEC* enteroaggregative *E. coli*, *EHEC* enterohemorrhagic *E. coli*, *EIEC* enteroinvasive *E. coli*, *EPEC* enteropathogenic *E. coli*, *ETEC* enterotoxigenic *E. coli*, *NMEC* neonatal meningitis causing *E. coli*, *UPEC* uropathogenic *E. coli*

[b]Data source: Virulence Factor Database (Liu et al. 2018)

12.3.3 Clinical Signs and Pathological Features

Avian colibacillosis produces varying morbidity and mortality along with wide ranging clinical signs depending on the nature of disease involved and organs affected. Whether generalized or localized, affected birds are generally underweight, with shabby appearance and lower productivity (Bryan et al. 2015; Collingwood et al. 2014; Dziva and Stevens 2008; Guabiraba and Schouler 2015; Nolan et al. 2013). Septicemic form of the disease results in moribund condition marked by lethargy and unresponsiveness to stimuli. Feces are diarrheic with greenish to yellowish white colorations. Anorexia and pronounced dehydration indicated by dark skin color or raised folds of skin are also observed (Bryan et al. 2015; Collingwood et al. 2014; Dziva and Stevens 2008; Guabiraba and Schouler 2015; Nolan et al. 2013). In the intestinal form of colibacillosis, the organisms enter the host through ingestion of contaminated feed and water followed by colonization in the intestinal tract and establishment of the disease leading to watery diarrhea resulting in fluid and electrolyte loss (Fig. 12.1). In case of localized infection, signs vary according to tissue affected. When joints or bones of legs are affected, difficulty in movement is apparent with typical hopping movement or partial to complete reluctance to stand or walk. Affection of spinal joints results in arched back appearance. If internal organ like yolk sac is affected, birds usually exhibit reluctance to move and may continue to sit at one place. Abdominal distention is also

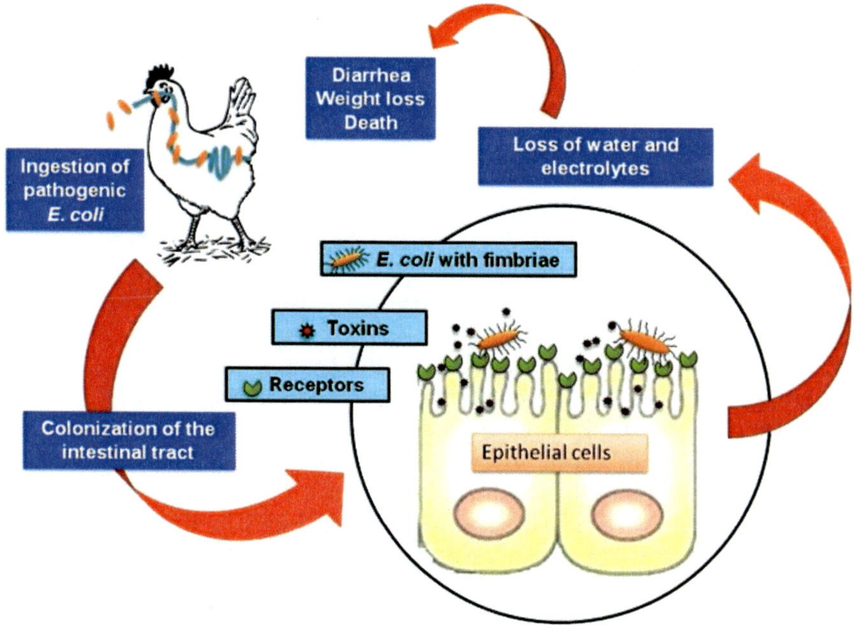

Fig. 12.1 Schematic diagram of gastrointestinal pathogenesis in colibacillosis

evident in case of affection of visceral organs (Dziva and Stevens 2008; Guabiraba and Schouler 2015; Nolan et al. 2013).

Pathological features of the colibacillosis in bird are diverse and wide ranging. Localized forms of the disease depending on the organs affected produce omphalitis, cellulitis, swollen head, salpingitis, and occasionally vaginitis. In case of omphalitis, cardinal signs include edematous swelling, abscess formation, abdominal distention, and hyperemia of associated blood vessels. Cellulitis caused by *E. coli* is characterized by the formation of serosanguinous exudation along the subcutaneous tissues of abdomen and thigh regions. The condition is often referred as "plaque." Similar to "plaque," swollen head is also a result of sub-acute cellulitis occurring around periorbital region of the head and is characterized by accumulation of inflammatory exudates under the dermis (Pattison et al. 2008; Nolan et al. 2013). Diarrheal form of colibacillosis is relatively uncommon in avian hosts and is characterized by dehydration, diarrhea, and pale discoloration of intestine and ceca which often are distended with fluid containing mucous and gas. Systemic forms of colibacillosis involve multiple organs and are manifested in many ways, including septicemia (hemorrhagic ones observed in turkeys), respiratory infection involving air sac and chronic respiratory disease, meningitis, panophthalmitis, osteomyelitis, synovitis, sternal bursitis, and fibrinous pericarditis (Pattison et al. 2008; Nolan et al. 2013).

12.4 Disease Management

12.4.1 Diagnosis

Primary diagnosis of colibacillosis in avian species is from clinical observation, epidemiological inputs, and history. For many small commercial units of poultry, usually this suffices. However, laboratory diagnosis is called for confirmation of the presumptive diagnosis.

Laboratory diagnosis of colibacillosis is attempted through bacteriological approach of isolation and identification of the agent. Various commercial media (e.g., Eosine Methylene Blue agar, McConkey agar, Tegritiol-7 agar) are available and are routinely employed in laboratories across the world. Following isolation, the bacteria is usually confirmed through a host of biochemical tests. To reduce the time requirement of manual identification, automated identification systems are commercially available and are also used. In addition, rapid biochemical testing systems such as API strips are also in use in many laboratories. For characterization of virulence genes, antimicrobial resistance genes, and epidemiological analysis, a wide variety of molecular and other methods are available. Serological identification is not commonly used though they are of real value in serotyping and epidemiological and surveillance purposes (Collingwood et al. 2014; Guabiraba and Schouler 2015; Gyles 2007; Nolan et al. 2013; Pattison et al. 2008).

12.4.2 Treatment and Control

Mainstay for the treatment of avian colibacillosis is administration of antibiotics to minimize the economic losses. Treatment with tetracyclines and apramycin has been proven helpful though the practices vary with farms, geography, prevailing husbandry norms, and others. With concerns rising over use and abuse of antibiotics in commercial poultry production, other non-antibiotic options are also being explored. These include use of essential oils, probiotics, use of bacteriophage, and vitamin supplementations (Guabiraba and Schouler 2015; Gyles 2007; Nolan et al. 2013; Projahn et al. 2018; Wernicki et al. 2017). Prevention of colibacillosis is primarily based on control of infection in breeding stock, exercising strict biosecurity measures and vaccination. Vaccination is however met with varying successes due to wide diversity of infecting strains. Usually vaccines containing live attenuated or killed bacteria provide protection against homologous infecting strains (Guabiraba and Schouler 2015; Nolan et al. 2013; Projahn et al. 2018; Wernicki et al. 2017).

12.5 *Escherichia coli*, Wild Birds, and the Public Health Concerns

Though colibacillosis is an important disease of birds and other mammals worldwide, the specific role of wild birds in the dissemination of this organism is poorly understood, and our current understanding is mostly based on studies with poultry birds. However, *E. coli* have been isolated from a wide variety of captive and non-captive birds, including pheasants, pigeons, guinea fowls, waterfowls, ostriches, emu, peacocks, gulls, starlings, crows, doves, lapwings, wild turkeys, hawks, songbirds, sparrows, swans, thrushes, and others (Collingwood et al. 2014; Espinosa et al. 2018; Guabiraba and Schouler 2015; Nolan et al. 2013; Farooq et al. 2009; Matin et al. 2017). High prevalence of *E. coli* had been detected in the feces of North America geese and up to 50% were found in the London parks at that time. It was noted that the North America goose posed a potential source for contamination of the urban and rural areas through their dropping during their migration journeys throughout the year (Borges et al. 2017; Espinosa et al. 2018). In a study on various avian hosts in India, urban pigeons were found to carry STEC isolates, and the authors concluded that pigeons might serve as vectors for transmission of STEC to humans and environment (Farooq et al. 2009).

Relation between avian pathogenic *E. coli* (APEC) and human infections with ExPEC (extraintestinal pathogenic *E. coli*) had been subject of considerable investigation exploring the possible link between the two. Wide range of infections, such as sepsis, urinary tract infection, and neonatal meningitis, are attributed to ExPEC in humans (Mellata 2013). However, the link between these two (APEC and ExPEC) are not always straightforward though evidences point toward such associations. In exploring the role of broiler chicken meat, pig, and pork with that of virulence genes of ExPEC in humans, Jakobsen and others concluded that the strains from meat and production animals did pose zoonotic risk (Jakobsen et al. 2010). Similarly, in a review of the linkages between poultry meat and human UTIs, it was observed that regardless of the quantum of the human ExPEC infections attributable to poultry meat, the public health consequence would be considerable (Manges 2016). A recent study from the Netherlands found that consumption of raw or undercooked meat was associated with poultry attributed STEC infections in humans (Mughini-Gras et al. 2018). Recently, in an extensive scoping review of STEC in wild animals, Espinosa and others found that much of the studies on *E. coli* in wild animals were limited to ruminants and urban birds, thus limiting our understanding of the role of wild birds (Espinosa et al. 2018). Nonetheless, a study from Brazil involving asymptomatic wild birds belonging to 15 orders (*Accipitriformes*, *Anseriformes*, *Cariamiformes*, *Cathartiformes*, *Charadriiformes*, *Columbiformes*, *Falconiformes*, *Galliformes*, *Gruiformes*, *Passeriformes*, *Pelecaniformes*, *Piciformes*, *Psittaciformes*, *Strigiformes*, and *Tinamiformes*), and free living urban pigeons revealed that they could act as carriers for STEC and EPEC strains for potential transmission to humans and environments (Borges et al. 2017). On the same line another study from Brazil identified captive wild birds of various Orders as putative reservoirs of EPEC and STEC (Sanches et al. 2017). Overall, these studies did indicate that wild and peri-domestic birds may play an important role in dissemination of pathogenic *E. coli* to humans.

References

Adesina T, Nwinyi O, De N, Akinnola O, Omonigbehin E (2019) First detection of carbapenem-resistant *Escherichia fergusonii* strains harbouring beta-lactamase genes from clinical samples. Pathogens 8. https://doi.org/10.3390/pathogens8040164

Awad A, Arafat N, Elhadidy M (2016) Genetic elements associated with antimicrobial resistance among avian pathogenic *Escherichia coli*. Ann Clin Microbiol Antimicrob 15:59. https://doi.org/10.1186/s12941-016-0174-9

Bhatt S, Egan M, Critelli B, Kouse A, Kalman D, Upreti C (2019) The evasive enemy: insights into the virulence and epidemiology of the emerging attaching and effacing pathogen *Escherichia albertii*. Infect Immun 87. https://doi.org/10.1128/IAI.00254-18

Borges CA, Cardozo MV, Beraldo LG, Oliveira ES, Maluta RP, Barboza KB, Werther K, Ávila FA (2017) Wild birds and urban pigeons as reservoirs for diarrheagenic *Escherichia coli* with zoonotic potential. J Microbiol 55:344–348. https://doi.org/10.1007/s12275-017-6523-3

Bryan A, Youngster I, McAdam AJ (2015) Shiga toxin producing *Escherichia coli*. In: Clinics in laboratory medicine, 1st edn. Elsevier Inc. https://doi.org/10.1016/j.cll.2015.02.004

Collingwood C, Kemmett K, Williams N, Wigley P (2014) Is the concept of avian pathogenic *Escherichia coli* as a single pathotype fundamentally flawed? Front Vet Sci 1:1–4. https://doi.org/10.3389/fvets.2014.00005

Dhaouadi S, Soufi L, Hamza A, Fedida D, Zied C, Awadhi E, Mtibaa M, Hassen B, Cherif A, Torres C, Abbassi MS, Landolsi RB (2020) Co-occurrence of mcr-1 mediated colistin resistance and β-lactamase-encoding genes in multidrug-resistant *Escherichia coli* from broiler chickens with colibacillosis in Tunisia. J Glob Antimicrob Resist 22:538–545. https://doi.org/10.1016/j.jgar.2020.03.017

Dziva F, Stevens MP (2008) Colibacillosis in poultry: unravelling the molecular basis of virulence of avian pathogenic *Escherichia coli* in their natural hosts. Avian Pathol 37:355–366. https://doi.org/10.1080/03079450802216652

Espinosa L, Gray A, Duffy G, Fanning S, McMahon BJ (2018) A scoping review on the prevalence of Shiga-toxigenic *Escherichia coli* in wild animal species. Zoonoses Public Health 65:911–920. https://doi.org/10.1111/zph.12508

Ewing WH (1986) Edwards and Ewing's identification of Enterobacteriaceae, 4th edn. Elsevier, Amsterdam, pp 1–536

Farooq S, Hussain I, Mir MA, Bhat MA, Wani SA (2009) Isolation of atypical enteropathogenic *Escherichia coli* and Shiga toxin 1 and 2f-producing *Escherichia coli* from avian species in India. Lett Appl Microbiol 48:692–697. https://doi.org/10.1111/j.1472-765X.2009.02594.x

Gaastra W, Kusters JG, van Duijkeren E, Lipman LJA (2014) Escherichia fergusonii. Vet Microbiol 172:7–12. https://doi.org/10.1016/j.vetmic.2014.04.016

Gao Q, Su S, Li X, Wang H, Liu J, Gao S (2020) Transcriptional analysis of RstA/RstB in avian pathogenic *Escherichia coli* identifies its role in the regulation of hdeD-mediated virulence and survival in chicken macrophages. Vet Microbiol 241:108555. https://doi.org/10.1016/j.vetmic.2019.108555

Guabiraba R, Schouler C (2015) Avian colibacillosis: still many black holes. FEMS Microbiol Lett 362:1–8. https://doi.org/10.1093/femsle/fnv118

Gyles CL (2007) Shiga toxin-producing *Escherichia coli*: an overview. J Anim Sci 85:E45–E62. https://doi.org/10.2527/jas.2006-508

Hu R, Li J, Zhao Y, Lin H, Liang L, Wang M, Liu H, Min Y, Gao Y, Yang M (2020) Exploiting bacterial outer membrane vesicles as a cross-protective vaccine candidate against avian pathogenic *Escherichia coli* (APEC). Microb Cell Factories 19:119. https://doi.org/10.1186/s12934-020-01372-7

Jakobsen L, Spangholm DJ, Pedersen K, Jensen LB, Emborg HD, Agersø Y, Aarestrup FM, Hammerum AM, Frimodt-Møller N (2010) Broiler chickens, broiler chicken meat, pigs and pork as sources of ExPEC related virulence genes and resistance in *Escherichia coli* isolates from community-dwelling humans and UTI patients. Int J Food Microbiol 142:264–272. https://doi.org/10.1016/j.ijfoodmicro.2010.06.025

Kim YB, Yoon MY, Ha JS, Seo KW, Noh EB, Son SH, Lee YJ (2020) Molecular characterization of avian pathogenic *Escherichia coli* from broiler chickens with colibacillosis. Poult Sci 99: 1088–1095. https://doi.org/10.1016/j.psj.2019.10.047

Kuznetsova MV, Gizatullina JS, Nesterova LY, Starčič Erjavec M (2020) *Escherichia coli* isolated from cases of colibacillosis in Russian poultry farms (Perm Krai): sensitivity to antibiotics and bacteriocins. Microorganisms 8. https://doi.org/10.3390/microorganisms8050741

Liu B, Zheng D, Jin Q, Chen L, Yang J (2018) VFDB 2019: a comparative pathogenomic platform with an interactive web interface. Nucleic Acids Res 47:D687–D692. https://doi.org/10.1093/nar/gky1080

Manges AR (2016) *Escherichia coli* and urinary tract infections: the role of poultry-meat. Clin Microbiol Infect 22:122–129. https://doi.org/10.1016/j.cmi.2015.11.010

Matin MA, Islam MA, Khatun MM (2017) Prevalence of colibacillosis in chickens in greater Mymensingh district of Bangladesh. Vet World 10:29–33. https://doi.org/10.14202/vetworld.2017.29-33

Mellata M (2013) Human and avian extraintestinal pathogenic *Escherichia coli*: infections, zoonotic risks, and antibiotic resistance trends. Foodborne Pathog Dis 10:916–932. https://doi.org/10.1089/fpd.2013.1533

Mu X, Gao R, Xiao W, Gao Q, Cao C, Xu H, Gao S, Liu X (2020) EntE, EntS and TolC synergistically contributed to the pathogenesis of APEC strain E058. Microb Pathog 141: 103990. https://doi.org/10.1016/j.micpath.2020.103990

Mughini-Gras L, van Pelt W, van der Voort M, Heck M, Friesema I, Franz E (2018) Attribution of human infections with Shiga toxin-producing *Escherichia coli* (STEC) to livestock sources and identification of source-specific risk factors, The Netherlands (2010–2014). Zoonoses Public Health 65:e8–e22. https://doi.org/10.1111/zph.12403

Nhung NT, Chansiripornchai N, Carrique-Mas JJ (2017) Antimicrobial resistance in bacterial poultry pathogens: a review. Front Vet Sci 4:126. https://doi.org/10.3389/fvets.2017.00126

Nielsen DW, Ricker N, Barbieri NL, Allen HK, Nolan LK, Logue CM (2020) Outer membrane protein A (OmpA) of extraintestinal pathogenic *Escherichia coli*. BMC Res Notes 13:51. https://doi.org/10.1186/s13104-020-4917-5

Nolan LK, Barnes HJ, Vaillancourt J, Abdul-aziz T, Logue CM (2013) Colibacillosis. In: Swayne DE (ed) Diseases of poultry. Wiley, pp 751–805

Ooka T, Seto K, Ogura Y, Nakamura K, Iguchi A, Gotoh Y, Honda M, Etoh Y, Ikeda T, Sugitani W, Konno T, Kawano K, Imuta N, Yoshiie K, Hara-Kudo Y, Murakami K, Hayashi T, Nishi J (2019) O-antigen biosynthesis gene clusters of *Escherichia albertii*: their diversity and similarity to *Escherichia coli* gene clusters and the development of an O-genotyping method. Microb Genomics 5. https://doi.org/10.1099/mgen.0.000314

Päivärinta M, Latvio S, Fredriksson-Ahomaa M, Heikinheimo A (2020) Whole genome sequence analysis of antimicrobial resistance genes, multilocus sequence types and plasmid sequences in ESBL/AmpC *Escherichia coli* isolated from broiler caecum and meat. Int J Food Microbiol 315: 108361. https://doi.org/10.1016/j.ijfoodmicro.2019.108361

Pattison M, McMullin P, Bradbury J, Alexander D (eds) (2008) Poultry diseases, 6th edn. https://doi.org/10.1016/B978-0-7020-2862-5.50031-3

Projahn M, Pacholewicz E, Becker E, Correia-Carreira G, Bandick N, Kaesbohrer A (2018) Reviewing interventions against Enterobacteriaceae in broiler processing: using old techniques for meeting the new challenges of ESBL E. coli? Biomed Res Int 2018:1–14. https://doi.org/10.1155/2018/7309346

Sanches LA, Gomes MS, Teixeira RHF, Cunha MPV, Oliveira MGX, Vieira MAM, Gomes TAT, Knobl T (2017) Captive wild birds as reservoirs of enteropathogenic *E. coli* (EPEC) and Shiga-toxin producing *E. coli* (STEC). Braz J Microbiol 48:760–763. https://doi.org/10.1016/j.bjm.2017.03.003

Sgariglia E, Aconiti Mandolini N, Napoleoni M, Medici L, Fraticelli R, Conquista M, Gianfelici P, Staffolani M, Fisichella S, Capuccella M, Sargenti M, Perugini G (2019) Antibiotic resistance pattern and virulence genesin avian pathogenic *Escherichia coli* (APEC) from different breeding systems. Vet Ital 55:26–33. https://doi.org/10.12834/VetIt.1617.8701.1

Simmons K, Islam MR, Rempel H, Block G, Topp E, Diarra MS (2016) Antimicrobial resistance of Escherichia fergusonii isolated from broiler chickens. J Food Prot 79:929–938. https://doi.org/10.4315/0362-028X.JFP-15-575

Song X, Jiang H, Qi Z, Shen X, Xue M, Hu J, Liu H, Zhou X, Tu J, Qi K (2020) APEC infection affects cytokine-cytokine receptor interaction and cell cycle pathways in chicken trachea. Res Vet Sci 130:144–152. https://doi.org/10.1016/j.rvsc.2020.03.016

Subedi M, Luitel H, Devkota B, Bhattarai RK, Phuyal S, Panthi P, Shrestha A, Chaudhary DK (2018) Antibiotic resistance pattern and virulence genes content in avian pathogenic *Escherichia coli* (APEC) from broiler chickens in Chitwan, Nepal. BMC Vet Res 14:113. https://doi.org/10.1186/s12917-018-1442-z

Thomrongsuwannakij T, Blackall PJ, Djordjevic SP, Cummins ML, Chansiripornchai N (2020) A comparison of virulence genes, antimicrobial resistance profiles and genetic diversity of avian pathogenic *Escherichia coli* (APEC) isolates from broilers and broiler breeders in Thailand and Australia. Avian Pathol:1–10. https://doi.org/10.1080/03079457.2020.1764493

Vounba P, Arsenault J, Bada-Alambédji R, Fairbrother JM (2019) Prevalence of antimicrobial resistance and potential pathogenicity, and possible spread of third generation cephalosporin resistance, in *Escherichia coli* isolated from healthy chicken farms in the region of Dakar, Senegal. PLoS One 14:e0214304. https://doi.org/10.1371/journal.pone.0214304

Wernicki A, Nowaczek A, Urban-Chmiel R (2017) Bacteriophage therapy to combat bacterial infections in poultry. Virol J 14:1–13. https://doi.org/10.1186/s12985-017-0849-7

Yoon MY, Kim YB, Ha JS, Seo KW, Noh EB, Son SH, Lee YJ (2020) Molecular characteristics of fluoroquinolone-resistant avian pathogenic *Escherichia coli* isolated from broiler chickens. Poult Sci 99:3628–3636. https://doi.org/10.1016/j.psj.2020.03.029

Yu L, Li W, Li Q, Chen X, Ni J, Shang F, Xue T (2020) Role of LsrR in the regulation of antibiotic sensitivity in avian pathogenic *Escherichia coli*. Poult Sci 99:3675–3687. https://doi.org/10.1016/j.psj.2020.03.064

Chapter 13
Avian Erysipelas

Abstract Avian erysipelas is an economically important zoonotic disease of birds caused by the Gram positive, rod-shaped, facultatively anaerobic, bacterium, *Erysipelothrix rhusiopathiae*. Many species of wild and domestic birds are affected by the disease, though the disease is of much commercial relevance in farmed turkeys. Due to ubiquitous distribution and saprophytic nature of the causative organism, infection is usually contracted from environment contaminated by previously infected birds. Pecking on carcasses of dead birds is particular risk for commercial flocks. Disease signs are often generalized in nature characterized by depression, diarrhea, and death, though marked cyanosis of head region in turkeys is an important feature. Diagnosis of the avian erysipelas largely depends on clinical signs, history, postmortem lesions, and laboratory tests. Penicillin and other antibiotics are commonly used for therapeutic management of the disease, though prevention largely depends on maintenance of farm hygiene and commercially available vaccines. Since *E. rhusiopathiae* is an important zoonotic agent for humans, avian infection poses particular risks to farm personnel. The organism was recovered from many species of captive and wild birds and may thus serve to disseminate the disease over large area.

Keywords *Erysipelothrix* · *E. rhusiopathiae* · Erysipelas · Zoonotic · Turkey · Erysipeloid

13.1 Introduction

Avian erysipelas is an infectious disease of birds caused by the organism *Erysipelothrix rhusiopathiae* with considerable economic importance especially in turkeys (Brooke and Riley 1999; Pattison et al. 2008; Reboli and Farrar 1989). The disease affects a wide range of hosts including humans in which the disease is mostly occupational in nature. The human disease is known as erysipeloid, particularly the cutaneous form of the infection (Brooke and Riley 1999; Pattison et al. 2008; Reboli and Farrar 1989). The organisms are ubiquitous in nature, and the diseases caused by them are known by many names such as erythema serpens, Rosenbach's erythema,

Y. S. Malik et al., *Role of Birds in Transmitting Zoonotic Pathogens*, Livestock Diseases and Management, https://doi.org/10.1007/978-981-16-4554-9_13

diamond disease in pigs, fish handler's disease, etc. (Brooke and Riley 1999; Reboli and Farrar 1989). The organism infects a wide range of birds including farmed chickens, turkey, and many wild and peri-domestic birds worldwide (Blyde and Woods 1999; Boerner et al. 2004; Griffiths and Buller 1991; Hennig et al. 2006; Mazaheri et al. 2006a; Opriessnig et al. 2005; Work et al. 1999).

13.2 Epidemiology

13.2.1 Causative Agent

Erysipelothrix rhusiopathiae, is a Gram positive, rod-shaped, facultatively anaerobic, nonmotile bacterium. The organism can grow at temperatures between 5 and 42 °C and is catalase and oxidase negative (Pattison et al. 2008). As per current listing, there are four species: *Erysipelothrix inopinata*, *Erysipelothrix larvae*, *Erysipelothrix rhusiopathiae* and *Erysipelothrix tonsillarum*. Of these, only *E. rhusiopathiae* is considered to be avian pathogenic species (Parte 2018; Pattison et al. 2008). However, taxonomic and other recent studies proposed several novel species for the genus (Bang et al. 2015; Opriessnig et al. 2020; Pomaranski et al. 2020; Takahashi et al. 1987, 2008; Verbarg et al. 2004) (Table 13.1).

13.2.2 Hosts and Distribution

The causative organism of *E. rhusiopathiae* is widely distributed in environment (soil, surface water, sewage effluent) and affects a wide variety of hosts. Pigs are considered to be natural reservoir of the organism. Many other hosts such as rodents, fishes, and birds may also harbor *E. rhusiopathiae*. In addition, amphibians, reptiles, insects, and humans are also infected. Among, domestic poultry birds, turkey is

Table 13.1 Members of the genus *Erysipelothrix* and their hosts

Erysipelothrix spp.	Hosts	References
E. rhusiopathiae	Mammals, birds, fishes	Opriessnig et al. (2020)
E. tonsillarum	Mammals, birds, fishes	Opriessnig et al. (2020), Takahashi et al. (1987)
E. species 1, 2, 3	Mammals, birds, fishes	Opriessnig et al. (2020), Takahashi et al. (2008)
E. inopinata	Vegetable peptone broth	Verbarg et al. (2004)
E. larvae	Rhinoceros beetle larvae	Bang et al. (2015)
E. piscisicarius sp. nov.	Fish	Pomaranski et al. (2020)

perhaps most susceptible, though infections have been recorded in other birds too. Humans, particularly animal handlers, fish processors, and butchers are at risk of contracting the infection (Apha et al. 2015; Eriksson et al. 2014; Pattison et al. 2008; Reboli and Farrar 1989; Wang et al. 2010). The disease has been reported from many parts of the world spanning all continents except Antarctica (Opriessnig et al. 2020; Pattison et al. 2008). Many farmed species of birds including chicken, turkey, and emu have been reportedly affected in many parts of the world (Eriksson et al. 2009, 2010, 2013, 2014; Griffiths and Buller 1991; Health and Agency 2013; Hollifield et al. 2006; Kurian et al. 2012; Mazaheri et al. 2006a, b; Morgan et al. 2011; Swan and Lindsey 1998). In addition, several cases have also been reported from wild, peri-domestic, and caged birds as well as from some of the endangered species of birds (Cousquer 2005; Gartrell et al. 2005; Hennig et al. 2006; Livingston et al. 2013; Opriessnig et al. 2005; Work et al. 1999).

13.2.3 Transmission

Despite *E. rhusiopathiae* being a well-known and widely reported pathogen affecting birds and many other species, the routes of transmission often remained obscure. This is particularly true in case of avian erysipelas (Pattison et al. 2008). In case of humans, the disease is considered an occupational zoonosis and is often reported to have gained entry in the host through breaks in the skin (Brooke and Riley 1999; Mutalib et al. 1993; Opriessnig et al. 2020; Reboli and Farrar 1989; Wang et al. 2010). In case of farmed poultry, often a history of indirect contact with pigs or sheep has been observed (Pattison et al. 2008). Moreover, recovered birds are also thought to be carriers for several weeks post infection and might contaminate their surroundings through fecal excretion of the pathogen (Pattison et al. 2008). Similarly, asymptomatic carriers might also play role in dissemination of the infection. In case of commercial poultry, contaminated feed, especially fish meal, may be an important source (Pattison et al. 2008). A number of researchers studied the risk factor of *E. rhusiopathiae* infection, and it has been observed that non-impervious earthen floors may be an influencing factor as is behavioral factors like fighting within the flock (Eriksson et al. 2013, 2014; Pattison et al. 2008). Pecking on carcasses of dead birds that died from infection of *E. rhusiopathiae* pose a particular risk for flock of poultry. Similarly, ingestion of compost material have been documented to have caused outbreak in racing pigeons indicating role of decaying materials in the spread and transmission of this soil-associated pathogen (Cousquer 2005; Pattison et al. 2008).

13.3 Disease Characteristics

13.3.1 Pathogenesis

Despite several decades of study, exact mechanisms of pathogenesis in wild birds are not well established. However, cues from domestic birds provide reasonable insights. A number of virulence factors are produced by the organism which cause cellular images in hosts and establish the infection (Table 13.2).

The enzyme neuraminidase usually cleaves alpha glycosidic linkages of the sialic acid which may lead to vascular damage as well as the formation of the hyaline thrombus. Most of the strains produce another virulence factors such as the hyaluronidase and coagulase. One more possible virulence factor is the bacterial capsule which is resistant to the phagocytic action of microphages. This phenomena is most commonly seen in the chronic infection (Mazaheri et al. 2006a, b). In addition, expression of heat shock proteins has been identified in the organism (Brooke and Riley 1999; Mazaheri et al. 2006a; Pattison et al. 2008).

13.3.2 Clinical Signs and Pathological Features

Clinical signs of the erysipelas are often not specific. In case of acute infection, the onset may be sudden. Death may ensue following a brief period of illness characterized by depression, diarrhea, and occasional darkening of the skin. In turkeys, marked cyanosis of the head region is noticed. Congestion of inter-digital areas is also observed in case of web-footed birds. Necropsy findings are also non-conspicuous with very mild or inapparent lesions. In the septicemic form of the disease, the carcasses of the dead birds may be congested, sometimes with petechial hemorrhage on the pericardial fat or pleura. Mottled appearance of liver

Table 13.2 Virulence factors of *E. rhusiopathiae*

Virulence factors	Major/putative role	References
Neuraminidase	Cleaves sialic acid moiety from sialo-glyco-conjugates	Opriessnig et al. (2020), Pattison et al. (2008), Shimoji (2000), Wang et al. (2010)
Hyalurinidase	Hydrolyzes matrix substance (hyaluronic acid) facilitating spread	Opriessnig et al. (2020), Pattison et al. (2008), Shimoji (2000), Wang et al. (2010)
Capsular antigens	Resistance to phagocytosis, intra-cellular survival	Shimoji (2000), Wang et al. (2010)
Adhesion proteins	Binding and adherence to cellular surfaces, biofilm formation	Shimoji (2000), Wang et al. (2010)
Additional surface proteins	Adhesion, unspecified role	Shimoji (2000)

along with enlargement of spleen and kidneys is also observed. In cases of enteritis, cecal ulceration yellowish circular lesion may be observed. Although arthritis is common in domestic birds, it is not very common in case of the wild birds (Mazaheri et al. 2006a, b; Pattison et al. 2008; Work et al. 1999).

13.4 Disease Management

13.4.1 Diagnosis

Diagnosis of avian erysipelas depends largely on history, clinical signs, and postmortem lesions. However, confirmatory diagnosis can be established only after reliable laboratory confirmation. There are many techniques for laboratory diagnosis of *E. rhusiopathiae* infection in birds, and they involve classical bacteriological approach to modern molecular methods.

Traditional bacteriological isolation and identification of the bacteria is usually done on one of the several commercially available media such as modified blood azide medium, Bohm's medium, Parker's medium, etc. Following selective plating and tentative identification, biochemical tests such as H_2S production are employed for further confirmation. Commercially available API systems have also been reportedly employed for identification of the organism. In addition to bacteriological identification and fluorescent antibody test, mouse protection assay has been employed for laboratory diagnosis. Among molecular methods, polymerase chain reaction (PCR) is widely used for the identification. Methods targeting 16S rRNA gene have been known to produce quick reproducible results. However, other PCR methodologies have also been reported (Bobrek and Gaweł 2015; Hennig et al. 2006; Kurian et al. 2012; Pattison et al. 2008; Shimoji et al. 2020; Wang et al. 2010).

13.4.2 Treatment and Control

Treatment of *E. rhusiopathiae* infection usually responds well to antibiotics such as penicillin, which is the preferred drug for therapy, and cephalosporins and erythromycins. For prevention of infection in birds, good hygienic practices and biosecurity measures are essential. In humans, care should be taken during handling of birds by wearing protective clothing and other necessary protective gears. Several commercial vaccines are available which are either live attenuated bacteria or bacterins offering immunity for 6–12 months in pigs and turkeys (Opriessnig et al. 2020; Pattison et al. 2008; Wang et al. 2010).

13.5 *Erysipelothrix*, Wild Birds, and the Public Health Concerns

It has been reported that *Erysipelothrix* infection is one of the low pathogenicity Zoonotic diseases. The main sources of contamination are the contact with the infected animals or even their products. Three forms of the disease had been reported in human. The erysipeloid form which is a mild cutaneous form differs from the diffuse cutaneous form and the systemic form which are rare and may result in septicemia and endocarditis. In animals, erysipelas is one of the most important disease affecting swines, domestic turkeys, chickens, and ducks with a great economic impact on those species (Apha et al. 2015; Eriksson et al. 2013, 2014; Health and Agency 2013; Hollifield et al. 2006; Kurian et al. 2012; Mazaheri et al. 2006a; Wang et al. 2010). The disease had also been reported from emus on multiple occasions (Eriksson et al. 2009; Griffiths and Buller 1991; Morgan et al. 2011; Pattison et al. 2008; Swan and Lindsey 1998). The transmission of the disease from the wild birds to the domestic animals still needs further investigations. *E. rhusiopathiae* was also isolated from various wild and captive birds including malleefowl (Blyde and Woods 1999), flocks of geese (Bobrek and Gaweł 2015), blue penguin (Boerner et al. 2004), racing pigeons (Cousquer 2005), parrots (Galindo-Cardiel et al. 2012), critically endangered kakapo (Gartrell et al. 2005), ring-necked pheasants (Hennig et al. 2006), laughing kookaburra (Opriessnig et al. 2005), takahe, kiwi, black stilts (Alley and Gartrell 2019), and crow (Work et al. 1999). However, according to the available data, *Erysipelothrix* infection in the wild birds is usually sporadic except in some cases there are some massive outbreaks in Eared Grebes as that reported in the Great Salt Lake (Jensen and Cotter 1976). Further studies are needed to determine the actual impact of the *Erysipelothrix* infection on the wild life population.

Erysipelas is obviously a zoonotic infection with occupational risks (Reboli and Farrar 1989; Wang et al. 2010), yet the causal links are not always easily established. However, a study investigating erysipelas infection of caged laying chicken documented that two attendants handling the dead birds developed localized lesion of erysipeloid in fingers eventually spreading to lymph nodes (Mutalib et al. 1993). Similarly, localized finger lesions of erysipeloid were detected in seven persons engaged in a commercial quail processing unit where an outbreak of erysipelas had been recorded among the quails (Mutalib et al. 2006). Though similar direct link between human infection and erysipelas in wild birds is rare, people in close contact to the wild birds should take extra precautions to avoid infection through the damaged or lacerated skin (Sheng et al. 2000).

References

Alley MR, Gartrell BD (2019) Wildlife diseases in New Zealand: recent findings and future challenges. N Z Vet J 67:1–11. https://doi.org/10.1080/00480169.2018.1520656

Apha F, Anglia E, Agency PH (2015) Erysipelas in turkeys, sheep and pigs. Vet Rec 176:302–305. https://doi.org/10.1136/vr.h1201

Bang B-H, Rhee M-S, Chang D-H, Park D-S, Kim B-C (2015) *Erysipelothrix larvae* sp. nov., isolated from the larval gut of the rhinoceros beetle, *Trypoxylus dichotomus* (Coleoptera: Scarabaeidae). Antonie Van Leeuwenhoek 107:443–451. https://doi.org/10.1007/s10482-014-0342-x

Blyde D, Woods R (1999) Erysipelas in malleefowl. Aust Vet J 77:434–435

Bobrek K, Gaweł A (2015) Erysipelas outbreaks in flocks of geese in Poland—biochemical and genetic analyses of the isolates. Avian Dis 59:436–439. https://doi.org/10.1637/11050-030115-case.1

Boerner L, Nevis KR, Hinckley LS, Weber ES, Frasca Jr S (2004) *Erysipelothrix septicemia* in a little blue penguin (*Eudyptula minor*). J Vet Diagnostic Investig 16:145–149

Brooke CJ, Riley TV (1999) *Erysipelothrix rhusiopathiae*: bacteriology, epidemiology and clinical manifestations of an occupational pathogen. J Med Microbiol 48:789–799. https://doi.org/10.1099/00222615-48-9-789

Cousquer G (2005) Erysipelas outbreak in racing pigeons following ingestion of compost [5]. Vet Rec 156:656. https://doi.org/10.1136/vr.156.20.656-a

Eriksson H, Jansson DS, Johansson KE, Båverud V, Chirico J, Aspán A (2009) Characterization of *Erysipelothrix rhusiopathiae* isolates from poultry, pigs, emus, the poultry red mite and other animals. Vet Microbiol 137:98–104. https://doi.org/10.1016/j.vetmic.2008.12.016

Eriksson H, Brännström S, Skarin H, Chirico J (2010) Characterization of *Erysipelothrix rhusiopathiae* isolates from laying hens and poultry red mites (*Dermanyssus gallinae*) from an outbreak of erysipelas. Avian Pathol 39:505–509. https://doi.org/10.1080/03079457.2010.518313

Eriksson H, Nyman AK, Fellström C, Wallgren P (2013) Erysipelas in laying hens is associated with housing system. Vet Rec 173:18. https://doi.org/10.1136/vr.101388

Eriksson H, Bagge E, Båverud V, Fellström C, Jansson DS (2014) *Erysipelothrix rhusiopathiae* contamination in the poultry house environment during erysipelas outbreaks in organic laying hen flocks. Avian Pathol 43:231–237. https://doi.org/10.1080/03079457.2014.907485

Galindo-Cardiel I, Opriessnig T, Molina L, Juan-Salles C (2012) Outbreak of mortality in psittacine birds in a mixed-species aviary associated with *Erysipelothrix rhusiopathiae* infection. Vet Pathol 49:498–502. https://doi.org/10.1177/0300985811417246

Gartrell BD, Alley MR, Mack H, Donald J, McInnes K, Jansen P (2005) Erysipelas in the critically endangered kakapo (*Strigops habroptilus*). Avian Pathol 34:383–387. https://doi.org/10.1080/03079450500268583

Griffiths G, Buller N (1991) *Erysipelothrix rhusiopathiae* infection in semi-intensively farmed emus. Aust Vet J 68:121–122

Health A, Agency VL (2013) Mortality in turkeys due to erysipelas. Vet Rec 172:62–65. https://doi.org/10.1136/vr.f116

Hennig GE, Goebel HD, Fabis JJ, Khan MI (2006) Diagnosis by polymerase chain reaction of erysipelas septicemia in a flock of ring-necked pheasants. Avian Dis 46:509–514. https://doi.org/10.1637/0005-2086(2002)046[0509:dbpcro]2.0.co;2

Hollifield JL, Cooper GL, Charlton BR (2006) An outbreak of erysipelas in 2-day-old poults. Avian Dis 44:721. https://doi.org/10.2307/1593119

Jensen WI, Cotter SE (1976) An outbreak of erysipelas in eared grebes (*Podiceps nigricollis*). J Wildl Dis 12:583–586

Kurian A, Neumann EJ, Hall WF, Marks D (2012) Serological survey of exposure to *Erysipelothrix rhusiopathiae* in poultry in New Zealand. N Z Vet J 60:106–109. https://doi.org/10.1080/00480169.2011.639058

Livingston M, Fidler A, Mellor D, de Kloet S, Eason D, Elliott G, Moorhouse R (2013) Prevalence of IgY antibodies against *Erysipelothrix rhusiopathiae* in a critically endangered parrot (kakapo,

Strigops habroptilus) and associated responses to vaccination. Avian Pathol 42:502–507. https://doi.org/10.1080/03079457.2013.832146

Mazaheri A, Lierz M, Hafez HM (2006a) Investigations on the pathogenicity of *Erysipelothrix rhusiopathiae* in laying hens. Avian Dis 49:574–576. https://doi.org/10.1637/7362-040805r.1

Mazaheri A, Philipp HC, Bonsack H, Voss M (2006b) Investigations of the vertical transmission of *Erysipelothrix rhusiopathiae* in laying hens. Avian Dis 50:306–308. https://doi.org/10.1637/7426-082605r.1

Morgan MJ, Britt JO, Cockrill JM, Eiten ML (2011) *Erysipelothrix Rhusiopathiae* infection in an emu (*Dromaius Novaehollandiae*). J Vet Diagnostic Investig 6:378–379. https://doi.org/10.1177/104063879400600319

Mutalib AA, King JM, Mcdonough PL (1993) Erysipelas in caged laying chickens and suspected erysipeloid in animal caretakers. J Vet Diagnostic Investig 5:198–201. https://doi.org/10.1177/104063879300500210

Mutalib A, Keirs R, Austin F (2006) Erysipelas in quail and suspected erysipeloid in processing plant employees. Avian Dis 39:191. https://doi.org/10.2307/1592002

Opriessnig T, Vance RK, Halbur PG (2005) *Erysipelothrix rhusiopathiae* septicemia in a Laughing kookaburra (*Dacelo novaeguineae*). J Vet Diagnostic Investig 17:497–499. https://doi.org/10.1177/104063870501700519

Opriessnig T, Forde T, Shimoji Y (2020) *Erysipelothrix* spp.: past, present, and future directions in vaccine research. Front Vet Sci 7:174. https://doi.org/10.3389/fvets.2020.00174

Parte AC (2018) LPSN—list of prokaryotic names with standing in nomenclature (Bacterio.net), 20 years on. Int J Syst Evol Microbiol 68:1825–1829. https://doi.org/10.1099/ijsem.0.002786

Pattison M, McMullin P, Bradbury J, Alexander D (eds) (2008) Poultry diseases, 6th edn. https://doi.org/10.1016/B978-0-7020-2862-5.50031-3

Pomaranski EK, Griffin MJ, Camus AC, Armwood AR, Shelley J, Waldbieser GC, LaFrentz BR, García JC, Yanong R, Soto E (2020) Description of *Erysipelothrix piscisicarius* sp. nov., an emergent fish pathogen, and assessment of virulence using a tiger barb (*Puntigrus tetrazona*) infection model. Int J Syst Evol Microbiol 70:857–867. https://doi.org/10.1099/ijsem.0.003838

Reboli AC, Farrar WE (1989) *Erysipelothrix rhusiopathiae*: an occupational pathogen. Clin Microbiol Rev 2:354–359. https://doi.org/10.1128/CMR.2.4.354

Sheng WH, Hsueh PR, Hung CC et al (2000) Fatal outcome of *Erysipelothrix rhusiopathiae* bacteremia in a patient with oropharyngeal cancer. J Formos Med Assoc 99:431–434

Shimoji Y (2000) Pathogenicity of *Erysipelothrix rhusiopathiae*: virulence factors and protective immunity. Microbes Infect 2:965–972. https://doi.org/10.1016/s1286-4579(00)00397-x

Shimoji Y, Shiraiwa K, Tominaga H, Nishikawa S, Eguchi M, Hikono H, Ogawa Y (2020) Development of a multiplex PCR-based assay for rapid serotyping of *Erysipelothrix* species. J Clin Microbiol 58. https://doi.org/10.1128/JCM.00315-20

Swan RA, Lindsey MJ (1998) Treatment and control by vaccination of erysipelas in farmed emus (*Dromaius novohollandiae*). Aust Vet J 76:325–327. https://doi.org/10.1111/j.1751-0813.1998.tb12356.x

Takahashi T, Fujisawa T, Benno Y, Tamura Y, Sawada T, Suzuki S, Muramatsu M, Mitsuoka T (1987) *Erysipelothrix tonsillarum* sp. nov. isolated from tonsils of apparently healthy pigs. Int J Syst Evol Microbiol 37:166–168. https://doi.org/10.1099/00207713-37-2-166

Takahashi T, Fujisawa T, Umeno A, Kozasa T, Yamamoto K, Sawada T (2008) A taxonomic study on *Erysipelothrix* by DNA-DNA hybridization experiments with numerous strains isolated from extensive origins. Microbiol Immunol 52:469–478. https://doi.org/10.1111/j.1348-0421.2008.00061.x

Verbarg S, Rheims H, Emus S, Frühling A, Kroppenstedt RM, Stackebrandt E, Schumann P (2004) *Erysipelothrix inopinata* sp. nov., isolated in the course of sterile filtration of vegetable peptone broth, and description of Erysipelotrichaceae fam. nov. Int J Syst Evol Microbiol 54:221–225. https://doi.org/10.1099/ijs.0.02898-0

Wang Q, Chang BJ, Riley TV (2010) Erysipelothrix rhusiopathiae. Vet Microbiol 140:405–417. https://doi.org/10.1016/j.vetmic.2009.08.012

Work TM, Ball D, Wolcott M (1999) Erysipelas in a free-ranging Hawaiian crow (*Corvus hawaiiensis*). Avian Dis 43:338–341

Chapter 14
Avian Mycoplasmosis

Abstract Avian mycoplasmosis is an economically important disease of birds including farmed and wild species. Among a number of pathogenic mycoplasmas, *Mycoplasma gallisepticum* and *M. synoviae* are most important causative agents of avian mycoplasmosis. Unlike other bacteria, mycoplasmas lack cell wall, lack rigidity, and appear pleomorphic. The infection spreads through direct and indirect contacts between infected and susceptible hosts. Vertical transmission of the disease is a serious concern in commercial hatcheries. The organisms enter the host through, nasal, oral, and conjunctival routes. Broken eggs from infected birds are particular risk factors within flocks. The disease in birds is manifested by respiratory signs of sneezing and coughing, conjunctivitis, ocular exudation, and others. Diagnosis of the disease in birds starts with clinical signs and often requires laboratory confirmation. In addition to microbiological methods of culturing the organism in suitable artificial medium, other tests based on immunological and nucleic acid techniques are available. Antibiotic treatment is usually successful in therapeutic management of the infection in birds. Adequate hygienic measures, avoiding overcrowding, and vaccination are mainstay of preventive measures. Though mycoplasmas are important human pathogens, the link between avian mycoplasmosis and corresponding human infections are not clear. However, rare cases of infection in occupationally exposed persons have been documented.

Keywords Avian mycoplasmosis · *Mycoplasma gallisepticum* · *Mycoplasma synoviae* · Zoonotic · Birds · Conjunctivitis

14.1 Introduction

Avian mycoplasmosis is an important disease of birds with considerable economic implications in poultry industry (OIE 2018). The disease is caused by a number of pathogenic mycoplasmas of which *Mycoplasma gallisepticum* and *M. synoviae* are most important (OIE 2018; Pattison et al. 2008). Apart from poultry birds a number of wild birds are also infected by *Mycoplasma* spp. (Anonymous 2018; OIE 2018; Pattison et al. 2008). In humans *Mycoplasma* infections are also common especially

Y. S. Malik et al., *Role of Birds in Transmitting Zoonotic Pathogens*, Livestock Diseases and Management, https://doi.org/10.1007/978-981-16-4554-9_14

infection with *M. pneumoniae* (Baseman and Tully 1997; Chaudhry et al. 2016). Avian disease with *Mycoplasma* has been reported from various parts of the world principally from poultry birds though reports from wild birds are not uncommon. However, the zoonotic potential of the organisms is not well known, and barring exceptions, *Mycoplasma* spp. are not considered a major pathogen that might possibly be transmitted from birds (Pattison et al. 2008).

14.2 Epidemiology

14.2.1 Etiology

The mycoplasmas belong to class *Mollicutes* which means organisms with soft skin (OIE 2018; Quinn et al. 2011, 2016; Waites and Talkington 2004). This is because unlike other bacteria, mycoplasmal cells are surrounded by soft flexible membrane devoid of peptidoglycan. The lack of rigidity of cellular structure allows mycoplasma to appear pleomorphic (spherical to filamentous). On the other hand, they are also the smallest of prokaryotic life forms capable of independent survival and self-replication. Usually the size of the organisms varies between 0.3 and 1.0 μm (OIE 2018; Quinn et al. 2011, 2016; Songer and Post 2005; Taylor-robinson 1996; Waites and Talkington 2004).

Mycoplasma spp. are important pathogens in human and veterinary medicine. The first mycoplasma that was isolated was the causative agent of bovine pleuropneumonia (Waites and Talkington 2004). This was followed by the description of a number of *Mycoplasma* spp. responsible for various disease conditions in humans, animals, and birds. In avian medicine, in terms of economic importance, the most important *Mycoplasma* spp. are *M. gallisepticum*, *M. synoviae*, *M. iowae*, and *M. meleagridis* (Pattison et al. 2008; Quinn et al. 2011, 2016; Songer and Post 2005). Apart from these a number of other *Mycoplasma* spp. also cause infections in birds (Table 14.1), but their importance is yet to be ascertained in detail. Very recently three novel species of *Mycoplasma* have been reported. Of these, two (*M. nasistruthionis* sp. nov. and *M. struthionis* sp. nov.) have been reported from ostriches suffering from respiratory illnesses, and another (*M. anserisalpingitidis* sp. nov.) was from European geese with reproductive infection (Spergser et al. 2020; Volokhov et al. 2020).

14.2.2 Host Range

Wide range of birds is affected by *Mycoplasma* spp. While most reports are from birds of direct economic importance such as chicken, turkey, ducks, and other species of birds are indeed infected from Mycoplasma spp. The topic of host specificity has been discussed in the literature, and it is believed that *Mycoplasma*

Table 14.1 Common avian hosts and associated *Mycoplasma* spp.

Host birds	*Mycoplasma* species
Chicken	*M. gallinaceum, M. gallinarum, M. gallisepticum, M. glycophilum, M. iners, M. iowae, M. lipofaciens, M. pullorum, M. synoviae*
Duck	*M. anatis, M. imitans*
Falcon	*M. falconis*
Goose	*M. anatis, M. anseris, M. cloacale, M. imitans, M. anserisalpingitidis* sp. nov
Partridge	*M. gallinaceum, M. gallisepticum, M. glycophilum, M. iners, M. imitans, M. pullorum*
Pheasant	*M. gallinaceum, M. gallisepticum, M. glycophilum, M. iners, M. pullorum*
Pigeon	*M. columbinasale, M. columbinum, M. columborale*
Songbird	*M. gallisepticum*
Starling	*M. sturni*
Turkey	*M. cloacale, M. gallinarum, M. gallisepticum, M. gallopavonis, M. iners, M. iowae, M. lipofaciens, M. meleagridis, M. synoviae*
Vulture	*M. corogypsi, M. gypis*
Ostrich	*M. nasistruthionis* sp. nov., *M. struthionis* sp. nov.

Data source: Various

spp. are capable of infecting multiple hosts (Pitcher and Nicholas 2005). Besides domestic birds *Mycoplasma* spp. are known to infect pheasants, partridges, quails, peafowl, parrots, flamingos, raptors, finches, woodpeckers, corvids, and sparrows (Anonymous 2018; Dhondt et al. 2005, 2014; Jordan 1985).

14.2.3 Transmission

Mycoplasma infection spreads through direct and indirect contacts. Vertical transmission of *Mycoplasma* infections, especially *M. gallisepticum* and *M. synoviae*, has also been documented. Usually the vertical transmissions occur in the hatchery of commercial poultry birds such as chicken and turkey (Pattison et al. 2008; Quinn et al. 2011, 2016; Songer and Post 2005).

Infection with *Mycoplasma* spp. usually result in respiratory symptoms of coughing and sneezing, which may facilitate spread of the organisms to susceptible individuals over short distances. In the susceptible uninfected individuals, the organisms enter the body through oral, respiratory, and conjunctival routes. Since the infection is acquired over short distances, commercial poultry birds maintained under high density pens may be at particular risk. While inanimate farm objects that have been contaminated by *Mycoplasma* spp. may play role in dissemination, one particular object of interest is broken eggs from infected birds. These eggs or their parts are known to readily spread infection among the healthy susceptible individuals (Anonymous 2018; OIE 2018; Pattison et al. 2008; Quinn et al. 2011, 2016; Songer and Post 2005; Ter Veen et al. 2020).

14.2.4 Mycoplasma in Wild Birds

It is believed that wild birds are important hosts of *Mycoplasma* spp., though limited information are available compared to birds of commercial interests. In fact they have been listed as one of the emerging bacterial pathogen in migratory birds (Parin et al. 2019). In a recent review, occurrences of *M. gallisepticum* were observed in 56 different species of birds under 11 different orders with or without current or past infection. The study indicated that wild birds might serve as putative reservoir of *M. gallisepticum* (Sawicka et al. 2020) (Fig. 14.1). A number of studies documented isolation or detection of infections with various species of *Mycoplasma* (Dhondt et al. 2005, 2014; Lecis et al. 2010, 2016; Lierz et al. 2000, 2007a, b; 2008a, b; Lierz and Hafez 2008; Loria et al. 2008; OIE 2018; Quinn et al. 2016; Stipkovits and Szathmary 2012; Ziegler et al. 2017).

In Germany, a study on birds of prey indicated isolation of *M. gypis*, *M. meleagridis*, *M. falconis*, and *M. buteonis* from tracheal swabs and air-sac samples of owls and raptors (Lierz et al. 2000). Similarly, *M. gypis*, *M. falconis*, and *M. buteonis* were identified from Eurasian Hobby, Western marsh harriers, and barn owl (Lierz et al. 2008a, b). Further, *M. sturni* was identified from tracheal swabs collected from free-ranging corvids in Germany (Ziegler et al. 2017). Mycoplasmal conjunctivitis are common in house finches and have been reportedly caused epidemics leading to population decline in finches (Anonymous 2018; Dhondt et al. 2005). In a study investigating the vertical transmission *Mycoplasma* in wild

Fig. 14.1 Red crested pochard ducks, member of the family Anatidae which are commonly infected by *M. anatis* and *M. imitans*. (Photo courtesy: Dr. Prosun Biswas, Veterinary Officer, West Bengal, India)

raptors, Lierz et al. (2007a, b) reported a low incidence of infection in eggs of raptors. The species of the *Mycoplasma* was ascertained as *M. lipofaciens*. Later, the same group of authors established the pathogenicity of the same isolate to turkey embryos (Lierz et al. 2007a, b). In an interesting study on role of wild birds semen in dissemination of *Mycoplasma* spp., it was observed that approximately 16% of raptors, including falcons and eagles, used for breeding as semen donors, were secreting *Mycoplasma* spp. thus potentially spreading the infection through genital route (Lierz and Hafez 2008).

Wide diversity of *Mycoplasma* spp. and their avian hosts underscores the fact that many of the *Mycoplasma* spp. might not have been identified till date. In 2010, a study from Italian wildlife recovery center in Sardinia reported two novel species of *Mycoplasma* of Hominis group from Eurasian griffon vultures (Bosnic et al. 2010). In another Italian study, Loria et al. (2008) reported identification of unusual *Mycoplasma* spp. from sick Eurasian griffons. Among the four isolates of *Mycoplasma* spp., only one could be identified as *M. gallinarum*, while another was identified as closely related to *M. glycophilum* and the rest two were related to *M. falconis* and *M. gateae* (Loria et al. 2008). A later study in 2016, again identified a number of *Mycoplasma* spp. from various raptors (common kestrel, common buzzard, honey buzzard, peregrine falcon, Eurasian sparrowhawk, western marsh harrier) and owls (little owl, barn owl, horn owl) (Lecis et al. 2016).

An extensive North American study involving wild bird hosts belonging to 19 families and 53 species documented that out of 53 species, 27 were positive for *M. gallisepticum* infection as detected by PCR test or Rapid Plate Agglutination test (Dhondt et al. 2014). Recently, several studies have reported detection of mycoplasmal infection in backyard poultry in Italy, swan goose in China, free-ranging pheasants in Germany, and in house finches (Felice et al. 2020; Gyuranecz et al. 2020; Liebing et al. 2020; Weitzman et al. 2020).

14.3 Disease Characteristics

14.3.1 Pathogenesis

The process of pathogenesis of mycoplasmal infections in birds is not well understood (Chaudhry et al. 2016; Lierz et al. 2007a, b; Pilo et al. 2005). This is especially true for possible mechanisms explaining development of clinical features, signs, and lesions arising out of infection with *Mycoplasma* spp. Nonetheless, the process begins with entry of the pathogen into the host body usually through inhalation or through conjunctiva. Entry is followed by colonization of the mucous membrane. For *M. gallisepticum*, the cell surface glycoproteins assist the colonization in conjunction with anti-ciliary property of the invading organism (Pattison et al. 2008; Quinn et al. 2011). Once adhered to host mucous surfaces, *Mycoplasma* spp. triggers various biochemical reactions that lead to local cellular damages. One important feature inflicting damage to cellular structure is H_2O_2 production. It is believed that the production of H_2O_2 is activated by the membrane bound enzyme, L-α-glycerophosphate oxidase (GlpO) of *Mycoplasma* (Pilo et al. 2005). Other virulence factors have also been described for *Mycoplasma* spp. In *M. pneumoniae*, which is the cause for community acquired pneumonia, a toxin named CARDS (community-acquired respiratory distress syndrome) has been described (Segovia et al. 2017). Interestingly, a neurotoxin, possibly causing arteritis and neural symptoms in turkeys, has also been described for *M. gallisepticum* (Pattison et al. 2008; Quinn et al. 2011).

In order to evade immune response from hosts, mycoplasmal organisms resort to a mechanism known as phenotype switching. The process involves varying surface antigen composition in the subpopulations which are not readily acted upon by host immune cells. Another mechanism deployed by mycoplasmal organisms to evade the host immune response involves similarity between host antigens and mycoplasmal antigens, thus avoiding detection by immune effector cells of the hosts. Besides these, *Mycoplasma* spp. are known to exert various modulating effects on host immune system including suppression/activation of lymphocytic activities and cytokine productions (Chaudhry et al. 2016; Pattison et al. 2008; Quinn et al. 2011, 2016; Songer and Post 2005; Waites and Talkington 2004).

14.3.2 Clinical Signs and Pathology

Infection of birds by *Mycoplasma* spp. might produce a variety of symptoms. Our current understanding of clinical signs of avian mycoplasmosis is predominantly based on observations recorded in commercial poultry and to a certain extent from other ornithological observations. In commercial poultry industry, four species of *Mycoplasma* (*M. gallisepticum*, *M. synoviae*, *M. meleagridis*, and *M. iowae*) are responsible for vast majority of disease incidence. Common clinical signs of *M. gallisepticum* infection include discharges from nasal orifices, cough, sneeze, and respiratory congestion giving rise to tracheal rales. Conjunctivitis with ocular exudation is also common. In case of turkeys, additional signs of sinusitis, encephalopathy, arthritis, and salpingitis may be observed. Infection with *M. synoviae* is usually less acute and assumes a more chronic course. Respiratory signs are usually milder, though reduction in growth and performance of commercial flock is noticed. In turkeys, infection with *M. meleagridis* and *M. iowae* are more commonly reported. Infections with these species usually result in reduced hatchability of eggs and higher embryo mortality in farm settings (Anonymous 2018; Pattison et al. 2008; Quinn et al. 2011, 2016; Songer and Post 2005; Waites and Talkington 2004).

Clinical signs of *Mycoplasma* infection in wild birds are reported to be more or less similar to poultry birds. In finches, outbreak of conjunctivitis has been reported with significant mortality (Anonymous 2018). Conjunctivitis, which is often bilateral, is usually accompanied by sinusitis and rhinitis. Conjunctivitis due to *Mycoplasma* infection have been observed in other passerine birds too (Anonymous 2018;

Kleven 1998; Levisohn and Kleven 2000; Ziegler et al. 2017). Birds of the Psittacine family too show signs of respiratory and ocular affections. However, reports from some corvids indicated that the infection may remain almost asymptomatic, possibly implying the communalistic nature of *Mycoplasma* spp. in these avian hosts (Ziegler et al. 2017). Less commonly observed features include neurological signs such as ataxia, tremor, paralysis, and torticollis (Anonymous 2018).

Postmortem lesions of mycoplasmal infections in birds are mostly seen in affected organs. Gross lesions are observed as tracheitis, sinusitis, airsacculitis, pneumonia, and salpingitis. Often, excess mucous is present in respiratory tract. Secondary infection with other pathogens, mostly bacteria, may cause the exudation to turn purulent. In turkeys, encephalitis and synovitis may be common. Among wild birds, conjunctivitis is observed almost always, especially in finches and pheasants (Anonymous 2018; Pattison et al. 2008; Quinn et al. 2011, 2016; Songer and Post 2005).

14.4 Disease Management

14.4.1 Diagnosis

Clinical signs and pathological lesions of *Mycoplasma* infection are not adequate to diagnose the infection. Laboratory diagnosis is therefore considered essential for correct inference (Pattison et al. 2008). There are three major approaches toward laboratory diagnosis of *Mycoplasma* infection in birds. These include isolation and identification of the agent, immunological or serological detection, and detection of nucleic acid of the infectious agent (OIE 2018; Quinn et al. 2016; Songer and Post 2005).

Isolation and identification of the agent is attempted on suitable culture medium usually containing serum followed by identification from studying colony morphology or biochemical reactions. Among serological and immunological methods rapid serum agglutination (RSA) is popular due to ease of performance and rapidity. Commercial antigen is available for the test. However, enzyme-linked immunosorbent assay (ELISA) is also employed worldwide for detection and surveillance purposes. There are many ELISA tests that are commercially available. In addition, there are other tests available under this category which include indirect fluorescent antibody test, indirect immuno-peroxidase test, and hemagglutination inhibition test, all of which are employed for laboratory diagnosis to varying degree depending on the available resources and conditions. Nucleic acid-based techniques for detection of *Mycoplasma* infection depend on detection of specific DNA fragment from suspected samples. Samples are usually collected as swabs from which DNA is extracted and then tested. Most popular among these tests is polymerase chain reaction (PCR) for which standardized primer sets are available. Many PCR assays have been commercialized too (Anonymous 2018; Ball et al. 2020; Felice et al. 2020; Kempf 1998; OIE 2018; Pattison et al. 2008; Quinn et al. 2016; Songer and

Post 2005). Molecular tools for detection of *Mycoplasma* spp. have become quite popular and are employed in many studies routinely (Ball et al. 2020; Felice et al. 2020). To enable pen-side diagnosis of the infection, improvised ELISA techniques and lateral flow assays have also been developed and reported (Elyazeed et al. 2020).

14.4.2 Treatment and Control

Treatment of birds affected with *Mycoplasma* infection is usually successful with appropriate antibiotic therapy. However, the unique biological feature of mycoplasmal cell structure needs to be factored in while deciding antimicrobial therapy. Since the *Mycoplasma* lack peptidoglycan in their cell membrane, antibiotics acting on cell wall synthesis will naturally be ineffective thus precluding a large section of antibiotics for use. Usually antibiotics such as tetracyclines, macrolides, fluoroquinolones, and aminoglycosides work well. Tiamulin has also been recommended for use against *Mycoplasma* infection (Anonymous 2018; Pattison et al. 2008).

Various control strategies are employed for *Mycoplasma* infection. Vaccination against *Mycoplasma* with live or killed vaccine is a proven strategy for prevention in commercial poultry flocks. Usually, vaccination is employed targeting two species of *Mycoplasma* (*M. gallisepticum* and *M. synoviae*), and separate vaccinations might be needed as the two strains are not cross protective against each other. Vaccination is impractical for wild birds. Among other measures employed in commercial farming, raising of flock from *Mycoplasma*-free parents is one. However, this approach needs to be complemented by a robust biosecurity plan to preempt introduction of infection from other sources. General sanitation and hygienic measures coupled with routine infection control protocols are usually practiced in commercial farms which may be extended to wild bird sanctuaries and supplementary feeding facilities. Since the disease is transmitted over close contact particularly during feeding, regular cleaning and disinfection of feeders are helpful in preventing spread of the diseases (Anonymous 2018; Gautier-Bouchardon 2018; Kleven 2008; OIE 2018; Pattison et al. 2008; Stipkovits and Szathmary 2012).

14.5 Public Health Concerns

Despite *Mycoplasma* spp. being an important pathogen of humans, they are generally considered non-zoonotic because of lack of solid evidence to suggest transmission between animals and humans. However, recent reports indicated transmission of *M. lipofaciens* (ML64 strain) infection to a veterinarian involved in laboratory experiments with the same pathogen in turkeys. The affected individual developed clinical signs of the disease characterized by sore throat, rhinitis, and nasal pain (Lierz et al. 2008a, b). Apart from this well-documented case of human infection with avian strains of *Mycoplasma*, few others have been reported also, though the evidences appear to be circumstantial and might not be related to avian strains. In 2013, a veterinarian was reportedly affected with a *Candidatus* Mycoplasma haematoparvum (Maggi et al. 2013) strain. Another case of human infection with a hemotropic mycoplasma was reported from Croatia in an immunocompromised patient suffering from systemic lupus erythematosus (SLE) (Bosnic et al. 2010).

References

Anonymous (2018) Avian mycoplasmosis (*Mycoplasma gallisepticum*). Iowa

Ball C, Felice V, Ding Y, Forrester A, Catelli E, Ganapathy K (2020) Influences of swab types and storage temperatures on isolation and molecular detection of *Mycoplasma gallisepticum* and *Mycoplasma synoviae*. Avian Pathol 49:106–110. https://doi.org/10.1080/03079457.2019.1675865

Baseman JB, Tully JG (1997) Mycoplasmas: sophisticated, reemerging, and burdened by their notoriety. Emerg Infect Dis 3:21–32. https://doi.org/10.3201/eid0301.970103

Bosnic D, Baresic M, Anic B, Sentic M, Cerovec M, Mayer M, Cikes N (2010) Rare zoonosis (hemotrophic mycoplasma infection) in a newly diagnosed systemic lupus erythematosus patient followed by a *Nocardia asteroides* pneumonia. Braz J Infect Dis 14:92–95. https://doi.org/10.1016/S1413-8670(10)70019-2

Chaudhry R, Ghosh A, Chandolia A (2016) Pathogenesis of *Mycoplasma pneumoniae*: an update. Indian J Med Microbiol 34:7–16. https://doi.org/10.4103/0255-0857.174112

Dhondt AA, Altizer S, Cooch EG, Davis AK, Dobson A, Driscoll MJL, Hartup BK, Hawley DM, Hochachka WM, Hosseini PR, Jennelle CS, Kollias GV, Ley DH, Swarthout ECH, Sydenstricker KV (2005) Dynamics of a novel pathogen in an avian host: mycoplasmal conjunctivitis in house finches. Acta Trop 94:77–93. https://doi.org/10.1016/j.actatropica.2005.01.009

Dhondt AA, DeCoste JC, Ley DH, Hochachka WM (2014) Diverse wild bird host range of *Mycoplasma gallisepticum* in Eastern North America. PLoS One 9. https://doi.org/10.1371/journal.pone.0103553

Elyazeed HA, Al-Atfeehy NM, Abotaleb R, Sayed R, Marouf S (2020) Preparation of ELISA and lateral flow kits for rapid diagnosis of *Mycoplasma gallisepticum* in poultry. Sci Rep 10:9056. https://doi.org/10.1038/s41598-020-65848-7

Felice V, Lupini C, Mescolini G, Silveira F, Guerrini A, Catelli E, Di Francesco A (2020) Molecular detection and characterization of *Mycoplasma gallisepticum* and *Mycoplasma synoviae* strains in backyard poultry in Italy. Poult Sci 99:719–724. https://doi.org/10.1016/j.psj.2019.12.020

Gautier-Bouchardon AV (2018) Antimicrobial resistance in *Mycoplasma* spp. Microbiol Spectr 6: 1–21. https://doi.org/10.1128/microbiolspec.arba-0030-2018

Gyuranecz M, Mitter A, Kovács ÁB, Grózner D, Kreizinger Z, Bali K, Bányai K, Morrow CJ (2020) Isolation of *Mycoplasma anserisalpingitidis* from swan goose (*Anser cygnoides*) in China. BMC Vet Res 16:178. https://doi.org/10.1186/s12917-020-02393-5

Jordan FTW (1985) People, poultry and pathogenic mycoplasmas. Br Poult Sci 26:1–15. https://doi.org/10.1080/00071668508416781

Kempf I (1998) DNA amplification methods for diagnosis and epidemiological investigations of avian mycoplasmosis. Avian Pathol 27:7–14. https://doi.org/10.1080/03079459808419268

Kleven SH (1998) Mycoplasmas in the etiology of multifactorial respiratory disease. Poult Sci 77: 1146–1149. https://doi.org/10.1093/ps/77.8.1146

Kleven SH (2008) Control of avian mycoplasma infections in commercial poultry. Avian Dis Dig 3: e1–e1. https://doi.org/10.1637/8424.1

Lecis R, Chessa B, Cacciotto C, Addis MF, Coradduzza E, Berlinguer F, Muzzeddu M, Lierz M, Carcangiu L, Pittau M, Alberti A (2010) Identification and characterization of novel *Mycoplasma* spp. belonging to the hominis group from griffon vultures. Res Vet Sci 89:58–64. https://doi.org/10.1016/j.rvsc.2009.12.016

Lecis R, Secci F, Mandas L, Muzzeddu M, Pittau M, Alberti A (2016) Molecular identification and sequence characterization of mycoplasmas in free-living birds of prey. J Zoo Wildl Med 47: 917–922. https://doi.org/10.1638/2015-0259.1

Levisohn S, Kleven SH (2000) Avian mycoplasmosis (*Mycoplasma gallisepticum*). Rev Sci Tech 19:425–442

Liebing J, Völker I, Curland N, Wohlsein P, Baumgärtner W, Braune S, Runge M, Moss A, Rautenschlein S, Jung A, Ryll M, Raue K, Strube C, Schulz J, Heffels-Redmann U, Fischer L, Gethöffer F, Voigt U, Lierz M, Siebert U (2020) Health status of free-ranging ring-necked pheasant chicks (*Phasianus colchicus*) in North-Western Germany. PLoS One 15: e0234044. https://doi.org/10.1371/journal.pone.0234044

Lierz M, Hafez HM (2008) Occurrence of mycoplasmas in semen samples of birds of prey. Avian Pathol 37:495–497. https://doi.org/10.1080/03079450802356961

Lierz M, Schmidt R, Brunnberg L, Runge M (2000) Isolation of *Mycoplasma meleagridis* from free-ranging birds of prey in Germany. J Vet Med Ser B 47:63–67. https://doi.org/10.1046/j.1439-0450.2000.00309.x

Lierz M, Deppenmeier S, Gruber AD, Brokat S, Hafez HM (2007a) Pathogenicity of *Mycoplasma lipofaciens* strain ML64 for turkey embryos. Avian Pathol 36:389–393. https://doi.org/10.1080/03079450701589126

Lierz M, Hagen N, Harcourt-Brown N, Hernandez-Divers SJ, Lüschow D, Hafez HM (2007b) Prevalence of mycoplasmas in eggs from birds of prey using culture and a genus-specific mycoplasma polymerase chain reaction. Avian Pathol 36:145–150. https://doi.org/10.1080/03079450701213347

Lierz M, Hagen N, Hernadez-Divers SJ, Hafez HM (2008a) Occurrence of mycoplasmas in free-ranging birds of prey in Germany. J Wildl Dis 44:845–850. https://doi.org/10.7589/0090-3558-44.4.845

Lierz M, Jansen A, Hafez HM (2008b) Avian *Mycoplasma lipofaciens* transmission to veterinarian. Emerg Infect Dis 14:1161–1163. https://doi.org/10.3201/eid1407.071703

Loria GR, Ferrantelli E, Giardina G, Vecchi LL, Sparacino L, Oliveri F, McAuliffe L, Nicholas RAJ (2008) Isolation and characterization of unusual *Mycoplasma* spp. from captive Eurasian griffon (*Gyps fulvus*) in Sicily. J Wildl Dis 44:159–163. https://doi.org/10.7589/0090-3558-44.1.159

Maggi RG, Mascarelli PE, Havenga LN, Naidoo V, Breitschwerdt EB (2013) Co-infection with *Anaplasma platys*, *Bartonella henselae* and *Candidatus* mycoplasma haematoparvum in a veterinarian. Parasit Vectors 6. https://doi.org/10.1186/1756-3305-6-103

OIE (2018) Avian mycoplasmosis. In: Manual of diagnostic tests and vaccines for terrestrial animals. OIE, pp 844–859

Parin U, Kirkan S, Erbas G (2019) Emerging bacterial zoonoses in migratory birds. In: Kideghesho J, Rija A (eds) Wildlife management—failures, successes and prospects. IntechOpen, London, pp 23–41. https://doi.org/10.5772/intechopen.72244

Pattison M, McMullin P, Bradbury J, Alexander D (eds) (2008) Poultry diseases, 6th edn. https://doi.org/10.1016/B978-0-7020-2862-5.50031-3

Pilo P, Vilei EM, Peterhans E, Bonvin-Klotz L, Stoffel MH, Dobbelaere D, Frey J (2005) A metabolic enzyme as a primary virulence factor of *Mycoplasma mycoides* subsp. mycoides small colony. J Bacteriol 187:6824–6831. https://doi.org/10.1128/JB.187.19.6824-6831.2005

Pitcher DG, Nicholas RAJ (2005) Mycoplasma host specificity: fact or fiction? Vet J 170:300–306. https://doi.org/10.1016/j.tvjl.2004.08.011

Quinn PJ, Markey BK, Leonard FC, FitzPatrick ES, Fanning S, Hartigan PJ (2011) Veterinary microbiology and microbial disease, 2nd edn, West Sussex

Quinn PJ, Markey BK, Leonard FC, FitzPatrick ES, Fanning S (2016) Concise review of veterinary microbiology. Wiley, West Sussex

Sawicka A, Durkalec M, Tomczyk G, Kursa O (2020) Occurrence of *Mycoplasma gallisepticum* in wild birds: a systematic review and meta-analysis. PLoS One 15:e0231545. https://doi.org/10.1371/journal.pone.0231545

Segovia JA, Chang T-H, Cagle MP, Winter VT, Coalson JJ, Kannan TR, Baseman JB (2017) *Mycoplasma pneumoniae* CARDS toxin is a unique virulence factor targeting the Th1/Th2 response for persistent infection. J Immunol 198:131.15 LP–131.15

Songer JG, Post KW (2005) Veterinary microbiology: bacterial and fungal agents of animal disease, 1st edn. Elsevier Saunders, St. Louis

Spergser J, Botes A, Nel T, Ruppitsch W, Lepuschitz S, Langer S, Ries S, Dinhopl N, Szostak M, Loncaric I, Busse H-J (2020) *Mycoplasma nasistruthionis* sp. nov. and *Mycoplasma struthionis* sp. nov. isolated from ostriches with respiratory disease. Syst Appl Microbiol 43:126047. https://doi.org/10.1016/j.syapm.2019.126047

Stipkovits L, Szathmary S (2012) Mycoplasma infection of ducks and geese. Poult Sci 91:2812–2819. https://doi.org/10.3382/ps.2012-02310

Taylor-robinson D (1996) Infections due to species of *Mycoplasma* and *Ureaplasma*: an update. Clin Infect Dis 23:671–684. https://doi.org/10.1093/clinids/23.4.671

Ter Veen C, de Wit JJ, Feberwee A (2020) Relative contribution of vertical, within-farm and between-farm transmission of *Mycoplasma synoviae* in layer pullet flocks. Avian Pathol 49:56–61. https://doi.org/10.1080/03079457.2019.1664725

Volokhov DV, Grózner D, Gyuranecz M, Ferguson-Noel N, Gao Y, Bradbury JM, Whittaker P, Chizhikov VE, Szathmary S, Stipkovits L (2020) *Mycoplasma anserisalpingitidis* sp. nov., isolated from European domestic geese (*Anser anser domesticus*) with reproductive pathology. Int J Syst Evol Microbiol 70:2369–2381. https://doi.org/10.1099/ijsem.0.004052

Waites KB, Talkington DF (2004) *Mycoplasma pneumoniae* and its role as a human pathogen. Clin Microbiol Rev 17:697–728. https://doi.org/10.1128/CMR.17.4.697-728.2004

Weitzman CL, Thomason C, Schuler EJA, Leon AE, Teemer SR, Hawley DM (2020) House finches with high coccidia burdens experience more severe experimental *Mycoplasma gallisepticum* infections. Parasitol Res. https://doi.org/10.1007/s00436-020-06814-0

Ziegler L, Palau-Ribes FM, Schmidt L, Lierz M (2017) Occurrence and relevance of *Mycoplasma sturni* in free-ranging corvids in Germany. J Wildl Dis 53:228–234. https://doi.org/10.7589/2015-12-350

Chapter 15
Avian Salmonellosis

Abstract Avian salmonellosis is an infectious disease affecting many species of domestic and wild birds causing heavy economic losses in commercial poultry. The disease is caused by various serovars of *Salmonella* spp., which are Gram negative, non-spore-forming, rod-shaped bacteria. Avian salmonellosis is reported from all over the world. Infected hosts shed the bacteria in the environment through their feces and new susceptible hosts contract the infection from contaminated environment. The infectious cycle is mainly feco-oral, though vertical transmissions are also reported. Birds are considered as the largest reservoir of various serovars of *Salmonella* spp. in nature. Wild birds are regarded as important disseminating agents for *Salmonella* in the environment, especially birds roosting around the commercial farms of poultry and other livestock. Clinical signs of the disease vary depending on the host factors. Usually, the disease is manifested by pasty vent diarrhea, depression, reluctance to move, occasional opacity of cornea, and marked loss of growth and production. Being one of the most studied pathogens of birds, a variety of laboratory detection methods ranging from traditional microbiological methods to modern molecular tools are available for diagnosis of the disease. In addition to therapeutic management of the disease by various antibiotics, preventive measures include strict farm hygiene and vaccination. Being an important foodborne zoonotic infection for humans, avian salmonellosis especially in commercially raised poultry is a particular risk, though the link between wild birds and human infections is relatively less understood.

Keywords Avian salmonellosis · *Salmonella* · Zoonotic · Foodborne · Wild birds · Human

15.1 Introduction

Salmonellosis is an infectious disease that affects birds, animals, and humans all over the world. The disease is usually manifested by septicemia and enteritis and may even remain without any clinical signs. Salmonellosis is caused by many serovars of the bacteria of the genus *Salmonella*. Infected hosts shed the bacteria in the

Y. S. Malik et al., *Role of Birds in Transmitting Zoonotic Pathogens*, Livestock Diseases and Management, https://doi.org/10.1007/978-981-16-4554-9_15

environment through their feces and new susceptible hosts contract the infection from contaminated environment. Salmonellosis causes heavy economic loss around the world affecting human and animal health including the commercial poultry industry (Barrow et al. 2012; OIE 2018a, b; Pattison et al. 2008). In humans, salmonellosis primarily takes two forms: Typhoid fever caused by *Salmonella* Typhi and infection by non-typhoidal salmonellae (NTS). In commercial poultry birds, salmonellosis is mainly caused by two host adapted biovars *Salmonella* Pullorum causing Pullorum disease and *Salmonella* Gallinarum causing fowl typhoid. However, other salmonellae also infects poultry and other birds (Barrow et al. 2012; OIE 2018a, b; Pattison et al. 2008).

15.2 Epidemiology

15.2.1 Etiology

Bacteriologically Salmonellae are Gram negative, non-spore-forming, rod-shaped bacteria that ferment glucose and reduce nitrates. Most salmonellae are motile with few exceptions, for example, *Salmonella* Pullorum and *Salmonella* Gallinarum. The taxonomy of *Salmonella* is complex and has undergone several changes and revisions. Currently the genus consists of two species *S. enterica* and *S. bongori*. Of the two species, the former contains all the salmonellae of medical and veterinary importance. The species is further subdivided into six subspecies (*S. enterica* subsp. *enterica*, *S. enterica* subsp. *salamae*, *S. enterica* subsp. *arizonae*, *S. enterica* subsp. *diarizonae*, *S. enterica* subsp. *houtenae*, and *S. enterica* subsp. *indica*). The serovars that cause disease in humans and birds and other mammals come under *S. enterica* subsp. *enterica*. Since it is inconvenient to use such long names, only the genus name (italicized) followed by serovar names (non-italicized) is used. Currently more than 2600 serovars of salmonellae are recognized (OIE 2018a; Pattison et al. 2008; Threlfall 2010).

Antigenic classification for serovar determination of salmonellae is done as per the White-Kauffmann-Le methodology based on lipopolysaccharide (O antigen) and flagella (H antigen). Serovars of *S. enterica* subsp. *enterica* are given names, while serovars of other subspecies and those belonging to *S. bongori* are designated by numbers (Brenner et al. 2000; OIE 2018a; Pattison et al. 2008; Threlfall 2010). Phage typing of *Salmonella* is also another important classification system which employs Felix O1 bacteriophage and has been used for the first time for typing of *S.* Typhimurium (OIE 2018a; Threlfall 2010).

15.2.2 Host Range

Infections with *Salmonella* are one of the widest spread infections in nature involving multiple hosts. As such the genus *Salmonella* has a very wide host range. Wild birds seem to be susceptible to most of *Salmonella* species and may play an important role as a carrier of *Salmonella* in their intestine. Salmonellae have been isolated from various species of farm animals such as cattle, buffalo, sheep, goat, horse, and pig as well as from poultry, duck, wild birds, reptiles, and amphibians. It is largely believed that domestic poultry birds constitute the largest reservoir of various *Salmonella* serovars (Barrow et al. 2012; Messens et al. 2013; OIE 2018a; Rajan et al. 2017; Threlfall 2010).

15.2.3 Transmission

The principal mode of transmission of *Salmonella* spp. is the fecal-oral route. Affected birds (and mammals) shed large number of salmonellae in their excreta and contaminate the environment including pastures, sheds, water, feeds, etc.. Susceptible hosts acquire the infection from such feed or water contaminated with the bacteria. Infection through either inhalation or conjunctival routes has also been reported. Vertical or transovarian transmission had been reported in several cases of salmonellosis due to the ova infection prior to the ovulation stage or even after deposition through contamination of the egg during the egg-laying process during passage through the cloaca or even after the egg deposition (Barrow et al. 2012; Chousalkar and Gole 2016; Denagamage et al. 2015; Magdy et al. 2020; OIE 2018a, b; Pattison et al. 2008; Quinn et al. 2016; Quinn et al. 2011; Songer and Post 2005; Threlfall 2010).

15.2.4 Wild Birds as Carriers of Salmonella

Wild birds are considered as important disseminator of salmonellosis in the environment and also to animals, humans, and other birds (Greig et al. 2015). A number of researchers from all continents except Antarctica expressed concerns regarding the potential role of wild birds in the spread of *Salmonella* (Blanco 2018; Dipineto et al. 2015; Elmberg et al. 2017; Fuentes-Castillo et al. 2019; Gargiulo et al. 2018; Hernandez et al. 2016; Janecko et al. 2015; Jurado-Tarifa et al. 2016; Konicek et al. 2016; Krawiec et al. 2015; Liakopoulos et al. 2016; Matias et al. 2016b; Pearson et al. 2016; Plaza et al. 2019; Tamamura et al. 2016).

In pig farms, wild birds were found to be harboring *Salmonella* spp. and were acting as potential introducer or receptor for existing infection in the pigs in Spain (Andrés-Barranco et al. 2014). However, a study on free-living waterfowls did not find any evidence of *Salmonella* infection in northeastern area of Spain (Antilles et al. 2015). Another study from Spain also indicated that waterfowls and raptors in southern Spain were potential carriers and disseminators of *Salmonella* in the environment (Jurado-Tarifa et al. 2016). The role of waterfowls in the dissemination of bacterial species including salmonellae was also documented recently in Canada where ducks, gerbes, and swans were detected to carry *Salmonella* infection (Vogt et al. 2019). In a separate yet similar study from Australia, it was observed that European Starlings inhabiting areas adjoining commercial piggeries were infected with salmonellae raising the possibility of transmission of pathogens to farmed pigs (Pearson et al. 2016). While studying the bacterial species richness, Benskin et al. (2015) isolated *Salmonella* spp. from wild blue tits in the UK. Similarly in Poland, wild birds, especially Eurasian siskins and green finches, were observed to carry *Salmonella* which were also prevalent in the environment (Krawiec et al. 2015).

In a study reported from southeast Texas, 17% of wild birds were found to be carrying *Salmonella* infection especially in suburban areas (Brobey et al. 2017). In Australia too, a One Health investigation of *Salmonella* Wangata over a period of 1 year revealed that indirect contact with wild birds was statistically associated with illnesses in humans (Collins et al. 2019).

Investigation into wild birds such as raptors, water birds, and passerines in a rescue center in Italy revealed that *Salmonella* Düsseldorf was one of the prominent enterobacteria that were isolated from these birds (Giacopello et al. 2016). Similar findings from wild birds of other rescue center was also reported with involvement of *Salmonella* Typhimurium (Dipineto et al. 2015). Other reports from Italy also indicated that birds of prey may carry *Salmonella* Typhimurium and *Salmonella* Napoli infections (Gargiulo et al. 2018). Wild birds appear to not only act as passive carriers of *Salmonella* infection, they may actively multiply the bacteria in their gut as was documented by De Lucia et al. (2018) who reported unusually high level (10^5 to 10^6 colony forming unit per gram) of *Salmonella* in wild bird droppings. Among wild birds, salmonellosis is known to be associated with a variety of species with both symptomatic and asymptomatic forms especially in gulls, and therefore, gulls have been suggested as a particular risk factor for occurrence of salmonellosis in domestic animals (Elmberg et al. 2017). In a study on wild birds in protected and human impacted ecological niches in Uganda, it was observed that strains of *Salmonella* were shared among birds and environment in the urban areas indicating dissemination and acquisition of bacteria by birds (Afema and Sischo 2016). Multidrug-resistant salmonellae were also isolated from Egyptian griffon vultures perhaps acquired from carcasses offered to them in supplementary feeding stations (Blanco 2018).

In Brazil, an interesting study on wild owls indicated that they may act as potential reservoirs international clones of *Salmonella* Infantis (Fuentes-Castillo et al. 2019). Likewise, wild Kelp Gulls in South America were identified as reservoirs of drug-resistant enterobacteria including *Salmonella* (Liakopoulos et al. 2016). Studies from Brazil further point out that wild birds held in rescue centers following confiscation from illegal trading are also potential sources for dissemination of *Salmonella* infection to other birds and environment (Matias et al. 2016a, b).

In the same line, data from wildlife rehabilitation center in Spain highlighted that asymptomatic wild birds (vultures, owls, hoopoe) were carriers of multidrug-resistant *Salmonella* Typhimurium (Molina-López et al. 2015).

While many of the wild birds act as asymptomatic carriers of *Salmonella* infection, they may suffer epidemic mortality as was evidenced in case of mass mortality in Eurasian Tree Sparrows in Hokkaido, Japan, due to infection with *Salmonella* Typhimurium DT40 (Fukui et al. 2014). Similar mortality in sparrows due to *Salmonella* Typhimurium and transmission of the bacteria to nearby dairy cattle farm was also reported from Japan (Tamamura et al. 2016). However, in New Zealand, other serotypes of *Salmonella* (Saintpaul) and other species (*Salmonella enterica* subsp. *houtenae*) were responsible for causing mortality in wild birds and other fauna (van Andel et al. 2015).

As wild birds often thrive around urban settlements, they tend to pick up *Salmonella* strains from the environments and as a result become colonized by the circulating strains. This assumption was substantiated by a study from Florida on white ibises, which indicated that the birds were carrying diverse serotypes of *Salmonella* and were capable of transmitting salmonellae to people (Hernandez et al. 2016). In India, Milton et al. (2018) isolated *Salmonella* Kentucky from captive golden pheasants and their caretakers and from the results, authors concluded that captive birds might play important role in the transmission of *Salmonella* through environmental contamination to other wildlife species, workers, and possibly visitors.

Migratory nature of the birds may also contribute to the transmission and dissemination of salmonellae across large distances. In order to assess the role of large corvids as distributor of salmonellae among urban, agricultural, and natural ecosystems, Janecko et al. (2015) concluded that the large corvids were harboring similar serovars of *Salmonella* as was in humans and were also capable of transmitting the same over great distances separating two continents (Europe and North America). In an interesting study on birds scavenging on rubbish dumps, it was observed that vultures were colonized by several *Salmonella* serotypes including Typhi and Paratyphi A (Plaza et al. 2019). The authors of the study contemplated that pathogens from the rubbish dumps could find another reservoir in birds who might in turn act as putative disseminators of the pathogens (Plaza et al. 2019). Nevertheless, researchers continue to document occurrence of *Salmonella* infection in various wild birds including psittacine birds, free-living birds, raptors and aquatic wild birds, and scavenging vultures (de Souza et al. 2020; Dos Santos et al. 2020; Naushad et al. 2020; Navarro-Gonzalez et al. 2020; Suárez-Pérez et al. 2020; Tardone et al. 2020; Tyson-Pello and Olsen 2020).

15.3 Disease Characteristics

15.3.1 Pathogenesis

A large number of hosts are infected by salmonellae, though host adapted *Salmonella* serotypes are not uncommon. As a general rule host adapted serotypes cause more severe diseases than other serotypes (Quinn et al. 2011; Songer and Post 2005; Threlfall 2010). Entry of *Salmonellae* into the host is usually through oral route. Survival in the harsh gastric environment is facilitated by food (proteins and fats) and acid-shock proteins secreted by the organisms. *Salmonella* attaches themselves to the intestinal mucosa with the help of fimbriae, especially long polar fimbrial adhesin which binds to Peyer's patches and M cells (Fig. 15.1). Following attachment *Salmonella* cells are internalized as membrane enclosed vacuoles and subsequent immuno-inflammatory events set in characterized by migration of neutrophils and secretion of inflammatory cytokines (Pattison et al. 2008; Quinn et al. 2011, 2016; Rostami et al. 2020; Songer and Post 2005). These events lead to onset of secretory diarrhea in hosts. Apart from local alimentary tract infection, salmonellae may invade blood stream and migrate to other organs, characterized by septicemia

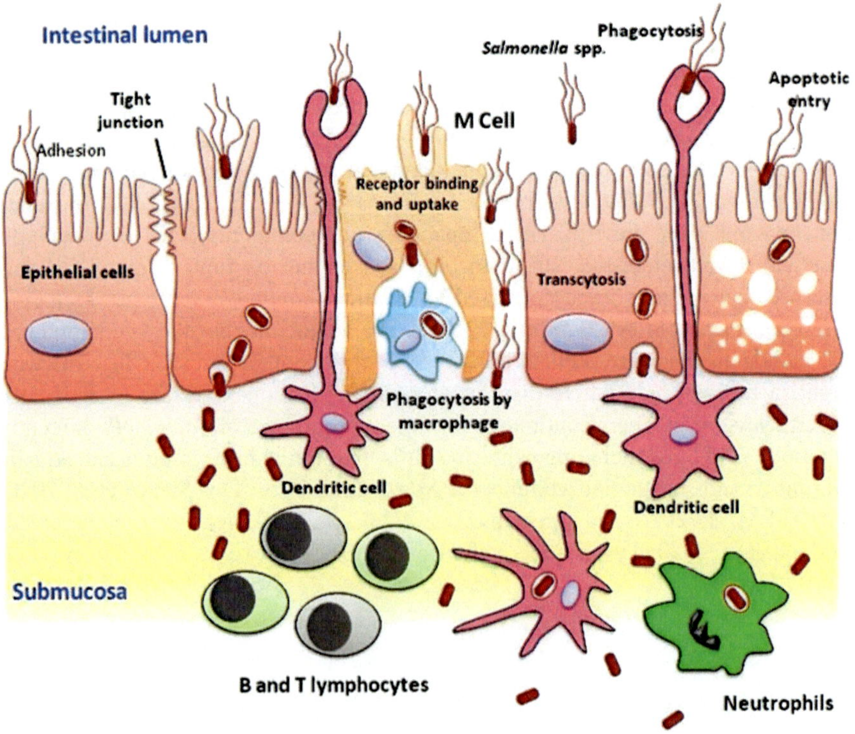

Fig. 15.1 Invasion and uptake of *Salmonella* spp. in the intestinal lumen

and establishment of foci of infection in other organs. It is believed that intraphagosomally survival of *Salmonellae* is a key factor for extra intestinal spread. However, on entry into blood stream and lymphoid system, the bacteria are countered by reticuloendothelial cells resulting in death of bacterial cells and release of lipopolysaccharide into the system which triggers cytokine releases leading to fever, shock, and other associated symptoms (Quinn et al. 2011, 2016; Songer and Post 2005; Threlfall 2010). In case of localized infection involving intestinal tract, a large number of salmonellae are shed through feces contaminating the environment and dissemination of the infection to other susceptible hosts. However, in case of systemic infection, reproductive tracts of birds may become affected enabling vertical transmission through eggs (Pattison et al. 2008; Quinn et al. 2011, 2016; Songer and Post 2005).

15.3.2 Clinical Signs and Pathology

Clinical signs of infections caused by *Salmonella* serovars in birds vary from inapparent infection to frank disease. Usually, younger birds are more susceptible than adults. In case of septicemic form of the infection, the disease is usually acute and may cause rapid deaths. In other forms of the disease, lethargy, ruffled feathers, droopiness, enteritis, difficulty in respiration, and shivering are frequently observed signs. In case of poultry birds, infections are caused usually by host adapted serovars and outbreaks are common with pasty vent diarrhea, depression, reluctance to move, occasional opacity of cornea, and marked loss of growth and production. However, in many cases the clinical signs remain nonspecific and may not aid in diagnosis (OIE 2018b; Pattison et al. 2008; Quinn et al. 2011, 2016).

Pathological lesions of salmonellosis may be variable and is not well documented specifically for wild birds. However, the septicemic form of the disease in poultry birds affects all vital organs, and on post mortem, these organs (spleen, liver, kidneys, and lungs) appear hyperemic and congested. There may be focal necrotic lesions in liver, heart, and enteritis. Inflammation of ceca is a commonly observed gross lesion along with swelling of pericardial sac and pericarditis. Ovarian lesions and deformities in egg formation stages are also reported (Barrow et al. 2012; OIE 2018a, b; Pattison et al. 2008; Quinn et al. 2011, 2016).

15.4 Disease Management

15.4.1 Diagnosis

As with other bacterial diseases of birds, diagnosis of salmonellosis starts with the clinical signs and symptoms which are substantiated by post-mortem lesions. Along with these, history and epidemiological observations also help in establishing a diagnosis. However, all presumptive diagnosis needs to be confirmed by results of laboratory investigation. For suspected *Salmonella* infection in birds, various laboratory techniques are utilized. Bacteriological investigations involve isolation of the organism by pre-enrichment and enrichment in broth (e.g., Rappaport-Vassiliadis broth) followed by selective plating on suitable media (e.g., XLD agar, *Salmonella-Shigella* agar). Isolation is usually followed by a battery of biochemical tests (e.g., Triple sugar iron agar test, H_2 production, sugar fermentation) to establish the identity of the isolated organism (Barrow et al. 2012; OIE 2018a, b; Pattison et al. 2008; Quinn et al. 2011, 2016; Songer and Post 2005; Tizard 2004). Apart from bacteriological tests, there a number of immunological tests that are also employed for identification of the infection by *Salmonella* spp. These tests include whole blood test, rapid slide agglutination test, serum agglutination test, enzyme-linked immunosorbent assay (ELISA), serotyping by specific antisera against O and H antigen, etc. (Barrow et al. 2012; OIE 2018a, b; Pattison et al. 2008; Quinn et al. 2011, 2016; Songer and Post 2005; Tizard 2004). Nucleic acid-based tests have also been put to use for rapid and sensitive detection of *Salmonella* in suspected cases of infection. Among these tests polymerase chain reaction (PCR) and real-time PCR have proved to be particularly useful, and many researchers documented various novel formats of assays. For epidemiological and typing purposes, multilocus sequence typing, restriction fragment length polymorphism, pulsed field gel electrophoresis, and other fingerprinting methods are usually employed (Crump et al. 2015; MacFadden et al. 2016; OIE 2018a; Quinn et al. 2011, 2016; Songer and Post 2005; Tizard 2004). With the advent of next-generation sequencing technology, whole genome sequencing (WGS) is being increasingly put to use for genomic epidemiological studies with much higher resolution (Crump et al. 2015; MacFadden et al. 2016; Naushad et al. 2020).

15.4.2 Treatment and Control

Treatment of avian salmonellosis in wild birds is a challenge as the infection often occurs in natural environment leaving minimal scope for therapeutic intervention. Infections are usually diagnosed post mortem or are detected only in captive wild birds. However, treatment of salmonellosis in commercial poultry birds is an economic necessity and is routinely undertaken. A number of antibiotics are available to control the morbidity and mortality in poultry birds including tetracyclines, fluoroquinolones (e.g. enrofloxacin), sulfonamides, and semisynthetic penicillin such as amoxycillin and others. Caution must be taken to avoid unnecessary usage of antimicrobials as salmonellae are notorious for harboring multiple antimicrobial resistance posing significant public health risks (Barrow et al. 2012; Desin et al. 2013; Massey 2003; OIE 2018a, b; Pattison et al. 2008; Quinn et al. 2011; Songer and Post 2005; Tizard 2004).

Control of *Salmonella* infection in wild birds is also difficult as birds tend to acquire infection in the wild, though reports are available indicating anthropogenic factors as predisposing causes (Afema and Sischo 2016; Blanco 2018; Plaza et al. 2019). For commercial poultry production, elaborate strategies for control of salmonellosis are practiced including implementation of biosecurity measures, litter management, control of pests (rodents, insects, etc.), safety of feed, hygienic management of hatcheries, cleaning and disinfection of farm premises, vaccination, and use of biological agents (e.g., probiotics for competitive exclusion for preventing colonization of GI tract, inclusion of small molecules of fatty acids in feed) among other measures. While these measures are not practicable, nor advisable for wild birds, ensuring safety of supplementary feed offered to wild birds and monitoring of wild life infection with *Salmonella* serovars are recommended (Blanco 2018; Collins et al. 2019; Elmberg et al. 2017; Pattison et al. 2008; Tizard 2004).

15.5 Public Health Concerns

Salmonellosis is perhaps the most commonly occurring foodborne zoonoses worldwide. Poultry meat, meat products, and eggs (and possibly products) are most important means of spread for foodborne salmonellosis (Antunes et al. 2016; Chousalkar and Gole 2016; Loharikar et al. 2012; Rajan et al. 2017; Threlfall 2010). Apart from the foodborne illnesses associated with salmonellosis, horizontal spread of multidrug resistance from salmonellae originating from poultry production chain is also of concern. Live poultry flocks especially that which are reared in the backyard also pose a significant threat. A number of outbreaks of *Salmonella* infections associated with live poultry have been reported (Basler et al. 2016; Behravesh et al. 2014; Threlfall 2010; Whiley and Ross 2015). While many of the outbreaks were attributable to backyard poultry rearing, few outbreaks were associated with high-risk practice of keeping poultry birds inside the house (Basler et al. 2016; Behravesh et al. 2014; Loharikar et al. 2012). Formation of biofilm by *Salmonella* serovars, especially on egg shell surfaces, is an important contributory factor for transmission through eggs (Whiley and Ross 2015). Biofilm formation poses a significant challenge in control of egg-borne salmonellosis in humans (Rajan et al. 2017; Whiley and Ross 2015).

Though poultry birds are most important in terms of salmonellosis acquired from birds, other wild birds are also known to be effective disseminator of *Salmonella* infection through fecal contamination of environment. Despite limitation of evidences directly connecting human infection with wild birds, a number of studies raised concerns regarding roles of wild birds, free roaming or caged, as a potential source of *Salmonella* infection in humans. Particular concerns have been raised for wild geese and water fowls (Elmberg et al. 2017; Vogt et al. 2019), urbanized white ibises (Hernandez et al. 2016), certain birds belonging to orders *Columbiformes* (Konicek et al. 2016), caged golden pheasants (Milton et al. 2018), owls (Fuentes-Castillo et al. 2019; Molina-López et al. 2015), decoys and raptors (Jurado-Tarifa et al. 2016), and wild Kelp gulls (Liakopoulos et al. 2016) among others. In addition to wild birds, several studies also highlighted the risks of transmission of *Salmonella* infection from domesticated avian species including commercially farmed birds (Antunes et al. 2016; Behravesh et al. 2014; Chousalkar and Gole 2016; Collins et al. 2019; De Lucia et al. 2018; Elmberg et al. 2017; Greig et al. 2015; Magdy et al. 2020; Messens et al. 2013; Sánchez-Salazar et al. 2020; Vogt et al. 2020; Whiley and Ross 2015). Another area of concern is the incursion of wild species of birds in animal farms and into live bird market. As wild birds are free living, this sort of interaction between farmed mammals and captive market birds provides opportunity for bidirectional spread and dissemination of *Salmonella* infection across long distances (De Lucia et al. 2018; Nabil et al. 2020). Though evidences suggest considerable risks of infection in and from wild and free-living birds, a pragmatic approach to risk assessment is necessary as there is often a chance for overestimating the associated risks (Smith et al. 2020).

References

Afema JA, Sischo WM (2016) Salmonella in wild birds utilizing protected and human impacted habitats, Uganda. EcoHealth 13:558–569. https://doi.org/10.1007/s10393-016-1149-1

Andrés-Barranco S, Vico JP, Garrido V, Samper S, Herrera-León S, de Frutos C, Mainar-Jaime RC (2014) Role of wild bird and rodents in the epidemiology of subclinical salmonellosis in finishing pigs. Foodborne Pathog Dis 11:689–697. https://doi.org/10.1089/fpd.2014.1755

Antilles N, Sanglas A, Cerdà-Cuéllar M (2015) Free-living waterfowl as a source of zoonotic bacteria in a dense wild bird population area in northeastern Spain. Transbound Emerg Dis 62: 516–521. https://doi.org/10.1111/tbed.12169

Antunes P, Mourão J, Campos J, Peixe L (2016) Salmonellosis: the role of poultry meat. Clin Microbiol Infect 22:110–121. https://doi.org/10.1016/j.cmi.2015.12.004

Barrow PA, Jones MA, Smith AL, Wigley P (2012) The long view: Salmonella—the last forty years. Avian Pathol 41:413–420. https://doi.org/10.1080/03079457.2012.718071

Basler C, Nguyen TA, Anderson TC, Hancock T, Behravesh CB (2016) Outbreaks of human *Salmonella* infections associated with live poultry, United States, 1990–2014. Emerg Infect Dis 22:1705–1711. https://doi.org/10.3201/eid2210.150765

Behravesh CB, Brinson D, Hopkins BA, Gomez TM (2014) Backyard poultry flocks and salmonellosis: a recurring, yet preventable public health challenge. Clin Infect Dis 58:1432–1438. https://doi.org/10.1093/cid/ciu067

Benskin CMH, Rhodes G, Pickup RW, Mainwaring MC, Wilson K, Hartley IR (2015) Life history correlates of fecal bacterial species richness in a wild population of the blue tit *Cyanistes caeruleus*. Ecol Evol 5:821–835. https://doi.org/10.1002/ece3.1384

Brenner FW, Villar RG, Angulo FJ, Tauxe R, Swaminathan B (2000) Salmonella nomenclature. J Clin Microbiol 38(7):2465–2467. https://doi.org/10.1128/JCM.38.7.2465-2467.2000. PMID: 10878026; PMCID: PMC86943

Blanco G (2018) Supplementary feeding as a source of multiresistant *Salmonella* in endangered Egyptian vultures. Transbound Emerg Dis 65:806–816. https://doi.org/10.1111/tbed.12806

Brobey B, Kucknoor A, Armacost J (2017) Prevalence of *Trichomonas*, *Salmonella*, and *Listeria* in wild birds from Southeast Texas. Avian Dis 61:347–352. https://doi.org/10.1637/11607-020617-regr

Chousalkar K, Gole VC (2016) Salmonellosis acquired from poultry. Curr Opin Infect Dis 29:514–519. https://doi.org/10.1097/QCO.0000000000000296

Collins J, Simpson KMJ, Bell G, Durrheim DN, Hill-Cawthorne GA, Hope K, Howard P, Kohlenberg T, Lawrence K, Lilly K, Porigneaux P, Sintchenko V, Wang Q, Ward MP, Wiethoelter A, Mor SM, Flint J (2019) A One Health investigation of *Salmonella enterica* serovar Wangata in North-Eastern New South Wales, Australia, 2016–2017. Epidemiol Infect 147:e150. https://doi.org/10.1017/S0950268819000475

Crump JA, Sjölund-Karlsson M, Gordon MA, Parry CM (2015) Epidemiology, clinical presentation, laboratory diagnosis, antimicrobial resistance, and antimicrobial management of invasive *Salmonella* infections. Clin Microbiol Rev 28:901–937. https://doi.org/10.1128/CMR.00002-15

De Lucia A, Rabie A, Smith RP, Davies R, Ostanello F, Ajayi D, Petrovska L, Martelli F (2018) Role of wild birds and environmental contamination in the epidemiology of *Salmonella* infection in an outdoor pig farm. Vet Microbiol 227:148–154. https://doi.org/10.1016/j.vetmic.2018.11.003

de Souza ML, Coelho ML, da Silva AO, da Silva Azuaga LB, Macedo Coutinho Netto CR, Galhardo JA, Brito Leal CR, do Nascimento Ramos CA (2020) *Salmonella* spp. infection in Psittacidae at a wildlife rehabilitation center in the state of Mato Grosso do Sul. Braz J Wildl Dis 56:288–293. https://doi.org/10.7589/2019-06-171

Denagamage T, Jayarao B, Patterson P, Wallner-Pendleton E, Kariyawasam S (2015) Risk factors associated with *Salmonella* in laying hen farms: systematic review of observational studies. Avian Dis 59:291–302. https://doi.org/10.1637/10997-120214-reg

Desin TS, Köster W, Potter AA (2013) Salmonella vaccines in poultry: past, present and future. Expert Rev Vaccines 12:87–96. https://doi.org/10.1586/erv.12.138

Dipineto L, Bossa LMDL, Pace A, Russo TP, Gargiulo A, Ciccarelli F, Raia P, Caputo V, Fioretti A (2015) Microbiological survey of birds of prey pellets. Comp Immunol Microbiol Infect Dis 41: 49–53. https://doi.org/10.1016/j.cimid.2015.05.001

Dos Santos EJE, Azevedo RP, Lopes ATS, Rocha JM, Albuquerque GR, Wenceslau AA, Miranda FR, Rodrigues DDP, Maciel BM (2020) *Salmonella* spp. in wild free-living birds from Atlantic forest fragments in Southern Bahia, Brazil. Biomed Res Int 2020:7594136. https://doi.org/10.1155/2020/7594136

Elmberg J, Berg C, Lerner H, Waldenstrom J, Hessel R (2017) Potential disease transmission from wild geese and swans to livestock, poultry and humans: a review of the scientific literature from a One Health perspective. Infect Ecol Epidemiol 7:1300450. https://doi.org/10.1080/20008686.2017.1300450

Fuentes-Castillo D, Farfán-López M, Esposito F, Moura Q, Fernandes MR, Lopes R, Cardoso B, Muñoz ME, Cerdeira L, Najle I, Muñoz PM, Catão-Dias JL, González-Acuña D, Lincopan N (2019) Wild owls colonized by international clones of extended-spectrum β-lactamase (CTX-M)-producing *Escherichia coli* and *Salmonella* Infantis in the Southern Cone of America. Sci Total Environ 674:554–562. https://doi.org/10.1016/j.scitotenv.2019.04.149

Fukui D, Takahashi K, Kubo M, Une Y, Kato Y, Izumiya H, Teraoka H, Asakawa M, Yanagida K, Bando G (2014) Mass mortality of Eurasian tree sparrows (*Passer montanus*) from *Salmonella* Typhimurium Dt40 in Japan, winter 2008–09. J Wildl Dis 50:484–495. https://doi.org/10.7589/2012-12-321

Gargiulo A, Fioretti A, Russo TP, Varriale L, Rampa L, Paone S, Bossa LMDL, Raia P, Dipineto L (2018) Occurrence of enteropathogenic bacteria in birds of prey in Italy. Lett Appl Microbiol 66: 202–206. https://doi.org/10.1111/lam.12836

Giacopello C, Foti M, Mascetti A, Grosso F, Ricciardi D, Fisichella V, Piccolo FL (2016) Antimicrobial resistance patterns of Enterobacteriaceae in European wild bird species admitted in a wildlife rescue centre. Vet Ital 52:139–144. https://doi.org/10.12834/VetIt.327.1374.2

Greig J, Rajić A, Young I, Mascarenhas M, Waddell L, Lejeune J (2015) A scoping review of the role of wildlife in the transmission of bacterial pathogens and antimicrobial resistance to the food chain. Zoonoses Public Health 62:269–284. https://doi.org/10.1111/zph.12147

Hernandez SM, Welch CN, Peters VE, Lipp EK, Curry S, Yabsley MJ, Sanchez S, Presotto A, Gerner-Smidt P, Hise KB, Hammond E, Kistler WM, Madden M, Conway AL, Kwan T, Maurer

JJ (2016) Urbanized White Ibises (*Eudocimus albus*) as carriers of *Salmonella enterica* of significance to public health and wildlife. PLoS One 11:1–22. https://doi.org/10.1371/journal.pone.0164402

Janecko N, Čížek A, Halová D, Karpíšková R, Myšková P, Literák I (2015) Prevalence, characterization and antibiotic resistance of salmonella isolates in large corvid species of Europe and North America between 2010 and 2013. Zoonoses Public Health 62:292–300. https://doi.org/10.1111/zph.12149

Jurado-Tarifa E, Torralbo A, Borge C, Cerdà-Cuéllar M, Ayats T, Carbonero A, García-Bocanegra I (2016) Genetic diversity and antimicrobial resistance of *Campylobacter* and *Salmonella* strains isolated from decoys and raptors. Comp Immunol Microbiol Infect Dis 48:14–21. https://doi.org/10.1016/j.cimid.2016.07.003

Konicek C, Vodrážka P, Barták P, Knotek Z, Hess C, Račka K, Hess M, Troxler S (2016) Detection of zoonotic pathogens in wild birds in the cross-border region Austria—Czech Republic. J Wildl Dis 52:850–861. https://doi.org/10.7589/2016-02-038

Krawiec M, Kuczkowski M, Kruszewicz AG, Wieliczko A (2015) Prevalence and genetic characteristics of *Salmonella* in free-living birds in Poland. BMC Vet Res 11. https://doi.org/10.1186/s12917-015-0332-x

Liakopoulos A, Olsen B, Geurts Y, Artursson K, Berg C, Mevius DJ, Bonnedahl J (2016) Molecular characterization of extended-spectrum-cephalosporin-resistant Enterobacteriaceae from wild kelp gulls in South America. Antimicrob Agents Chemother 60:6924–6927. https://doi.org/10.1128/aac.01120-16

Loharikar A, Briere E, Schwensohn C, Weninger S, Wagendorf J, Scheftel J, Garvey A, Warren K, Villamil E, Rudroff JA, Kurkjian K, Levine S, Colby K, Morrison B, May A, Anderson S, Daly E, Marsden-Haug N, Erdman MM, Gomez T, Rhorer A, Castleman J, Adams JK, Theobald L, Lafon P, Trees E, Mitchell J, Sotir MJ, Behravesh CB (2012) Four multistate outbreaks of human Salmonella infections associated with live poultry contact, United States, 2009. Zoonoses Public Health 59:347–354. https://doi.org/10.1111/j.1863-2378.2012.01461.x

MacFadden DR, Bogoch II, Andrews JR (2016) Advances in diagnosis, treatment, and prevention of invasive *Salmonella* infections. Curr Opin Infect Dis 29:453–458. https://doi.org/10.1097/QCO.0000000000000302

Magdy OS, Moussa IM, Hussein HA, El-Hariri MD, Ghareeb A, Hemeg HA, Al-Maary KS, Mubarak AS, Alwarhi WK, Eljakee JK, Kabli SA (2020) Genetic diversity of *Salmonella enterica* recovered from chickens farms and its potential transmission to human. J Infect Public Health 13:571–576. https://doi.org/10.1016/j.jiph.2019.09.007

Massey JG (2003) Diseases and medical management of wild Passeriformes. Semin Avian Exot Pet Med 12:29–36. https://doi.org/10.1053/saep.2003.127876

Matias CAR, Pereira IA, de Araújo MS, Santos AFM, Lopes RP, Christakis S, Rodrigues DP, Siciliano S (2016a) Characteristics of *Salmonella* spp. isolated from wild birds confiscated in illegal trade markets, Rio de Janeiro, Brazil. Biomed Res Int 2016:1–7. https://doi.org/10.1155/2016/3416864

Matias CAR, Pereira IA, dos Reis EMF, Rodrigues DP, Siciliano S (2016b) Frequency of zoonotic bacteria among illegally traded wild birds in Rio de Janeiro. Braz J Microbiol 47:882–888. https://doi.org/10.1016/j.bjm.2016.07.012

Messens W, Vivas-Alegre L, Bashir S, Amore G, Romero-Barrios P, Hugas M (2013) Estimating the public health impact of setting targets at the European level for the reduction of zoonotic *Salmonella* in certain poultry populations. Int J Environ Res Public Health 10:4836–4850. https://doi.org/10.3390/ijerph10104836

Milton AAP, Agarwal RK, Priya GB, Athira CK, Saminathan M, Reddy A, Aravind M, Kumar A (2018) Occurrence, antimicrobial susceptibility patterns and genotypic relatedness of *Salmonella* spp. isolates from captive wildlife, their caretakers, feed and water in India. Epidemiol Infect 146:1543–1549. https://doi.org/10.1017/s0950268818001553

Molina-López RA, Vidal A, Obón E, Martín M, Darwich L (2015) Multidrug-resistant *Salmonella enterica* serovar Typhimurium monophasic variant 4,12:i:- isolated from asymptomatic wildlife in a Catalonian Wildlife Rehabilitation Center, Spain. J Wildl Dis 51:759–763. https://doi.org/10.7589/2015-01-019

Nabil NM, Erfan AM, Tawakol MM, Haggag NM, Naguib MM, Samy A (2020) Wild birds in live birds markets: potential reservoirs of enzootic avian influenza viruses and antimicrobial resistant Enterobacteriaceae in Northern Egypt. Pathogens 9. https://doi.org/10.3390/pathogens9030196

Naushad S, Duceppe M-O, Dupras AA, Gao R, Ogunremi D (2020) Closed genome sequences and antimicrobial resistance profiles of eight wild bird *Salmonella* isolates obtained with MinION and Illumina MiSeq sequencing. Microbiol Resour Announc 9. https://doi.org/10.1128/MRA.00228-20

Navarro-Gonzalez N, Wright S, Aminabadi P, Gwinn A, Suslow TV, Jay-Russell MT (2020) Carriage and subtypes of foodborne pathogens identified in wild birds residing near agricultural lands in California: a repeated cross-sectional study. Appl Environ Microbiol 86. https://doi.org/10.1128/AEM.01678-19

OIE (2018a) Salmonellosis. In: Manual of diagnostic tests and vaccines for terrestrial animals. OIE, pp 1735–1752

OIE (2018b) Fowl typhoid and Pullorum disease. In: Manual of diagnostic tests and vaccines for terrestrial animals. OIE, pp 914–930

Pattison M, McMullin P, Bradbury J, Alexander D (eds) (2008) Poultry diseases, 6th edn. https://doi.org/10.1016/B978-0-7020-2862-5.50031-3

Pearson HE, Lapidge SJ, Hernández-Jover M, Toribio J-ALML (2016) Pathogen presence in European starlings inhabiting commercial piggeries in South Australia. Avian Dis 60:430–436. https://doi.org/10.1637/11304-101815-reg

Plaza PI, Blanco G, Madariaga MJ, Boeri E, Teijeiro ML, Bianco G, Lambertucci SA (2019) Scavenger birds exploiting rubbish dumps: pathogens at the gates. Transbound Emerg Dis 66: 873–881. https://doi.org/10.1111/tbed.13097

Quinn PJ, Markey BK, Leonard FC, FitzPatrick ES, Fanning S, Hartigan PJ (2011) Veterinary microbiology and microbial disease, 2nd edn, West Sussex

Quinn PJ, Markey BK, Leonard FC, FitzPatrick ES, Fanning S (2016) Concise review of veterinary microbiology. Wiley, West Sussex

Rajan K, Shi Z, Ricke SC (2017) Current aspects of *Salmonella* contamination in the US poultry production chain and the potential application of risk strategies in understanding emerging hazards. Crit Rev Microbiol 43:370–392. https://doi.org/10.1080/1040841X.2016.1223600

Rostami S, Mehrzad J, Yahyaraeyat R, Salehi TZ (2020) Pathogenic *Salmonella* weakens avian enriched blood monocytes through ATP depletion, apoptosis induction and phagocytosis inefficiency. Vet Microbiol 240:108505. https://doi.org/10.1016/j.vetmic.2019.108505

Sánchez-Salazar E, Gudiño ME, Sevillano G, Zurita J, Guerrero-López R, Jaramillo K, Calero-Cáceres W (2020) Antibiotic resistance of *Salmonella* strains from layer poultry farms in central Ecuador. J Appl Microbiol 128:1347–1354. https://doi.org/10.1111/jam.14562

Smith OM, Snyder WE, Owen JP (2020) Are we overestimating risk of enteric pathogen spillover from wild birds to humans? Biol Rev Camb Philos Soc 95:652–679. https://doi.org/10.1111/brv.12581

Songer JG, Post KW (2005) Veterinary microbiology: bacterial and fungal agents of animal disease, 1st edn. Elsevier Saunders, St. Louis

Suárez-Pérez A, Corbera JA, González-Martín M, Donázar JA, Rosales RS, Morales M, Tejedor-Junco MT (2020) Microorganisms resistant to antimicrobials in wild Canarian Egyptian vultures (*Neophron percnopterus majorensis*). Animals 10. https://doi.org/10.3390/ani10060970

Tamamura Y, Uchida I, Tanaka K, Nakano Y, Izumiya H, Takahashi T, Kikuchi N (2016) A case study on *Salmonella enterica* serovar Typhimurium at a dairy farm associated with massive sparrow death. Acta Vet Scand 58:4–7. https://doi.org/10.1186/s13028-016-0205-8

Tardone R, Rivera D, Dueñas F, Sallaberry-Pincheira N, Hamilton-West C, Adell AD, Moreno-Switt AI (2020) *Salmonella* in raptors and aquatic wild birds in Chile. J Wildl Dis 56:707–712. https://doi.org/10.7589/2019-08-198

Threlfall EJ (2010) Salmonella. In: Topley & Wilson's microbiology and microbial infections. Major reference works. https://doi.org/10.1002/9780470688618.taw0054

Tizard I (2004) Salmonellosis in wild birds. Semin Avian Exot Pet Med 13:50–66. https://doi.org/10.1053/j.saep.2004.01.008

Tyson-Pello SJ, Olsen GH (2020) Emerging diseases of avian wildlife. Vet Clin North Am Exot Anim Pract 23:383–395. https://doi.org/10.1016/j.cvex.2020.01.002

van Andel M, Jackson BH, Midwinter AC, Alley MR, Ewen JG, McInnes K, Jakob Hoff R, Reynolds AD, French N (2015) Investigation of mortalities associated with *Salmonella* spp. infection in wildlife on Tiritiri Matangi Island in the Hauraki Gulf of New Zealand. N Z Vet J 63:235–239. https://doi.org/10.1080/00480169.2014.990065

Vogt N, Pearl D, Taboada E, Mutschall S, Janecko N, Reid-Smith R, Jardine C (2019) Carriage of *Campylobacter*, *Salmonella*, and antimicrobial-resistant, non-specific *Escherichia coli* by waterfowl species collected from three sources in Southern Ontario, Canada. J Wildl Dis 55. https://doi.org/10.7589/2018-12-288

Vogt NA, Stevens CPG, Pearl DL, Taboada EN, Jardine CM (2020) Generalizability and comparability of prevalence estimates in the wild bird literature: methodological and epidemiological considerations. Anim Health Res Rev 18:1–7. https://doi.org/10.1017/S1466252320000043

Whiley H, Ross K (2015) *Salmonella* and eggs: from production to plate. Int J Environ Res Public Health 12:2543–2556. https://doi.org/10.3390/ijerph120302543

Chapter 16
Avian Tuberculosis

Abstract Avian tuberculosis also known as avian mycobacteriosis is a chronic disease affecting wild and domestic birds including captive and pet birds. The disease is primarily caused by acid fast mycobacterial species *Mycobacterium avium* and *M. genavense*. Avian tuberculosis is worldwide in distribution. Various species of wild and captive birds are affected by the disease. Infected birds shed the causative agent in their fecal droppings and the transmission usually occurs through contaminated environment especially water, soil, litter, pens, farm equipment, and implements. Due to chronic nature of the infection, clinical signs are usually slow to appear and generally include loss of condition, emaciation, loss of skeletal muscle volume, comb discoloration, and persistent diarrhea. Postmortem lesions are characterized by marked emaciation and occurrence of nodules in affected organs. Clinico-epidemiological observations complemented by laboratory methods are important for diagnosis of the disease in birds. Due to lack of effective antimicrobial therapy and vaccine, disease management remains a challenge. Avian tuberculosis is an important zoonotic disease in humans and is manifested as disseminated mycobacteriosis. Fecal matter of the infected birds is important source of infections for humans. As avian tuberculosis is affecting many domestic animals, the avian infection may also be indirectly transmitted to humans through non-avian hosts.

Keywords Avian tuberculosis · Avian mycobacteriosis · *Mycobacterium* · *M. avium* complex · Zoonotic · Acid fast · Captive birds

16.1 Introduction

Avian tuberculosis is a zoonotic disease of chronic nature that affects wild and domestic birds including captive and pet birds. The disease is worldwide in occurrence and is mostly caused by *Mycobacterium avium* (serotypes 1–3) and *M. genavense* (Daley 2017; Dhama et al. 2011; Eslami et al. 2019; Lande et al. 2018; OIE 2018; Srivastava et al. 2017). However, other species of *Mycobacterium* have also been reported from birds (OIE 2018). A number of bird species are affected by the disease including domestic poultry, pheasants, quails, partridges,

Y. S. Malik et al., *Role of Birds in Transmitting Zoonotic Pathogens*, Livestock Diseases and Management, https://doi.org/10.1007/978-981-16-4554-9_16

and captive pet birds (Dahlhausen et al. 2012; Daley 2017; Dhama et al. 2011; Eslami et al. 2019; Lennox 2007; OIE 2018; Slany et al. 2016). The disease runs a chronic course in affected birds causing loss of growth and production in farmed birds. The affected birds usually suffer from debilitating disease with emaciation, decrease in the performance of the bird, and finally death after couple of months. Although the causative agent had been identified since long, there is no effective vaccine or medication available so far (OIE 2018; Pattison et al. 2008; Slany et al. 2016).

16.2 Epidemiology

16.2.1 Causative Agent

Avian tuberculosis or avian mycobacteriosis is caused by mycobacteria which are acid fast, Gram positive, rod-shaped bacteria thriving intracellularly inside hosts. Mycobacteria are highly resistant to most of the environmental conditions such as ultraviolet light, freezing, and thawing. Environmental resistance and persistence of these bacteria is attributable to the presence of the very rigid cell wall rich in lipids and waxy material. In birds, the disease is caused by members of *M. avium* group. Of these, *M. avium* subsp. *avium*, *M. avium* subsp. *hominissuis*, *M. intracellulare*, and *M. avium* subsp. *silvaticum* are particularly relevant for avian infections though the group comprises of a number of species (Table 16.1) (OIE 2018; Pattison et al. 2008; Slany et al. 2016). However, other species such as *M. genavense*, *M. fortuitum*, *M. intracellulare*, *M. scrofulaceum, M. bovis*, and *M. tuberculosis* have also been reported to cause avian mycobacteriosis. These species, except, *M. genavense*, are rather rare in causing the disease in birds.

Table 16.1 Current members and subspecies of *M. avium* group

Species	Sub-species
M. avium	*M. avium* subsp. *avium*
	M. avium subsp. *paratuberculosis*
	M. avium subsp. *silvaticum*
M. intracellulare	
M. chimaera	
M. colombiense	
M. arosiense	
M. bouchedurhonense	
M. marseillense	
M. timonense	
M. vulneris	
M. yongenense	

Data sources: (Daley 2017; Eslami et al. 2019; OIE 2018; Parte 2018; Pattison et al. 2008; Slany et al. 2016)

16.2.2 Hosts

Avian mycobacteriosis has been reported to occur in all species of avian. Common domestic chickens and game birds such as partridges and pheasants are quite susceptible to the disease. Turkey and water birds like duck and geese are relatively less susceptible though diseases have been reported in these species too (Aranaz et al. 1997; Chen et al. 2019; Dahlhausen et al. 2012; Dhama et al. 2011; Dvorska et al. 2007; González et al. 2006; Hejlícek and Treml 1995; Massey 2003; OIE 2018; Pattison et al. 2008; Salamatian et al. 2020; Shivaprasad and Palmieri 2012; VanDerHeyden 1997a; Zhu et al. 2016). In addition, *M. avium* causes infection in a number of wild, captive, and peri-domestic birds including pigeons, vultures, egrets, ibis, captive brolgas, flamingoes, raptors, falcon, buzzards, finches, various psittacine birds, parakeets, macaws, owls, and others (Bougiouklis et al. 2006; Carrasco-Garcia et al. 2018; Dvorska et al. 2007; Hodge et al. 2019; Kock et al. 2013; Kriz et al. 2011, 2013; Ledwoń et al. 2018; Lennox 2007; Massey 2003; Millán et al. 2010; Patiño et al. 2018; Soler et al. 2009; VanDerHeyden 1997a). Moreover, *M. avium* and *M. genavense* are capable of causing disease in many mammals including domesticated animals and humans (Coelho et al. 2013; Daley 2017; OIE 2018; Slany et al. 2016).

16.2.3 Transmission

The source of infection is usually infected birds which shed the organisms with their droppings and other excrements. These infected birds contaminate the environment, especially water, soil, litter, pens, farm equipment, and implements, which in turn act as source of infection for susceptible hosts (OIE 2018; Pattison et al. 2008). Spread of the bacteria is also facilitated by high environmental resistance of the bacteria enabling them to thrive in the environment for long. Tubercular nodules formed in different organs of the hosts may release the bacilli for a long time and play an important role in the spreading the disease in certain area where the infected birds reside. Although vertical transmission is uncommon in wild birds, it has been reported in domestic birds. Experimental study with inoculated chicken eggs revealed that hatched birds shed the bacteria in their feces indicating infection through vertical route. Mechanical transmission through the ticks had been also reported in some cases in some poultry yards that have the ticks which can harbor the bacteria and transmit them to the birds.

There are several influencing factors for avian tuberculosis. Important predisposing factors include stress such as temperature, malnutrition, overcrowding, and extreme environmental conditions (OIE 2018; Pattison et al. 2008).

16.2.4 Avian Mycobacteriosis in Wild Birds

Reports of avian tuberculosis in wild and non-farmed birds are relatively rare in scientific literatures, yet a number of authors reported occurrence of avian mycobacterioses in a variety of species of birds. An outbreak of avian mycobacterioses in Greece was reported in a flock of pigeons which were possibly in contact with many wild and peri-domestic birds such as sparrows, partridges, and pheasants (Bougiouklis et al. 2006). Similar outbreak of avian tuberculosis in a flock of domestic pigeons has also been reported from elsewhere (Kriz et al. 2011). In a study from Czech Republic, occurrence of avian tuberculosis was documented in captive water birds, namely little egret, buff-backed heron, great white egret, bittern, and sacred ibis (Dvorska et al. 2007). In another report from Czech Republic, raptors were found to be infected with *M. avium* subsp. *avium* and the infection was perhaps acquired from infected domestic fowls which were in close contact with the raptors (Kriz et al. 2013). Similarly, in a wildlife rehabilitation center in Spain, a 3-year study of avian mycobacteriosis in raptors indicated an overall prevalence rate of 2.4% (Millán et al. 2010). Though rare, the disease is also known to affect psittacine birds too and the infection have been recorded in brotogerid parakeets, Amazon parrots, budgerigars, and pionus parrots (Lennox 2007). In Bogota, Colombia, free ranging stygian owls were found infected with avian tuberculosis (Soler et al. 2009). An outbreak of the disease in commercially reared Pekin Ducks was recorded in China with high mortality rate of 20%. The report is rather unusual as avian tuberculosis has not been reported in commercial duck farms and perhaps constituted the first report (Zhu et al. 2016). However, in recent reports outbreaks of avian mycobacterioses in pekin ducks and commercial turkey flocks have been documented (Chen et al. 2019; Salamatian et al. 2020). Further, a study from Australia documented the occurrence of disease outbreak in captive brolgas (Hodge et al. 2019).

16.3 Disease Characteristics

16.3.1 Pathogenesis

Avian tuberculosis is slowly progressive chronic disease in birds. The mycobacterium pathogenesis depends on various factors—bacterial survival inside host, replication, as well as induction of the immune response. Infection is usually initiated after ingestion of the bacteria. Owing to the protective cell wall, the bacteria survive the acidic environment of the stomach until it passes to the intestine. In the intestine, the bacteria attach to the intestinal mucosa and submucosa and then enter circulation to reach liver and spleen. Mycobacterial cell wall consisting of complex hydrocarbon (arabinogalactan-peptidoglycan) and mycolic acid core and other glycopeptidolipids such as phospholipids and sulfolipids enables the bacteria to downregulate the normal phagocytic killing by delaying fusion of phagosome and lysosomes as well as inhibition of cytokine production (Shivaprasad and Palmieri 2012). The immune response to mycobacterial infection is predominantly cell mediated with delayed type hypersensitivity reaction appearing as early as 2 days post infection (Shivaprasad and Palmieri 2012). Mycobacterium can persist in the granulomatous nodules inactively for several years until immunosuppression in hosts reactivates the bacteria to spread again to other organs in the body and may be excreted outside to infect another host (Hejlícek and Treml 1995; Lande et al. 2018; Lennox 2007; OIE 2018; Pattison et al. 2008; Shivaprasad and Palmieri 2012; Srivastava et al. 2017).

16.3.2 Clinical Signs and Pathology

Due to chronic nature of the disease the clinical signs develop over a relatively long period. Usually there is slow progressive loss of condition in birds, with unthriftiness and lethargy. Over time emaciation becomes evident with significant loss of muscle especially the sternal muscles. The keel bone appears sharply prominent. Other associated signs include discoloration of comb and persistent diarrhea often soiling the tail feathers. Altered gait and posture may be observed in birds with bone marrow affected in the long bones and death is usually noticed following long periods of illness (OIE 2018; Pattison et al. 2008; Shivaprasad and Palmieri 2012).

Gross pathological lesions of avian tuberculosis vary depending on the host species, host age, immune status, stage of the infection, and other concurrent infections (Shivaprasad and Palmieri 2012). Poor emaciated body condition and atrophy of sternal muscles with prominent keel bone are usually observed. Most commonly affected internal organs in birds are liver, lungs, air sacs, spleen, intestine, thoracic, and abdominal chambers. In these organs, nodules of pale white to tan color may be observed. Nodules vary in sizes measuring from few millimeters to two centimeters. Larger nodules may appear caseous but not calceated. However, in certain species of birds such as canaries, song birds, and finches gross lesions are not always prominent to make a diagnosis (OIE 2018; Pattison et al. 2008; Shivaprasad and Palmieri 2012). Microscopic examination of nodules may reveal tubercle bacilli following acid fast staining. Histopathological examination of the affected tissue reveal centrally necrosed area with cellular debris surrounded by polynucleated giant cells (OIE 2018; Shivaprasad and Palmieri 2012).

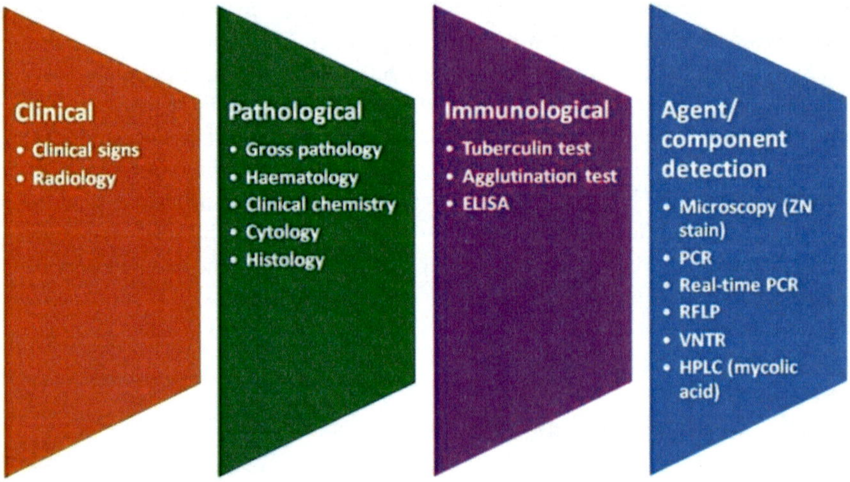

Fig. 16.1 Diagnostic framework of avian mycobacteriosis

16.4 Disease Management

16.4.1 Diagnosis

Diagnosis of avian mycobacteriosis is primarily based on four cardinal approaches (Fig. 16.1). While the first impression is always formed from clinical or epidemiological observations, these need to be fortified with laboratory findings. There are a host of techniques and methodologies available for the laboratory diagnosis of avian tuberculosis. All these techniques come with inherent pros and cons and find application in appropriate settings. Generally, it is believed that confirmation of clinical diagnosis is best supported by microscopy (Ziehl-Nielsen staining), polymerase chain reaction (PCR), and culture of agent on suitable artificial medium. On the other hand, population level surveillance for the disease or screening of individual birds are best achieved by tests that are easy to perform, easy to interpret, and affordable, such as tuberculin test and hemagglutination with stained antigen. Depending on the availability of advanced laboratories several other tests can also be employed including histopathology, enzyme-linked immunosorbent assay (ELISA), real-time PCR, and high-performance liquid chromatography for mycolic acid analysis. Further, for molecular epidemiological analysis and typing of isolates RFLP (restriction fragment length polymorphism) and VNTR (variable number tandem repeats) have been employed (Aranaz et al. 1997; Dahlhausen et al. 2012; Daley 2017; Dhama et al. 2011; Kasperbauer and Daley 2008; OIE 2018; Pattison et al. 2008; Srivastava et al. 2017). Many researchers also attempted direct DNA detection from avian fecal samples for shortening the time to results and improvised novel methodologies for detection of *M. avium* from samples. Recently a loop-mediated isothermal assay has been developed which has shown promise for field applicability without the need for sophisticated equipment (Dahlhausen et al. 2012; OIE 2018; Patiño et al. 2018; Rindi and Garzelli 2014; Sattar et al. 2018; Yashiki et al. 2019; Zhu et al. 2017).

16.4.2 Treatment and Control

Like other mycobacterial infection, treatment of *M. avium* complex is difficult and time consuming. That members of MAC are inherently resistant to many antibiotics and antitubercular drugs poses further therapeutic challenges. Therefore, treatment is neither economically feasible nor advisable in commercial poultry. Vaccination is also not economically viable in poultry birds (Pattison et al. 2008; Slany et al. 2016). However, in pet and companion birds various antibiotics are used depending on the clinician's choice as have been used in captive and zoo birds. Antibiotics that have been reportedly used include isoniazid, rifamycin, ethambutol, aminoglycosides, fluoroquinolones, and macrolides (Buur and Saggese 2012). Control of infection in birds mainly hinges on keeping premises clean and hygienic, with regular monitoring of infection in new birds, proper disposal of infected materials, and awareness about the infection (Dhama et al. 2011; Griffith 2018; Kasperbauer and Daley 2008; Kwon et al. 2019; Lande et al. 2018; Ledwoń et al. 2018; Lennox 2007; OIE 2018; Pattison et al. 2008; VanDerHeyden 1997b).

16.5 Public Health Concerns

Avian tuberculosis in a zoonotic disease with potential to infect humans. In humans the disease is manifested as disseminated mycobacteriosis characterized by prolonged fever, weight loss, weakness, and anemia (Lennox 2007). As with other mycobacterial infection, immune status of the human host plays important role in establishing an infection with avian mycobacteria. Thus, immunocompromised people, especially persons with HIV infection, are particularly at risk of acquiring infection with *M. avium*. The most important sources of human infection are the fecal material of the infected chickens and wild birds; however, there is no strong evidence for the direct infection of the human due to the wild birds but the most possible role is through the indirect transmission through the contaminated fecal matter from the infected birds. Many other animals are infected by *M. avium* complex and develop clinical to subclinical disease. Available reports suggest that almost all domestic and companion animal are susceptible to the infection and thus in turn may spread the disease to humans (Biet et al. 2005; Daley 2017; Hoop 1997; OIE 2018; Pattison et al. 2008). Based on our current understanding Slany et al. (2016) summarized that environment is the most probable source of *M. avium* infection for humans with no convincing evidence on human to human transfer till date (Slany et al. 2016).

References

Aranaz A, Liébana E, Mateos A, Dominguez L (1997) Laboratory diagnosis of avian mycobacteriosis. Semin Avian Exot Pet Med 6:9–17. https://doi.org/10.1016/s1055-937x(97)80036-9

Biet F, Boschiroli ML, Thorel MF, Guilloteau LA (2005) Zoonotic aspects of *Mycobacterium bovis* and *Mycobacterium avium-intracellulare* complex (MAC). Vet Res 36:411–436. https://doi.org/10.1051/vetres:2005001

Bougiouklis P, Brellou G, Fragkiadaki E, Iordanidis P, Vlemmas I, Georgopoulou I (2006) Outbreak of avian mycobacteriosis in a flock of two-year-old domestic pigeons (Columba livia f. domestica). Avian Dis 49:442–445. https://doi.org/10.1637/7325-011005r.1

Buur J, Saggese MD (2012) Taking a rational approach in the treatment of avian mycobacteriosis. Vet Clin North Am Exot Anim Pract 15:57–70. https://doi.org/10.1016/j.cvex.2011.12.001

Carrasco-Garcia R, Barroso P, Perez-Olivares J, Montoro V, Vicente J (2018) Consumption of big game remains by scavengers: a potential risk as regards disease transmission in Central Spain. Front Vet Sci 5:4. https://doi.org/10.3389/fvets.2018.00004

Chen H, Zhu D, Wang M, Jia R, Chen S, Liu M, Zhao X, Yang Q, Wu Y, Zhang S, Liu Y, Zhang L, Yu Y, Chen X, Cheng A (2019) Amyloid A amyloidosis secondary to avian tuberculosis in naturally infected domestic Pekin ducks (*Anas platyrhynchos domestica*). Comp Immunol Microbiol Infect Dis 63:136–141. https://doi.org/10.1016/j.cimid.2019.01.016

Coelho A, Pinto M, Matos A, Matos M, Pires M (2013) *Mycobacterium avium* complex in domestic and wild animals. pp 91–128. https://doi.org/10.5772/54323

Dahlhausen B, Tovar DS, Saggese MD (2012) Diagnosis of mycobacterial infections in the exotic pet patient with emphasis on birds. Vet Clin North Am Exot Anim Pract 15:71–83. https://doi.org/10.1016/j.cvex.2011.11.003

Daley CL (2017) *Mycobacterium avium* complex disease. Microbiol Spectr 5. https://doi.org/10.1128/microbiolspec.TNMI7-0045-2017

Dhama K, Mahendran M, Tiwari R, Dayal Singh S, Kumar D, Singh S, Sawant PM (2011) Tuberculosis in birds: insights into the *Mycobacterium avium* infections. Vet Med Int 2011: 712369. https://doi.org/10.4061/2011/712369

Dvorska L, Matlova L, Ayele WY, Fischer OA, Amemori T, Weston RT, Alvarez J, Beran V, Moravkova M, Pavlik I (2007) Avian tuberculosis in naturally infected captive water birds of the Ardeideae and Threskiornithidae families studied by serotyping, IS901 RFLP typing, and virulence for poultry. Vet Microbiol 119:366–374. https://doi.org/10.1016/j.vetmic.2006.09.010

Eslami M, Shafiei M, Ghasemian A, Valizadeh S, Al-Marzoqi AH, Shokouhi Mostafavi SK, Nojoomi F, Mirforughi SA (2019) *Mycobacterium avium* paratuberculosis and *Mycobacterium avium* complex and related subspecies as causative agents of zoonotic and occupational diseases. J Cell Physiol 234:12415–12421. https://doi.org/10.1002/jcp.28076

González M, Rodriguez-Bertos A, Gimeno I, Flores JM, Pizarro M (2006) Outbreak of avian tuberculosis in 48-week-old commercial layer hen flock. Avian Dis 46:1055–1061. https://doi.org/10.1637/0005-2086(2002)046[1055:ooatiw]2.0.co;2

Griffith DE (2018) Treatment of *Mycobacterium avium* complex (MAC). Semin Respir Crit Care Med 39:351–361. https://doi.org/10.1055/s-0038-1660472

Hejlícek K, Treml F (1995) Pathogenesis of avian mycobacteriosis in the domestic goose (*Anser anser* f. *domestica*) and duck (*Anas platyrhynchos* f. *domestica*). Vet Med (Praha) 40:117–121

Hodge PJ, Sandy JR, Noormohammadi AH (2019) Avian mycobacteriosis in captive brolgas (*Antigone rubicunda*). Aust Vet J. https://doi.org/10.1111/avj.12784

Hoop RK (1997) Public health implications of exotic pet mycobacteriosis. Semin Avian Exot Pet Med 6:3–8. https://doi.org/10.1016/s1055-937x(97)80035-7

Kasperbauer SH, Daley CL (2008) Diagnosis and treatment of infections due to *Mycobacterium avium* complex. Semin Respir Crit Care Med 29:569–576. https://doi.org/10.1055/s-0028-1085708

Kock ND, Kock RA, Wambua J, Kamau GJ, Mohan K (2013) *Mycobacterium avium*-related epizootic in free-ranging lesser flamingos in Kenya. J Wildl Dis 35:297–300. https://doi.org/10.7589/0090-3558-35.2.297

Kriz P, Slana I, Kralik P, Babak V, Skoric M, Fictum P, Docekal J, Pavlik I (2011) Outbreak of *Mycobacterium avium* subsp. *avium* infection in one flock of domestic pigeons. Avian Dis Dig 6:e59–e60. https://doi.org/10.1637/9815-963811-digest.1

Kriz P, Kaevska M, Bartejsova I, Pavlik I (2013) *Mycobacterium avium* subsp. *avium* found in raptors exposed to infected domestic fowl. Avian Dis 57:688–692. https://doi.org/10.1637/10446-110612-case.1

Kwon YS, Koh WJ, Daley CL (2019) Treatment of *Mycobacterium avium* complex pulmonary disease. Tuberc Respir Dis (Seoul) 82:15–26. https://doi.org/10.4046/trd.2018.0060

Lande L, George J, Plush T (2018) *Mycobacterium avium* complex pulmonary disease: new epidemiology and management concepts. Curr Opin Infect Dis 31:199–207. https://doi.org/10.1097/QCO.0000000000000437

Ledwoń A, Napiórkowska A, Augustynowicz-Kopeć E, Szeleszczuk P (2018) Drug susceptibility of non-tuberculous strains of *Mycobacterium* isolated from birds from Poland. Polish J Microbiol 67:487–492. https://doi.org/10.21307/pjm-2018-057

Lennox AM (2007) Mycobacteriosis in companion psittacine birds: a review. J Avian Med Surg 21: 181–187. https://doi.org/10.1647/1082-6742(2007)21[181:micpba]2.0.co;2

Massey JG (2003) Diseases and medical management of wild Passeriformes. Semin Avian Exot Pet Med 12:29–36. https://doi.org/10.1053/saep.2003.127876

Millán J, Negre N, Castellanos E, de Juan L, Mateos A, Parpal L, Aranaz A (2010) Avian mycobacteriosis in free-living raptors in Majorca island, Spain. Avian Pathol 39:1–6. https://doi.org/10.1080/03079450903389945

OIE (2018) Avian tuberculosis. In: Manual of diagnostic tests and vaccines for terrestrial animals. OIE, pp 860–870

Parte AC (2018) LPSN—list of prokaryotic names with standing in nomenclature (Bacterio.net), 20 years on. Int J Syst Evol Microbiol 68:1825–1829. https://doi.org/10.1099/ijsem.0.002786

Patiño WLC, Monge O, Suzán G, Gutiérrez-Espeleta G, Chaves A (2018) Molecular detection of *Mycobacterium avium avium* and *Mycobacterium genavense* in feces of free-living scarlet macaws (Ara macao) in Costa Rica. J Wildl Dis 54:357–361. https://doi.org/10.7589/2017-05-124

Pattison M, McMullin P, Bradbury J, Alexander D (eds) (2008) Poultry diseases, 6th edn. https://doi.org/10.1016/B978-0-7020-2862-5.50031-3

Rindi L, Garzelli C (2014) Genetic diversity and phylogeny of *Mycobacterium avium*. Infect Genet Evol 21:375–383. https://doi.org/10.1016/j.meegid.2013.12.007

Salamatian I, Ghaniei A, Mosavari N, Nourani H, Keshavarz R, Eslampanah M (2020) Outbreak of avian mycobacteriosis in a commercial turkey breeder flock. Avian Pathol 49:296–304. https://doi.org/10.1080/03079457.2020.1740167

Sattar A, Zakaria Z, Abu J, Aziz SA, Gabriel RP (2018) Evaluation of six decontamination procedures for isolation of *Mycobacterium avium* complex from avian feces. PLoS One 13:1–16. https://doi.org/10.1371/journal.pone.0202034

Shivaprasad HL, Palmieri C (2012) Pathology of mycobacteriosis in birds. Vet Clin North Am Exot Anim Pract 15:41–55. https://doi.org/10.1016/j.cvex.2011.11.004

Slany M, Ulmann V, Slana I (2016) Avian mycobacteriosis: still existing threat to humans. Biomed Res Int 2016:1–12. https://doi.org/10.1155/2016/4387461

Soler D, Brieva C, Ribón W (2009) Avian mycobacteriosis in two free-ranging Stygian owls (*Asio stygius*). In: AAZV Annu. Conf. II, pp 212–213

Srivastava V, Dahiya A, Singh SV, Kulshreshtha S (2017) Diagnostic approaches to avian tuberculosis. Worlds Poult Sci J 73:857–871. https://doi.org/10.1017/S0043933917000836

VanDerHeyden N (1997a) Clinical manifestations of mycobacteriosis in pet birds. Semin Avian Exot Pet Med 6:18–24. https://doi.org/10.1016/S1055-937X(97)80037-0

VanDerHeyden N (1997b) New strategies in the treatment of avian mycobacteriosis. Semin Avian Exot Pet Med 6:25–33. https://doi.org/10.1016/s1055-937x(97)80038-2

Yashiki N, Yamazaki Y, Subangkit M, Okabayashi T, Yamazaki W, Goto Y (2019) Development of a LAMP assay for rapid and sensitive detection and differentiation of *Mycobacterium avium* subsp. *avium* and subsp. *hominissuis*. Lett Appl Microbiol 69:155–160. https://doi.org/10.1111/lam.13188

Zhu D-K, Song X-H, Wang J-B, Zhou W-S, Ou X-M, Chen H-X, Liu M-F, Wang M-S, Jia R-Y, Chen S, Sun K-F, Yang Q, Wu Y, Chen X-Y, Cheng A-C (2016) Outbreak of avian tuberculosis in commercial domestic Pekin ducks (*Anas platyrhynchos domestica*). Avian Dis 60:677–680. https://doi.org/10.1637/11396-021916-resnote.1

Zhu L, Peng Y, Ye J, Wang T, Bian Z, Qin Y, Zhang H, Ding J (2017) Isolation, identification, and characterization of a new highly pathogenic field isolate of *Mycobacterium avium* spp. *avium*. Front Vet Sci 4:243. https://doi.org/10.3389/fvets.2017.00243

Part IV
Parasitic and Mycotic Infections

Chapter 17
Cryptosporidiosis

Abstract Cryptosporidiosis is a highly contagious intestinal infection caused by *Cryptosporidium*, a protozoan parasite. Infections caused by *Cryptosporidium baileyi*, *C. galli*, *C. meleagridis*, and *C. avium* are regularly identified among pet and wild avian species. In our context, *C. meleagridis* is the most significant one, since it is the only *Cryptosporidium* species known to cause natural infection in humans and avian (turkeys) species and is the third most common reason for human cryptosporidiosis. *Cryptosporidium* species associated with mammals rarely infect birds; however, birds act as carrier of oocysts of zoonotic species like *C. hominis* and *C. parvum*. Children below 5 years are at higher risk of this infection; however, adults in developed countries can also get infected through contaminated food and water. The most at-risk group is the immunocompromised persons such as AIDS and cancer patients. Sporulated oocysts are the infective stage of *Cryptosporidium* spp. and may be excreted by infected humans or birds/animals into the environment. Cryptosporidiosis typically causes watery diarrhea that may be profuse and prolonged, increase in body temperature, and dehydration with subsequent bodyweight loss. The diagnosis of cryptosporidiosis in the laboratory is accomplished by demonstration of oocysts in fecal/intestinal fluid/bowel biopsy samples and molecular methods. Presently, nitazoxanide is the only proven antiparasitic drug effective for treatment of *Cryptosporidium* infections. The prevention and control of avian cryptosporidiosis mostly depends on nutritional and sanitary management to counteract the introduction of oocysts and the prophylaxis of accompanying maladies that are usually connected with the avian cryptosporidiosis.

Keywords Cryptosporidiosis · Zoonotic · Diarrhea · Avian species · *C. meleagridis* · Oocysts

17.1 Introduction

Cryptosporidiosis is among the most common parasitic infections in domestic as well as wild birds around the world (Wang et al. 2014; Zahedi et al. 2015). Around 30 species of birds have been accounted for infection with *Cryptosporidium* (Ryan

Y. S. Malik et al., *Role of Birds in Transmitting Zoonotic Pathogens*, Livestock Diseases and Management, https://doi.org/10.1007/978-981-16-4554-9_17

2010). In pet and wild avian species, cryptosporidiosis is regularly identified with infections of *Cryptosporidium baileyi*, *C. galli*, *C. meleagridis*, and *C. avium* (Holubová et al. 2016). Each can infect numerous avian species, yet they vary in the host, site of infection, and symptomatology. Furthermore, 13 other *Cryptosporidium* genotypes have been depicted in winged animals around the world, including avian genotypes I–VI, goose genotypes I–V, dark duck genotype, and Eurasian woodcock genotype (Nakamura and Meireles 2015; Chelladurai et al. 2016). *C. meleagridis* is the only known species to cause infection in humans and has public health importance (Xiao 2010).

Cryptosporidiosis is caused by coccidian parasite belonging to the genus *Cryptosporidium* in the family *Apicomplexa*. The most favorite site for replication is respiratory and gastrointestinal tract's epithelial cells and it leads to respiratory and gastrointestinal diseases (Aubert and Favennec 2017). The *Cryptosporidium* infection in birds was first reported by Tyzzer (1929) in the cecal epithelium of the chicken. Other types of *Cryptosporidium* causing mortality in young turkeys were described by Slavin (1955), and he recommended the name *C. meleagridis*. Almost two decades later, cryptosporidiosis was detected in domestic geese (Proctor and Kemp 1974) and chickens (Fletcher et al. 1975). Current et al. (1986) explained the life cycle of *Cryptosporidium* in chickens and named the species *Cryptosporidium baileyi*. The third authentic types of this parasite, *C. galli*, were described by Pavlásek (1999) from the proventriculi of chickens and were later reviewed by Ryan et al. (2003).

Cryptosporidiosis has been reported for the first time as a causative agent of diarrhea in humans in 1976. It is one of the primary causes of diarrhea in children in developing countries. Children below 5 years are at higher risk; however, adults in developed countries can also get infected through contaminated food and water. The most at-risk group is the immunocompromised persons having low CD4+ lymphocytes count (200 cells/μL) (Vanathy et al. 2017; Mohebali et al. 2020). Cryptosporidiosis is among the primary protozoan diseases in the birds. It occurs as either a respiratory or a digestive illness, and it influences a substantial number of avian species in all the continents except in Antarctica. *Cryptosporidium* cause infection in the epithelial cells of the respiratory and gastrointestinal tracts of infected hosts (Huber et al. 2007; Gomes et al. 2012). Over 90% of the human cryptosporidiosis is brought about by *C. parvum* and *C. hominis*. But, different other species like *C. canis*, *C. muris*, *C. meleagridis*, *C. felis*, *C. andersoni*, and *C. suis* also cause infection in humans. *Cryptosporidium* species have been described in different avian species, like quails, turkeys, peacocks, geese, chickens, ducks, fowls, including different pet and wild birds (Morgan et al. 2000; Fayer 2010).

17.2 Epizootiology and Modes of Transmission

There are many species of *Cryptosporidium* involved in the infection of animals. However, the main species causing infection in humans are *C. parvum* and *C. hominis*. Earlier *C. hominis* was known as *C. parvum* or genotype 1, and later it was renamed as *C. hominis*. They are usually transmitted from humans to humans, while *C. parvum* can also infect ruminants and is considered as a zoonotic pathogen (Danišová et al. 2017; Feng et al. 2018). The life cycle begins with secretion of the infectious oocyst in the feces of an infected host. These oocysts can infect other hosts through contaminated food and water. In the new host ingested oocyst's wall gets ruptured with the favorable body temperature and acidity of the gastric and bile secretions of the host. Sporozoites are produced which infect the intestinal epithelial cells. In the epithelial cells, sporozoites undergo sexual maturation and produce oocysts. In the infectious stage, *Cryptosporidium* oocyst is 4–6 μm in diameter with either thick or thin walls. These oocysts may either get released in feces or may rupture in the same host and continue the infection, which may lead to severe disease (Pumipuntu and Piratae 2018). *C. baileyi* is most commonly detected in birds of the 12 avian orders, with reports of clinical or subclinical infection. Additionally, this is the most common species in the order Galliformes. *C. galli* has also been reported in five different avian orders, most commonly in Passeriformes and Psittaciformes; however, *C. meleagridis* has been identified in four avian orders, with disease happening especially among the Galliformes. The only avian species which can infect mammals both naturally and experimentally have been reported as *C. meleagridis* (Darabus 1997; Sréter et al. 2000; Darabus and Olariu 2003).

The data on the host specificity of *Cryptosporidium* avian genotypes is scarce or less. Avian genotype I is reported in Indian peafowl (*Pavo cristatus*) and canaries (*Serinus canaria*), whereas genotype III has been found in a few species of Psittaciformes and Passeriformes (Ng et al. 2006; Gomes et al. 2012; Nakamura et al. 2014). The avian genotype II has been depicted in Psittaciformes species and ostriches (Nguyen et al. 2013). Wang et al. (2010) reported that 3 out of 385 fecal samples of chickens have avian genotype II, while Meireles et al. (2006) did not find infection in the chickens experimentally tainted with avian genotype II and *Cryptosporidium* infection was screened employing histology, cytology, and oocyst screening in the feces.

Transmission of the *Cryptosporidium* is through the fecal-oral route by the ingestion of oocysts through the utilization of fecally contaminated food or water or contact with a tainted individual or animal. After the ingestion, the incubation time is 7 days (1–12 days) (Jokipii and Jokipii 1986). *Cryptosporidium* is highly chlorine-tolerant, and inactivation takes 3.5–10.6 days at free chlorine levels (Shields et al. 2008). So, the *Cryptosporidium* transmission can happen even in treated recreational waters, and an outbreak in such water can expand into a community-wide flare-up related to other recreational waters or other settings like daycare (Centers for Disease Control and Prevention 2008). It has been reported that *Cryptosporidium* is mainly transmitted person to person by direct contact as documented in a daycare center. In the center, there was a widespread disease among children and attendants (Fayer and Unger 1986). Indirectly, the disease is transmitted through contaminated food and water or swimming pools contaminated with oocysts, which acts as the zoonotic mode of disease transmission (Ryan et al. 2017).

17.3 Cryptosporidiosis in Canadian Geese

Cryptosporidium is omnipresent genera of protozoa and is present in the feces of Canadian geese (Zhou et al. 2004). The Canadian goose plays an important role in the dissemination of Cryptosporidia pathogens (Graczyk et al. 1998; Kassa et al. 2004). *C. parvum* oocysts retain its viability and infectivity after passing through the Canadian geese. The Canadian migratory geese further may disperse oocysts of *C. parvum* into the open water sources through their feces (Graczyk et al. 1998). This is of specific importance as Canadian geese have a preference for the aquatic habitats, and contaminated water is the primary source of human *Cryptosporidium* disease in the United States (Dietz and Roberts 2000). The role of the geese in the dissemination of the human pathogenic *Cryptosporidium* is unclear. Graczyk et al. (1998) reported that seven of nine areas have Cryptosporidium-positive samples in the United States. Kassa et al. (2001) revealed a comparative commonness of *Cryptosporidium* spp. in geese feces (14/18 samples) in Ohio, yet did not distinguish these isolates at the species level.

Although prevalence at the genus level was reported high, Zhou et al. (2004) found that just 5 out of 49 fecal samples of Canada geese were positive for either *C. parvum* or *C. hominis*. This low abundance at the species level gives rise to the conclusion that Canada geese are the carriers of pathogenic *Cryptosporidium* species and assume a minor role in the dissemination to the humans (Zhou et al. 2004; Graczyk et al. 2007). This issue is additionally entangled by the thought of host-specific *C. parvum* genotypes and whether these extra, nonhuman-specific genotypes are equipped for causing human sickness (Zhou et al. 2004). Indeed, even with the profound change in the prevalence of geese populaces, the stability of oocysts in the environment (Robertson et al. 1992) and the low infectious dose of *Cryptosporidium* species (Hlavsa et al. 2005) permit further thought of the danger of contamination brought about by *Cryptosporidium*. This thought is relative in perspective on the report that there are around 370 *Cryptosporidium* oocysts in a gram of feces, and a mean single dropping weighs 17.2 g; this means 6363 oocysts per positive Canada goose fecal sample (Graczyk et al. 1998).

Cryptosporidium causes approximately 748,000 cases of cryptosporidiosis in humans in the United States every year (Scallan et al. 2011). The significance of this *Cryptosporidium* spp., concerning waterborne maladies, has been mounting in the United States in recent decades, including both drinking water-related flare-ups and those related to recreational water (U.S. Centers for Disease Control and Prevention 2010). With low (10–30 oocysts) dose (DuPont et al. 1995; Okhuysen et al. 1999) and the capacity of infected (human) people to discharge a large number of oocysts every day (Chappel et al. 1996), it is not unusual that cryptosporidiosis episodes are often connected with recreational water (Hlavsa et al. 2005, 2014).

17.4 Pathogenicity

Different field studies have been performed on human volunteers to study the pathogenesis of *Cryptosporidium*. These studies suggest a direct relationship between the inoculum (number of ingested oocysts) and the possibility of infection (Certad et al. 2017). The possibility of infection was 20% for the dose of 30 oocysts, while possibility was 100% when the dose increased to 1000 oocysts. The equation derived from these studies suggests that even if the inoculum is a single oocyst, the possibility of infection will be 0.5%. So, the clinical symptoms are dose dependent (Messner and Berger 2016).

Albeit many recently revealed *Cryptosporidium* infections in the intestinal and respiratory tracts and the bursa of Fabricius in birds had shown the presence of *C. baileyi* and *C. meleagridis* (Ryan 2010), while the role of other species of *Cryptosporidium* causing diseases not characterized molecularly cannot be neglected.

Cryptosporidiosis typically causes watery diarrhea that may be profuse and prolonged, increase in body temperature, and dehydration with subsequent bodyweight loss (Ungar et al. 1990). The common symptoms are diarrhea and abdominal pain in which patients need medicinal consideration, prompting to the laboratory diagnosis of cryptosporidiosis. Other symptoms include vomiting, nausea, and low-grade fever. Once in a while, nonspecific manifestations, for example, weakness, myalgia, headache, malaise, and anorexia, can happen (Current and Garcia 1991). The parasite characteristics and host factors play an essential role in the persistence, severity, and extremeness of the infection. Host factors are the immune status and the exposure frequency of the infected individual; though, little is known in regard to the pathogenic attributes of *Cryptosporidium* spp. (Meinhardt et al. 1996). In severe cases of *Cryptosporidium* infection, symptoms may include asymptomatic shedding of oocysts to an extremely severe disease leading to death. Immunocompetent people only have a transient self-constraining disease, which can go up to 2–3 weeks. Although, in immunocompromised patients, cryptosporidiosis might be a severe disease with tenacious symptoms prompting dehydration and squandering (Blackburn et al. 2004; Chen et al. 2002; O'Donoghue 1995) and was related to noteworthy death rates (Juranek 1995; Manabe et al. 1998). Likewise, *Cryptosporidium* infection might cause atypical signs in immunocompromised patients, for example, biliary tract infection, respiratory tract infection, gastrointestinal infection, and pancreatitis (Hunter and Nichols 2002; Bouzid et al. 2013).

The disease is more severe in immunocompromised persons having congenital immunodeficiency, HIV, cancer (under chemotherapeutic treatment), and congenital gammaglobulinemia. The disease may be life-threatening (Mahmoudi et al. 2017). A study from the UK shows that 19% of patients who had a dual infection (HIV and cryptosporidiosis) died (Connolly et al. 1988). However, another study reported 46 and 29% mortality in HIV and other immunosuppressive syndromes patients (Fayer and Unger 1986). However, in general cases, the mortality is very low but may have severe diarrhea and dehydration (Ungar et al. 1990). The infection is usually in the entire intestinal epithelium which may transmit to other body parts like respiratory tract, gall bladder, and pancreas (Framm and Soave 1997).

17.5 Public Health Significance of Cryptosporidiosis

In a survey of stool analysis, cryptosporidium was reported in 1–4% of patients in Europe and North America and 3–20% in Asia, Australia, Africa, and Central America (Current and Garcia 1991). Several asymptomatic carriers in daycare centers are reported (Lacroix et al. 1987). Similarly, a seroprevalence analysis revealed the presence of cryptosporidiosis as 25–35% in the Western countries and 95% in South America (Casemore et al. 1997).

Over 90% of human *Cryptosporidium* diseases are identified with *C. hominis* or *C. parvum*. However, contaminations with other *Cryptosporidium* species or genotypes have been reported. *C. meleagridis* has a broader host range among the *Cryptosporidium* species and genotypes of avian hosts, which can infect people; in fact, it is the third most common reason for human cryptosporidiosis. In certain countries, for example, Peru and Thailand, *C. meleagridis* causes 10–20% of human *Cryptosporidium* infections, with a recurrence like that of disease by *C. parvum* (Xiao and Feng 2008; Elwin et al. 2012; Insulander et al. 2013). Silverlås et al. (2012) reported that the *C. meleagridis* found in the birds might be associated with species from humans, and the birds may act as a source of human diseases by *C. meleagridis* (Stensvold et al. 2014; Wang et al. 2014). The *C. meleagridis* have been reported by different epidemiological investigations in pet birds and environmental samples. Baroudi et al. (2013) reported a high predominance of *C. meleagridis* in Algeria: 34% in chickens and 44% in turkeys. While in China, a low number was obtained for *C. meleagridis* in chickens (0.52%), hens (0.19%), and quails (0.22%) (Wang et al. 2010, 2012, 2014). Li et al. (2012) discovered *C. meleagridis* in 24.4% of the wastewater samples gathered from four different cities in China. The fecal samples of asymptomatic pet birds or the birds in zoos sporadically show the presence of *C. parvum* oocysts. As a rule, the birds act as a mechanical carrier of oocysts (Nakamura et al. 2009; Quah et al. 2011). *Cryptosporidium* species associated with mammals rarely infect birds. However, the birds act as carrier of oocysts of zoonotic species, for example, *C. hominis* and *C. parvum*, and may take part in the epidemiological chain of human cryptosporidiosis through environmental contamination (Graczyk et al. 1998; Zhou et al. 2004; Graczyk et al. 2008; Plutzer and Tomor 2009).

17.6 Diagnosis, Treatment, and Control Measures

The different diagnostic methods are used in the diagnosis of cryptosporidiosis. The most common and least expensive are microscopy involving oocysts screening after centrifugal flotation in Sheather's solution, followed by bright-field microscopy or phase-contrast microscopy (Cardozo et al. 2008) and different other staining techniques like negative malachite green staining (Elliot et al. 1999) and acid-fast staining (Ortolani 2000; Cardozo et al. 2008). For the screening of the developmental stages of *Cryptosporidium*, staining techniques are used in histological parts and smears of the mucosa, including hematoxylin and eosin, acid-fast, and safranin-methylene blue stains. Further, *Cryptosporidium* DNA can also be identified in tissue sections by using FISH (fluorescent in situ hybridization) (Chvala et al. 2006; Jex et al. 2008). For *Cryptosporidium* spp. diagnosis, immunological methods have been widely used (Jex et al. 2008; Chalmers and Katzer 2013). The capture ELISA (enzyme-linked immunosorbent assay) or DFA (direct fluorescent antibody) assay has been used for the detection of *Cryptosporidium* using commercially available antibodies in fecal and environmental samples; generally speaking, they have higher sensitivity and higher specificity than oocyst-staining systems. The capture ELISA and DFA targeting different antigens show cross-reactivity between different species of *Cryptosporidium*, so it cannot be used in species-specific diagnosis (Jex et al. 2008). Albeit the two strategies are normally used to identify *C. parvum antigens* (Jex et al. 2008), they might be helpful for the detection of avian cryptosporidiosis (Pagès-Manté et al. 2007).

PCR is used for the molecular characterization of *Cryptosporidium* which is followed by sequencing or restriction fragment length polymorphism (RFLP) of the amplified segments. The 18S rRNA sequencing is mostly used for the identification of the species or genotype. Only a few sequences of avian *Cryptosporidium* have been published in comparison to the mammals in the GenBank. At the point when more resolution is required to distinguish genetically similar genotypes or species, the related sequences like HSP-70 (heat shock protein), COWP (*Cryptosporidium* oocyst wall protein), and actin gene of avian *Cryptosporidium* are used. The subtyping of *C. meleagridis* is performed using GP60 (60-kDa glycoprotein) gene in the molecular examination of disease transmission (Stensvold et al. 2014). In the 18S rRNA sequencing, the avian genotypes II and V show 99.9% genetic similarity. The sequences of avian genotypes II and V are differentiated by the substitution of G by A only at the positions 329 and 378, respectively. For this genetic closeness, classification of these two avian genotypes is suggested simply when one gene presents more noteworthy interspecies polymorphism, for example, the HSP-70 or the actin gene (Abe and Makino 2010; Ng et al. 2006; Meireles et al. 2006) has been broken down. A real-time PCR was developed by Nakamura et al. (2014) for diagnosing avian genotype III and *C. galli.*

Numerous medications have been tried for the treatment of cryptosporidiosis. However, the US Food and Drug Administration has recommended the use of nitazoxanide in humans (Striepen 2013). The halofuginone has also indicated variable efficiency; a compelling medication for the treatment of cryptosporidiosis in animals is still lacking (Sréter et al. 2000; Shahiduzzaman and Daugschies 2012). *Cryptosporidium* oocysts are generally resistant to environmental factors and the commonly used disinfectants in avian facilities. The anticipation and control of avian cryptosporidiosis mostly depend on nutritional and sanitary management to counteract the introduction of oocysts and the prophylaxis of accompanying maladies that are usually connected with the avian cryptosporidiosis (Santos et al. 2005; Silva et al. 2010; Shahiduzzaman and Daugschies 2012).

References

Abe N, Makino I (2010) Multilocus genotypic analysis of *Cryptosporidium* isolates from cockatiels, Japan. Parasitol Res 106(6):1491–1497

Aubert D, Favennec L (2017) *Eimeria* and *Cryptosporidium*: recent advances in the therapeutic field. In: Antimicrobial drug resistance. Springer, Cham, pp 685–688

Baroudi D, Khelef D, Goucem R, Adjou KT, Adamu H, Zhang H et al (2013) Common occurrence of zoonotic pathogen *Cryptosporidium meleagridis* in broiler chickens and turkeys in Algeria. Vet Parasitol 196(3–4):334–340

Blackburn BG, Craun GF, Yoder JS, Hill V, Calderon RL, Chen N, Lee SH, Levy DA, Beach MJ (2004) Surveillance for waterborne-disease outbreaks associated with drinking water—United States, 2001-2002. MMWR Surveill Summ 53:23–45

Bouzid M, Hunter PR, Chalmers RM, Tyler KM (2013) *Cryptosporidium* pathogenicity and virulence. Clin Microbiol Rev 26(1):115–134

Cardozo SV, Teixeira WLF, CWG L (2008) Avaliação das técnicas de rotina no diagnóstico de oocistos de *Cryptosporidium baileyi* em amostras de fezes de frangos de corte (*Gallus gallus domesticus*). Rev Bras Parasitol Vet 17(S1 Suppl 1):351–353

Casemore DP, Wright SE, Coop RL (1997) Cryptosporidiosis—human and animal epidemiology. CRC Press, Boca Raton

Centers for Disease Control and Prevention (2008) Communitywide cryptosporidiosis outbreak—Utah, 2007. MMWR Morb Mortal Wkly Rep 57:989–993

Certad G, Viscogliosi E, Chabé M, Cacciò SM (2017) Pathogenic mechanisms of *Cryptosporidium* and *Giardia*. Trends Parasitol 33(7):561–576

Chalmers RM, Katzer F (2013) Looking for *Cryptosporidium*: the application of advances in detection and diagnosis. Trends Parasitol 29(5):237–251

Chappel C, Okhuysen P, Sterling C, DuPont H (1996) *Cryptosporidium parvum*: intensity of infection and oocyst excretion patterns in healthy volunteers. J Infect Dis 1996:232–236

Chelladurai JJ, Clark ME, Kváč M, Holubová N, Khan E, Stenger BL, Giddings CW, McEvoy J (2016) *Cryptosporidium galli* and novel *Cryptosporidium avian* genotype VI in North American red-winged blackbirds (*Agelaius phoeniceus*). Parasitol Res 115:1901–1906

Chen X, Keithly JS, Paya CV, LaRusso NF (2002) Cryptosporidiosis. N Engl J Med 346:1723–1731

Chvala S, Fragner K, Hackl R, Hess M, Weissenböck H (2006) *Cryptosporidium* infection in domestic geese (*Anser anser f. domestica*) detected by in-situ hybridization. J Comp Pathol 134(2–3):211–218

Connolly GM, Dryden MS, Shanson DC, Gazzard BG (1988) Cryptosporidial diarrhoea in AIDS and its treatment. Gut 29(5):593–597

Current WL, Garcia LS (1991) Cryptosporidiosis. Clin Microbiol Rev 4(3):325–358

Current WL, Upton SJ, Haynes TB (1986) The life cycle of *Cryptosporidium baileyi* n. sp. (Apicomplexa, Cryptosporidiidae) infecting chickens. J Protozool 33(2):289–296

Danišová O, Valenčáková A, Stanko M, Luptáková L, Hatalová E, Čanády A (2017) Rodents as a reservoir of infection caused by multiple zoonotic species/genotypes of *C. parvum*, *C. hominis*, *C. suis*, *C. scrofarum*, and the first evidence of *C. muskrat* genotypes I and II of rodents in Europe. Acta Trop 172:29–35

Darabus G (1997) Experimental studies of inter- and intraspecific transmission of *Cryptosporidium parvum* and *C. meleagridis*. Rev Rom Med Vet 7:155–160

Darabus G, Olariu R (2003) The homologous and interspecies transmission of *Cryptosporidium parvum* and *Cryptosporidium meleagridis*. Pol J Vet Sci 6(3):225–228. PMID:14510055

Dietz VJ, Roberts JM (2000) National surveillance for infection with *Cryptosporidium parvum*, 1995-1998: what have we learned? Public Health Rep 115:358–363

DuPont H, Chappel C, Sterling C, Okhuysen P, Rose J, Jakubowski W (1995) The infectivity of *Cryptosporidium parvum* in healthy volunteers. N Engl J Med 332:855–859

Elliot A, Morgan UM, Thompson RCA (1999) Improved staining method for detecting *Cryptosporidium* oocysts in stools using malachite green. J Gen Appl Microbiol 45(3):139–142

Elwin K, Hadfield SJ, Robinson G, Chalmers RM (2012) The epidemiology of sporadic human infections with unusual cryptosporidia detected during routine typing in England and Wales, 2000-2008. Epidemiol Infect 140(4):673–683

Fayer R (2010) Taxonomy and species delimitation in *Cryptosporidium*. Exp Parasitol 124(1):90–97

Fayer R, Unger BL (1986) *Cryptosporidium* and cryptosporidiosis. Microbial Rev 50:458–483

Feng Y, Ryan UM, Xiao L (2018) Genetic diversity and population structure of *Cryptosporidium*. Trends Parasitol 34(11):997–1011

Fletcher OJ, Munnell JF, Page RK (1975) Cryptosporidiosis of the bursa of Fabricius of chickens. Avian Dis 19(3):630–639

Framm SR, Soave R (1997) Agents of diarrhea. Med Clin N Am 81(2):427–447

Gomes RS, Huber F, da Silva S, do Bomfim TC (2012) *Cryptosporidium* spp. parasitize exotic birds that are commercialized in markets, commercial aviaries, and pet shops. Parasitol Res 110(4):1363–1370

Graczyk TK, Fayer R, Trout JM, Lewis EJ, Farley CA, Sulaiman I et al (1998) Giardia sp. cysts and infectious *Cryptosporidium parvum* oocysts in the feces of migratory Canada geese (*Branta canadensis*). Appl Environ Microbiol 64(7):2736–2738

Graczyk TK, Cranfield MR, Fayer R, Trout J, Goodale HJ (2007) Infectivity of *Cryptosporidium parvum* oocysts is retained upon intestinal passage through a migratory waterfowl species (Canada goose, *Branta canadensis*). Tropical Med Int Health 2:341–347

Graczyk TK, Majewska AC, Schwab KJ (2008) The role of birds in dissemination of human waterborne enteropathogens. Trends Parasitol 24(2):55–59

Hlavsa MC, Watson JC, Beach MJ (2005) Cryptosporidiosis surveillance—United States 1999–2002. Morb Mortal Wkly Rep 54:1–8

Hlavsa MC, Roberts VA, Kahler AM, Hilborn ED, Wade TJ, Backer LC, Yoder JS (2014) Recreational water-associated disease outbreaks—United States, 2009–2010. Morb Mortal Wkly Rep 63:6–10

Holubová N, Sak B, Horčičková M, Hlásková L, Květoňová D, Menchaca S, McEvoy J, Kváč M (2016) *Cryptosporidium avium* n. sp. (Apicomplexa: Cryptosporidiidae) in birds. Parasitol Res 115:2243–2251

Huber F, da Silva S, Bomfim TCB, Teixeira KRS, Bello AR (2007) Genotypic characterization and phylogenetic analysis of sp. from domestic animals in Brazil, *Cryptosporidium*. Vet Parasitol 150(1–2):65–74

Hunter P, Nichols G (2002) Epidemiology and clinical features of *Cryptosporidium* infection in immunocompromised patients. Clin Microbiol Rev 15:145–154

Insulander M, Silverlås C, Lebbad M, Karlsson L, Mattsson JG, Svenungsson B (2013) Molecular epidemiology and clinical manifestations of human cryptosporidiosis in Sweden. Epidemiol Infect 141(5):1009–1020

Jex AR, Smith HV, Monis PT, Campbell BE, Gasser RB (2008) *Cryptosporidium*—biotechnological advances in the detection, diagnosis and analysis of genetic variation. Biotechnol Adv 26(4):304–317

Jokipii L, Jokipii A (1986) Timing of symptoms and oocyst excretion in human cryptosporidiosis. N Engl J Med 1986:1643–1647

Juranek D (1995) Cryptosporidiosis: sources of infection and guidelines for prevention. Clin Infect Dis 22:S57–S61

Kassa H, Harrington BJ, Bisesi MS (2001) Risk of occupational exposure to *Cryptosporidium*, *Giardia*, and *Campylobacter* associated with the feces of giant Canada geese. Appl Occup Environ Hyg 16:905–909

Kassa H, Harrington BJ, Bisesi MS (2004) Cryptosporidiosis: a brief literature review and update regarding *Cryptosporidium* in feces of Canada geese (*Branta canadensis*). J Environ Health 66: 34–39

Lacroix C, Berthier M, Agius G, Bonneau D, Pallu B, Jacquemin JL (1987) *Cryptosporidium* oocysts in immunocompetent children: epidemiologic investigations in the day-care centers of Poitiers, France. Eur J Epidemiol 3(4):381–385

Li N, Xiao L, Wang L, Zhao S, Zhao X, Duan L et al (2012) Molecular surveillance of *Cryptosporidium* spp., *Giardia duodenalis*, and *Enterocytozoon bieneusi* by genotyping and subtyping parasites in wastewater. PLoS Negl Trop Dis 6(9):e1809

Mahmoudi MR, Ongerth JE, Karanis P (2017) *Cryptosporidium* and cryptosporidiosis: the Asian perspective. Int J Hyg Environ Health 220(7):1098–1099

Manabe YC, Clark DP, Moore RD, Lumadue JA, Dahlman HR, Belitsos PC, Chaisson RE, Sears CL (1998) Cryptosporidiosis in patients with AIDS: correlates of disease and survival. Clin Infect Dis 27:536–542

Meinhardt P, Casemore DP, Miller KB (1996) Epidemiologic aspects of human cryptosporidiosis and the role of waterborne transmission. Epidemiol Rev 18:118–136

Meireles MV, Soares RM, Santos MM, Gennari SM (2006) Biological studies and molecular characterization of a *Cryptosporidium* isolate from ostriches (*Struthio camelus*). J Parasitol 92(3):623–626

Messner MJ, Berger P (2016) *Cryptosporidium* infection risk: results of new dose-response modeling. Risk Anal 36(10):1969–1082

Mohebali M, Yimam Y, Woreta A (2020) *Cryptosporidium* infection among people living with HIV/AIDS in Ethiopia: a systematic review and meta-analysis. Pathog Global Health 4:1

Morgan UM, Xiao L, Limor J, Gelis S, Raidal SR, Fayer R et al (2000) *Cryptosporidium meleagridis* in an Indian ring-necked parrot (*Psittacula krameri*). Aust Vet J 78(3):182–183

Nakamura AA, Meireles MV (2015) *Cryptosporidium* in birds—a review. Braz J Vet Parasitol 24: 253–267

Nakamura AA, Simões DC, Antunes RG, Silva DC, Meireles MV (2009) Molecular characterization of *Cryptosporidium* spp. from fecal samples of birds kept in captivity in Brazil. Vet Parasitol 166(1–2):47–51

Nakamura AA, Homem CG, Silva AMJ, Meireles MV (2014) Diagnosis of gastric cryptosporidiosis in birds using a duplex real-time PCR assay. Vet Parasitol 205(1–2):7–13

Ng J, Pavlasek I, Ryan U (2006) Identification of novel *Cryptosporidium* genotypes from avian hosts. Appl Environ Microbiol 72(12):7548–7553

Nguyen ST, Fukuda Y, Tada C, Huynh VV, Nguyen DT, Nakai Y (2013) Prevalence and molecular characterization of *Cryptosporidium* in ostriches (*Struthio camelus*) on a farm in Central Vietnam. Exp Parasitol 133(1):8–11

O'Donoghue P (1995) *Cryptosporidium* and cryptosporidiosis in man and animals. Int J Parasitol 25:139–195

Okhuysen PC, Chappel C, Crabb J, Sterling C, DuPont H (1999) Virulence of three distinct *Cryptosporidium parvum* isolates for healthy adults. J Infect Dis 180:1275–1281

Ortolani EL (2000) Standardization of the modified Ziehl-Neelsen technique to stain oocysts of *Cryptosporidium*. Rev Bras Parasitol Vet 9(1):29–31

Pagès-Manté A, Pagès-Bosch M, Majó-Masferrer N, Gómez-Couso H, Ares-Mazás E (2007) An outbreak of disease associated with cryptosporidia on a red-legged partridge (*Alectoris rufa*) game farm. Avian Pathol 36(4):275–278

Pavlásek I (1999) Cryptosporidia: biology, diagnosis, host spectrum, specificity, and the environment. Remed Klinicka Mikrobiol 3:290–301

Plutzer J, Tomor B (2009) The role of aquatic birds in the environmental dissemination of human pathogenic *Giardia duodenalis* cysts and *Cryptosporidium* oocysts in Hungary. Parasitol Int 58(3):227–231

Proctor SJ, Kemp RL (1974) *Cryptosporidium anserinum* sp. n. (Sporozoa) in a domestic goose *Anser anser* L., from Iowa. J Protozool 21(5):664–666

Pumipuntu N, Piratae S (2018) Cryptosporidiosis: a zoonotic disease concern. Vet World 11(5):681

Quah JX, Ambu S, Lim YAL, Mahdy MAK, Mak JW (2011) Molecular identification of *Cryptosporidium parvum* from avian hosts. Parasitology 138(5):573–577

Robertson LJ, Campbell AT, Smith HV (1992) Survival of *Cryptosporidium parvum* oocysts under various environmental pressures. Appl Environ Microbiol 58:3494–3500

Ryan U (2010) *Cryptosporidium* in birds, fish and amphibians. Exp Parasitol 124:113–120

Ryan UM, Xiao L, Read C, Sulaiman IM, Monis P, Lal AA et al (2003) A redescription of *Cryptosporidium galli* Pavlasek, 1999 (Apicomplexa: Cryptosporidiidae) from birds. J Parasitol 89(4):809–813

Ryan U, Lawler S, Reid S (2017) Limiting swimming pool outbreaks of cryptosporidiosis—the roles of regulations, staff, patrons and research. J Water Health 15(1):1–6

Santos MMAB, Peiró JR, Meireles MV (2005) *Cryptosporidium* infection in ostriches (*Struthio camelus*) in Brazil: clinical, morphological and molecular studies. Braz J Poult Sci 7(2):113–117

Scallan E, Hoekstra RM, Angulo FJ, Tauxe RV, Widdowson M-A, Roy SL, Jones JL, Griffin PM (2011) Foodborne illness acquired in the United States—major pathogens. Emerg Infect Dis 17: 7–15

Shahiduzzaman M, Daugschies A (2012) Therapy and prevention of cryptosporidiosis in animals. Vet Parasitol 188(3–4):203–214

Shields JM et al (2008) Inactivation of *Cryptosporidium parvum* under chlorinated recreational water conditions. J Water Health 6:513–520

Silva DC, Homem CG, Nakamura AA, Teixeira WF, Perri SH, Meireles MV (2010) Physical, epidemiological, and molecular evaluation of infection by *Cryptosporidium galli* in Passeriformes. Parasitol Res 107(2):271–277

Silverlås C, Mattsson JG, Insulander M, Lebbad M (2012) Zoonotic transmission of *Cryptosporidium meleagridis* on an organic Swedish farm. Int J Parasitol 42(11):963–967

Slavin D (1955) *Cryptosporidium* (sp. nov.). *meleagridis*. J Comp Pathol 65(3):262–266

Sréter T, Kovács G, Silva AJ, Pieniazek NJ, Széll Z, Dobos-Kovács M et al (2000) Morphologic, host specificity, and molecular characterization of a Hungarian *Cryptosporidium meleagridis* isolate. Appl Environ Microbiol 66(2):735–738

Stensvold CR, Beser J, Axén C, Lebbad M (2014) High applicability of a novel method for gp60-based subtyping of *Cryptosporidium meleagridis*. J Clin Microbiol 52(7):2311–2319

Striepen B (2013) Parasitic infections: time to tackle cryptosporidiosis. Nature 503(7475):189–191

Tyzzer EE (1929) Coccidiosis in gallinaceous birds. Am J Epidemiol 10(2):269–383

U.S. Centers for Disease Control and Prevention (2010) Parasites—*Cryptosporidium* (also known as "crypto"). http://www.cdc.gov/parasites/crypto/gen_info/index.html

Ungar BL, Ward DJ, Fayer R, Quinn CA (1990) Cessation of *Cryptosporidium*-associated diarrhea in an acquired immunodeficiency syndrome patient after treatment with hyperimmune bovine colostrum. Gastroenterology 98(2):486–489

Vanathy K, Parija SC, Mandal J, Hamide A, Krishnamurthy S (2017) Detection of *Cryptosporidium* in stool samples of immunocompromised patients. Trop Parasitol 7(1):41

Wang R, Jian F, Sun Y, Hu Q, Zhu J, Wang F et al (2010) Large-scale survey of *Cryptosporidium* spp. in chickens and Pekin ducks (*Anas platyrhynchos*) in Henan, China: prevalence and molecular characterization. Avian Pathol 39(6):447–451

Wang R, Wang F, Zhao J, Qi M, Ning C, Zhang L et al (2012) *Cryptosporidium* spp. in quails (*Coturnix coturnix japonica*) in Henan, China: molecular characterization and public health significance. Vet Parasitol 187(3–4):534–537

Wang L, Xue X, Li J, Zhou Q, Yu Y, Du A (2014) Cryptosporidiosis in broiler chickens in Zhejiang Province, China: molecular characterization of oocysts detected in fecal samples. Parasite 21:36

Xiao L (2010) Molecular epidemiology of cryptosporidiosis: an update. Exp Parasitol 124:80–88

Xiao L, Feng Y (2008) Zoonotic cryptosporidiosis. FEMS Immunol Med Microbiol 52(3):309–323

Zahedi A, Paparini A, Jian F, Robertson I, Ryan U (2015) Public health significance of zoonotic *Cryptosporidium* species in wildlife: critical insights into better drinking water management. Int J Parasitol Parasit Wildl 5(1):88–109

Zhou L, Kassa H, Tischler ML, Xiao L (2004) Host-adapted *Cryptosporidium* spp. in Canada geese (*Branta canadensis*). Appl Environ Microbiol 70:4211–4215

Chapter 18
Giardiasis

Abstract Giardiasis is an important protozoan disease-causing diarrhea in both children and adults and transmitted mainly through fecal-oral route and distributed throughout the world. *Giardia* spp. also causes disease in pet and wild birds which may act as asymptomatic mechanical carriers of Giardia cysts to humans and other mammals. Among the six *Giardia* spp. *G. lamblia* (syn. *G. intestinalis*) is a complex species with numerous assemblages of which assemblage A and B are mainly associated with human illnesses. Drinking water contaminated with cyst may lead to the development of disease in humans. An infectious dose of as few as 10 cysts is enough to establish the infection in humans. Of note, an infected patient can excrete a very high shedding rate of up to one billion cysts for every single day. Chronic cases in children may lead to malabsorption of vit A and B12. Proper and timely diagnosis of infection is important in order to devise appropriate preventive and control measures. Laboratory identification of *Giardia* spp. mainly involves identification of cyst from human stool samples. Further, antigen-antibody based assays are available for more specific detection of *Giardia* spp. Metronidazole is the drug of choice for the treatment of Giardiasis in humans. Proper hygienic measures including hand hygiene, toilet hygiene, and also hygiene in food environment are necessary to prevent giardiasis in humans.

Keywords Giardiasis · *G. intestinalis* · Birds · Mechanical carrier · Water-borne · Protozoa

18.1 Introduction

Giardia is an important water-borne protozoon parasite found in different animal species (Meyer 1994), causing around 10,000 cases per year (Warren 1989). Giardia has been accounted for disease in wild and pet birds around the world (Reboredo-Fernández et al. 2015). Avian giardiasis is caused by generally two species of *Giardia*, *G. ardeae* and *G. psittaci* (Ryan and Caccio 2013). *G. duodenalis* has additionally been accounted for birds. So far, A and B zoonotic assemblages, as well as D and F non-zoonotic assemblages, have been reported in birds

Y. S. Malik et al., *Role of Birds in Transmitting Zoonotic Pathogens*, Livestock Diseases and Management, https://doi.org/10.1007/978-981-16-4554-9_18

(Reboredo-Fernández et al. 2015; Majewska et al. 2009). Giardiasis commonly occurs in cockatiels and parakeets; but in most cases it is asymptomatic. The two species which infect birds are *G. psittaci* (Erlandsen and Bemrick 1987) and *G. ardeae* (Erlandsen et al. 1990; Ichikawa et al. 2019).

Human giardiasis is brought about by *G. duodenalis* assemblages A and B, and B has a higher predominance than assemblage A (Ryan and Caccio 2013). Giardiasis shows retarded growth, weight reduction, dehydration, repetitive diarrhea, culling of plumes, dry skin, and depression in birds (Greiner and Ritchie 1994; Filippich et al. 1998). Sometime, healthy birds can also shed *Giardia* cysts without any clinical signs (Fernandes et al. 2014). Albeit some of them are host specific; however *G. duodenalis* A and B can infect both animals and humans (Feng and Xiao 2011). Moreover, there are numerous reports of finding assemblages A, B, D, F, and H in fecal samples of different birds (Reboredo-Fernández et al. 2015; Cano et al. 2016; Cunha et al. 2017). Birds are important link of the epidemiological chain of giardiasis in mammals as mechanical transporters of Giardia cysts through the feces (Plutzer and Tomor 2009; Graczyk et al. 2008).

18.2 Epizootiology and Mode of Transmission

Currently, six *Giardia* species have been identified based on host predilection and morphology: *Gracilinanus agilis* (amphibians), *G. lamblia* (mammals), *Giardia muris* (rodents), *Giardia psittaci*, and *Giardia ardeae* (both mainly in birds) (Gutiérrez 2017). There is a slight difference between human and animal species of *Giardia*, and the role of animals and birds in disease transmission is not well defined (Erlandsen et al. 1988). *G. lamblia*, a parasitic protozoan in the order Retortomonadida, shifts back and forth among trophozoites and cysts shapes during its life cycle, stages involved in the clinical disease, and the transmission of the sickness, respectively. *Giardia* growths look oval shaped, measuring around 11–14 by 7–10 μm, having four nuclei (cysts), generally arranged toward one side, and bent middle bodies and linear axonemes. Amid the procedure of encystment, which can be seen under a microscope, trophozoites at first turn out to be latent, rounded, and progressively refractile as encystment starts. At that point, the quadri-nucleate infectious cyst is produced after nuclear division. The thick hyaline wall in the *Giardia* cysts shields them from ecological stressors, for example, the basic condition in the small intestinal, extreme temperatures, water chlorination, etc. This extra protection makes cysts to survive in water as long as 3 months (Gutiérrez 2017). Upon excystation, every cyst discharges two trophozoites, which proceed with the life cycle.

Giardia is mainly transmitted through fecal-oral route in underdeveloped countries with a lack of proper hygiene. Contaminated water having *Giardia* oocyst generally leads to the infection. The daycare centers lacking good hygienic practices are the primary source of infection and secondary infection occurs when people attend the daycare centers (Islam 1990: van de Bosch 1991).

The water-borne giardiasis had been reported 30 years ago in different countries like the United States, North America, New Zealand, Sweden, and the United Kingdom (Cráun 1990).

18.3 Pathobiology

The studies on the pathogenicity of giardiasis in humans revealed a direct relationship between the dose of infectious oocyst and the possibility of infection (Rose et al. 1991), and asymptomatic carriers play an essential role in the transmission of disease in a particular environment (Farthing 1993). *Giardia* infection causes damage to intestinal villi which leads to malabsorption. However, no toxin production is reported (Buret 1994).

Giardia has a similar life cycle to that of the *Cryptosporidium*. The oocysts are released through the feces of the infected person and are transmitted by contaminated food and water. The ingested oocyst reaches the small intestine and gets ruptured to release the trophozoites. These trophozoites attach and feed on the wall of the intestine. The trophozoites are fragile, and if excreted through feces, they do not survive outside the body. In contrast, the remaining trophozoites undergo encystation and are excreted with the fecal matter (Meyer 1994).

The *Giardia* pathogenicity has been debated for quite some time; be that as it may, the capacity of this parasite to cause intense or determined diarrhea and the potential for long haul sequelae, including impeded kid development and psychological advancement, have been perceived (Bartelt and Sartor 2015). *Giardia* predominantly infects the small intestine and invades the lumen and epithelial surface; however, it does not attack further deep in the mucosal layers. Ongoing examinations recommend that changeability among parasite strains, host dietary status, the organization of gut microbiota, and coinfection with different enteropathogens, immune modulation, and mucosal immune responses are important variables that impact sickness symptoms (Miyamoto and Eckmann 2015). The contribution of genetic variability and host factors of the *Giardia* parasites in the development of clinical symptoms of giardiasis has been the subject of various investigations, yet their results are conflicting. For sure, assemblage A has been related to diarrhea in investigations from India, Turkey, and Spain.

In contrast, a relationship with assemblage B was accounted for in The Netherlands, Malaysia, and Ethiopia. No association with any of the assemblage was found in the United Kingdom, Brazil, Cuba, and Albania (Cacciò et al. 2011). Regardless of whether those clashing outcomes are because of contrasts in the investigation plan, in the populace considered (grown-ups versus kids), in the meaning of symptoms, or the geographical location of assemblages or its variations, is unknown. The idea that variability in the strains or subtypes to a limited extent explains the pathogenicity of the host has additionally been proposed for another similar intestinal parasite, *Blastocystis* sp. (Wu et al. 2014), whose job as a pathogen is disputable (Lukeš et al. 2015). In this specific circumstance, it is essential to review that, at times, certain parasite species/genotypes/assemblages are able to infect a specific host (Schmid-Hempel 2009).

18.4 Giardiasis and Canada Goose

The life cycle of *Giardia* has a cyst stage similar to *Cryptosporidium*, which has high resistance to environmental stresses. Different authors have recovered *Giardia* from the feces of Canada geese. Graczyk et al. (2007) found that all the fecal samples were positive for *Giardia* collected from geese fecal dropping sites. Similarly, Kassa et al. (2001) reported 16.7% of *Giardia* cyst prevalence in Canada geese feces.

Despite the fact that both humans and geese are influenced by this same species of *Giardia*, however, molecular evidence proposes that wildlife plays a negligible role in spreading *Giardia* to humans (Yoder et al. 2010) in light of the fact that the species *G. intestinalis* seems to be generally species-specific (Yoder et al. 2010). Until further illumination is given, it is prescribed that studies be preceded for giardiasis fecal samples of humans and environmental samples of *G. intestinalis* to understand the assemblage types to comprehend the importance of animal-to-human transmission of *Giardia* (Yoder et al. 2010). Regardless of the role played by Canada geese in the transfer of *Giardia* cysts to humans, the CDC reported that 6–8% of children and 2% of adults in developed nations, and 33% of individuals in developing countries are infected with *Giardia* (U.S. Centers for Disease Control and Prevention 2011). The most common symptoms of giardiasis are diarrhea, oily stools, gas, and stomach throbs, among other basic signs and manifestations of gastrointestinal disease, yet may likewise prompt "weight reduction and inability to assimilate fat, lactose, vitamin A and B12" (U.S. Centers for Disease Control and Prevention 2011). The most popular transmission course is human-to-human contact with just only an infectious dose of 10 cysts and a very high shedding rate in the patient of up to one billion cysts every day per infected people (Yoder et al. 2010). However, the main transmission course is through human-to-human contact, and wildlife may act as a reservoir of *Giardia* cysts (U.S. Centers for Disease Control and Prevention 2011). It would thus have the capacity for human transmission by affecting the quality of the surface water.

18.5 Public Health Significance of Giardiasis

The *Giardia* cysts usually take 12–19 days from its ingestion to shedding of the oocysts. The most important clinical signs are diarrhea, abdominal cramps, vomiting, nausea, and hyperthermia (Farthing 1993). Around 30–50% of cases become chronic with the shedding of the cysts. These chronic cases may be due to genetic changes in the protozoan as the patients were found to have changed the trophozoite epitopes. Giardiasis, particularly in young children, is of a significant health concern as it may result in malabsorption of some essential nutrients such as vitamin A and B12 (Farthing 1993).

Molecular methods have resolved the taxonomy of *Cryptosporidium* and *Giardia*, further and have increased our knowledge in understanding the host range of the diverse species and genotypes. Specifically, the capacity to describe cysts straightforwardly from environmental samples or feces utilizing PCR-based methods has been valuable in deciding the risk factors (Hunter and Thompson 2005), and also has a high potential in determining the sources of infection in the outbreak and the transmission procedure of the parasites in the endemic host. Different strains of *Cryptosporidium* and *Giardia* collected from various types of hosts across multiple land areas have been genotyped (Monis and Thompson 2003). Such information is characteristic of zoonotic potential, and most experts would concur that *C. parvum* and *G. duodenalis* are zoonotic. It is well known that dogs can possess infections of either zoonotic or host-specific assemblage of *Giardia* (Caccio et al. 2005), and this explained in various studies of Brazil, Mexico, Italy, Japan, and Poland (Itagaki et al. 2005; Berrilli et al. 2004; Eligio-Garcia et al. 2005; Lalle et al. 2005; Zygner et al. 2006; Volotão et al. 2007). In a study from Germany, it was reported that from 60 *Giardia* positive samples gathered arbitrarily from dogs in urban zones, 60% were tainted with zoonotic *Giardia* of the assemblage A, 12% were infected with dog-specific assemblages C and D, and the rest 28% have mixed infections (Leonhard et al. 2007). Vasilopulos et al. (2007) found 17 positive samples for Giardia out of the 250 cats analyzed from Mississippi and Alabama, USA. They discovered six tainted with assemblage AI and 11 with assemblage F. All things considered, the finding of *Giardia* in the excretion of companion animals is the legitimization for treatment.

Although different studies highlighted the potential public health risk from the diverse genotypes of *Giardia* in cats and dogs, the information on the recurrence of zoonotic *Giardia* transmission is deficient. Such data can be acquired from the studies on molecular epidemiological examinations of the parasites from susceptible hosts in restricted foci of transmission or because of genotyping of positive cases and longitudinal observation. In the previous, ongoing exploration in limited endemic foci of transmission has given proof for the role of dogs in cycles of zoonotic *Giardia* transmission including people and dogs from the tea growing territories of Assam, India, and in communities of the temple in Thailand and Bangkok (Inpankaew et al. 2007; Traub et al. 2004). In the two investigations, a few dogs and their possessors having a similar living region were found to have *G. duodenalia* from a similar assemblage.

18.6 Diagnosis, Treatment, and Control Measures

Giardia is mainly transmitted by means of the fecal-oral course and is of financial significance around the world. The exact identification and characterization of the diverse species and assemblages and sub-assemblages of *Giardia* is essential in understanding the pattern of transmission and host range. For the diagnosis purpose, the human fecal samples are routinely fixed in polyvinyl liquor or formalin-based fixatives before analysis (Ballweber et al. 2010), while in canine and fresh cat samples are commonly analyzed. *Giardia* trophozoites are visible in fresh new fecal samples prepared in saline solutions and also on slides having stained polyvinyl liquor fixed samples. For further microscopic examination, the *Giardia* cysts are concentrated using floatation or sedimentation strategies. Tragically, trophozoites and cysts obtained from polyvinyl liquor cannot be utilized for molecular analysis.

Further at the molecular level, immunofluorescent antibody assay and antigen detection tests are more specific than floatation methodology, albeit each of the three methods shows comparable sensitivity (Ballweber et al. 2010). Immunofluorescent antibody assay is a more specific and sensitive method in the case of specimens from symptomatic patients for detecting Giardia. At the same time, fecal antigen identification is a good alternative where no specialized laboratory is required. Youngsters and grown-ups who have *Giardia* disease without symptoms or side effects usually require no treatment except if they have the possibility of spreading the parasites. Mostly people having *Giardia* get well in a few weeks on their own without any treatment. In most of the cases, giardiasis usually disappears on its own. However, in severe or persistent cases, doctors may prescribe some medications. Doctors mostly recommend antiparasitic drugs rather than treating them on their own. Metronidazole (Flagyl) is the most commonly used antibiotic and is recommended for 5–7 days. It may cause side effects like nausea and a metallic taste in the mouth. Alcohol should be avoided while on this medication. Another antibiotic, tinidazole (Tindamax) is also as effective as metronidazole and is also used to treat giardiasis in a single dose. The side effects are almost the same as in metronidazole. Nitazoxanide (Alinia) is available in liquid form and is commonly used to treat giardiasis in children (Soliman et al. 2016).

References

Ballweber LR, Xiao L, Bowman DD, Kahn G, Cama VA (2010) Giardiasis in dogs and cats: update on epidemiology and public health significance. Trends Parasitol 26:180–189

Bartelt LA, Sartor RB (2015) Advances in understanding *Giardia*: determinants and mechanisms of chronic sequelae. F1000Prime Rep 7:62

Berrilli F, Di Cave D, De Liberato C, Franco A, Scaramozzino P, Orecchia P (2004) Genotype characterization of *Giardia duodenalis* isolates from domestic and farm animals by ssu-rRNA gene sequencing. Vet Parasitol 122:193–199

Buret A (1994) Pathogenesis-how does *Giardia* cause disease? In: Giardia: from molecules to disease, pp 293–315

Caccio SM, Thompson RCA, McLauchlin J, Smith HV (2005) Unravelling *Cryptosporidium* and *Giardia* epidemiology. Trends Parasitol 21:430–437

Cacciò SM et al (2011) Epidemiology of giardiasis in humans. In Lujab HD, Svard S (eds) *Giardia*: a model organism. Springer, pp 17–28

Cano L, Lucio A, Bailo B, Cardona GA, Muadica ASO, Lobo L, Carmena D (2016) Identification and genotyping of *Giardia* spp. and *Cryptosporidium* spp. isolates in aquatic birds in the Salburua wetlands, Álava, Northern Spain. Vet Parasitol 221:144–148

Cráun GF (1990) Waterborne giardiasis. Giardiasis:267–293

Cunha MJR, Cury MC, Santín M (2017) Molecular identification of *Enterocytozoon bieneusi*, *Cryptosporidium*, and *Giardia* in Brazilian captive birds. Parasitol Res 116:487–493

Eligio-Garcia L, Cortes-Campos A, Jiminez-Cardoso E (2005) Genotype of *Giardia intestinalis* isolates from children and dogs and its relationship to host origin. Parasitol Res 97:1–6

Erlandsen SL, Bemrick WJ (1987) SEM evidence for a new species, *Giardia psittaci*. J Parasitol 73: 623–629

Erlandsen SL, Sherlock LA, Januschka M, Schupp DG, Schaefer FW, Jakubowski W, Bemrick WJ (1988) Cross-species transmission of *Giardia* spp.: inoculation of beavers and muskrats with cysts of human, beaver, mouse, and muskrat origin. Appl Environ Microbiol 54(11):2777–2785

Erlandsen SL, Bemrick WJ, Wells CL, Feely DL, Knudson L, Campbell SR, Van Keulen H, Jarroll EL (1990) Axenic culture and characterization of *Giardia ardeae* from the great blue heron (*Ardea herodias*) J. Parasitology 76:717–724

Farthing MJ (1993) Giardiasis as a disease. In: *Giardia*: from molecules to disease. pp 15–37

Feng Y, Xiao L (2011) Zoonotic potential and molecular epidemiology of *Giardia* species and giardiasis Clin. Microbiol Rev 24:110–140

Fernandes CA, Grespan A, Knöbl T (2014) Pesquisa de cistos de *Giardia* spp. em fezes de psitacídeos cativos Asa 2:25–32

Filippich LJ, Mcdonnell PA, Munoz E, Upcroft JA (1998) *Giardia* infection in budgerigars. Aust Vet J 76:246–249

Graczyk TK, Cranfield MR, Fayer R, Trout J, Goodale HJ (2007) Infectivity of *Cryptosporidium parvum* oocysts is retained upon intestinal passage through a migratory waterfowl species (Canada goose, *Branta canadensis*). Tropical Med Int Health 2:341–347

Graczyk TK, Majewska AC, Schwab KJ (2008) The role of birds in dissemination of human waterborne enteropathogens. Trends Parasitol 24(2):55–59

Greiner EC, Ritchie BW (1994) Parasites. In: Ritchie BW, Harrison GJ, Harrison LR (eds) Avian medicine: principles and application, 1st edn. Wingers Publishing, Florida, pp 1007–1029

Gutiérrez AMQ (2017) Giardiasis epidemiology. In: Rodriguez-Morales AJ (ed) Current topics in giardiasis. IntechOpen. https://doi.org/10.5772/intechopen.70338

Hunter PR, Thompson RCA (2005) The zoonotic transmission of *Giardia* and *Cryptosporidium*. Int J Parasitol 35:1181–1190

Ichikawa RS, Santana BN, Ferrari ED, do Nascimento IG, Nakamura AA, Nardi ARM, Meireles MV (2019) Detection and molecular characterization of *Giardia* spp. in captive Psittaciformes in Brazil. Prev Vet Med 164:10–12

Inpankaew T, Traub R, Thompson RCA, Sukthana Y (2007) Canine parasitic zoonoses and temple communities in Thailand. Southeast Asian J Trop Med Public Health 38:247–255

Islam AS (1990) Giardiasis in developing countries. Giardiasis:235–266

Itagaki T, Kinoshita S, Aoki M, Itoh N, Saeki H, Sato N, Uetsuki J, Izumiyama S, Yagita K, Endo T (2005) Genotyping of *Giardia intestinalis* from domestic and wild animals in Japan using glutamate dehydrogenase gene sequencing. Vet Parasitol 133:283–287

Kassa H, Harrington BJ, Bisesi MS (2001) Risk of occupational exposure to *Cryptosporidium*, *Giardia*, and *Campylobacter* associated with the feces of giant Canada geese. Appl Occup Environ Hyg 16:905–909

Lalle M, Jiminez-Cardosa E, Caccio SM, Pozio E (2005) Genotyping of *Giardia duodenalis* from humans and dogs from Mexico using a beta-giardin nested polymerase chain reaction assay. J Parasitol 91:203–205

Leonhard S, Pfister K, Beelitz P, Wielinga C, Thompson RCA (2007) The molecular characterisation of *Giardia* from dogs in Southern Germany. Vet Parasitol 150:33–38

Lukeš J et al (2015) Are human intestinal eukaryotes beneficial or commensals? PLoS Pathog 11: e1005039

Majewska AC, Taddeus KG, Słodkowicz-Kowalska A, Tamang L, Jędrzejewski S, Zduniak P, Solarczyk P, Nowosad A, Nowosad P (2009) The role of free-ranging, captive, and domestic birds of Western Poland in environmental contamination with *Cryptosporidium parvum* oocysts and *Giardia lamblia* cysts. Parasitol Res 104:1093–1099

Meyer EA (1994) *Giardia* as an organism. In: Giardia: from molecules to disease. pp 3–13

Miyamoto Y, Eckmann L (2015) Drug development against the major diarrhea-causing parasites of the small intestine, *Cryptosporidium* and *Giardia*. Front Microbiol 6:1208

Monis PT, Thompson RCA (2003) *Cryptosporidium* and *Giardia*-zoonoses: fact or fiction? Infect Genet Evol 3:233–244

Plutzer J, Tomor B (2009) The role of aquatic birds in the environmental dissemination of human pathogenic *Giardia duodenalis* cysts and *Cryptosporidium* oocysts in Hungary. Parasitol Int 58(3):227–231

Reboredo-Fernández A, Ares-Mazás E, Cacciò SM, Gómez-Couso H (2015) Occurrence of *Giardia* and *Cryptosporidium* in wild birds in Galicia (Northwest Spain). Parasitology 142(7):917–925

Rose JB, Haas CN, Regli S (1991) Risk assessment and control of waterborne giardiasis. Am J Public Health 81(6):709–713

Ryan U, Caccio SM (2013) Zoonotic potential of *Giardia*. Int J Parasitol 43:943–956

Schmid-Hempel P (2009) Immune defence, parasite evasion strategies and their relevance for 'macroscopic phenomena' such as virulence. Philos Trans R Soc Lond Ser B Biol Sci 364: 85–98

Soliman A, Aufy S, Ezzat Moussa H, Saber M, Zaki M, El Akkad D (2016) A broad spectrum antiparasitic effect of nitazoxanide: an important advancement. Kasr Al Ainy Med J 22:56–62

Traub RJ, Monis PT, Robertson I, Irwin P, Mencke N, Thompson RCA (2004) Epidemiological and molecular evidence supports the zoonotic transmission of *Giardia* among humans and dogs living in the same community. Parasitology 128:258–262

U.S. Centers for Disease Control and Prevention (2011) Parasites—*Giardia* [WWW Document]. http://www.cdc.gov/parasites/giardia/

van de Bosch DA (1991) Prevalentie van Giardia lamblia in peuterspeelzalen en kinderdagverblijven in Tilburg. Inf Dent 2(8):2–6

Vasilopulos RJ, Rickard LG, Mackin AJ, Pharr GT, Huston CL (2007) Genotypic analysis of *Giardia duodenalis* in domestic cats. J Vet Intern Med 21:352–355

Volotão AC, Costa-Macedo LM, Haddad FSM, Brandão A, Peralta JM, Fernandes O (2007) Genotyping of Giardia duodenalis from human and animal samples from Brazil using β-giardin gene: a phylogenetic analysis. Acta Trop 102:10–19

Warren KS (1989) Selective primary health care and parasitic diseases. In: New strategies in parasitology. Churchill Livingstone, Edinburgh, pp 217–231

Wu Z et al (2014) Intra-subtype variation in enteroadhesion accounts for differences in epithelial barrier disruption and is associated with metronidazole resistance in Blastocystis subtype-7. PLoS Negl Trop Dis 8:e2885

Yoder JS, Harral C, Beach MJ (2010) Cryptosporidiosis surveillance 2006–2008 and giardiasis surveillance—United States, 2006–2008. Morb Mortal Wkly Rep 59:15–24

Zygner W, Jaros D, Skowronska M, Bogdanowicz-Kamirska M, Wedrychowicz H (2006) Prevalence of *Giardia intestinalis* in domestic dogs in Warsaw. Wiadomosci Parasitologica 52:311–315

Chapter 19
Role of Birds in Tick-Borne Diseases

Abstract Ticks are considered as the second most potential source of vector-borne diseases to humans. Migratory birds are long-distance transporters of ticks and have been accounted for carrying different human pathogens such as tick-borne encephalitis virus, Crimean Congo Hemorrhagic Fever (CCHF) virus, *Anaplasma marginale, Babesia divergens, Anaplasma phagocytophilium, Ehrichia,* and *Borrelia burgdorferi*. The majority of the cases of human parasitism are identified with hard ticks compared to soft ticks. *Ixodes persulcatus* and *Ixodes ricinus* are the most important vectors for tick-borne pathogens in Asia and Europe, respectively. CCHF is a tick-borne zoonotic viral endemic in Asia, Africa, the Middle East, and Eastern Europe. CCHF virus was identified around 31 species of ticks in seven genera of the Ixodidae family. Amid the different genera of Ixodidae, the most proficient and basic vectors for CCHF are the members of the genus *Hyalomma* in which transovarial and transstadial transmission of infection can be seen. CCHF causes no disease in animals but causes a hemorrhagic disease in humans. *Dermanyssus gallinae*, also known as the Poultry Red Mite, is a hematophagous parasite that infects many bird species and is associated with the transmission of various poultry pathogens, including zoonotic pathogens like *Salmonella enteritidis, Borrelia burgerdorferi*, and Avian Influenza virus. Birds also serve as vectors for *Trichobilharzia szidati*, a lung fluke which usually lodges in the lungs of birds and can cause severe parasitic pneumonia, followed by lymphatic lesions and additionally death of the animal in extreme cases. The most commonly used serologic tests for TBD diagnoses are enzyme-linked immunosorbent assay (ELISA), indirect immunofluorescent assay (IFA), and western blot. Microscopy and PCR offer good choices and use of immunodominant epitopes can improve protein-based diagnostic methods. Treatment modalities, such as doxycline, are available for bacterial and parasitic infections are no specific antiviral treatment available but not for viral infections.

Keywords Tick-borne diseases · Migratory birds · Hard ticks · Zoonoses · Vector-borne · Transmission

Y. S. Malik et al., *Role of Birds in Transmitting Zoonotic Pathogens*, Livestock Diseases and Management, https://doi.org/10.1007/978-981-16-4554-9_19

19.1 Introduction

Ticks are responsible for the highest incidence of transmission of pathogens among the arthropod vectors and are considered second to mosquitoes as a potential source of vector-borne diseases to humans (Kuo et al. 2017). The primary hosts for immature ticks are the small mammals (Horak et al. 2002); however, many studies have given importance to the birds as hosts and the role of migratory birds in the transmission of ticks and tick-borne pathogens (Hasle 2013; Loss et al. 2016). Birds fly over long distances throughout a couple of days, especially amid their seasonal migration. Also, referred for their role in epidemiology of ticks, inferring that tick-borne pathogens present in the ticks on the avian hosts are distributed to far off places (Hornok et al. 2014). Even though birds are important for the survival of a few ticks and tick-borne pathogens, their significance varies in different species (Kuo et al. 2017). For instance, birds scrounging principally on the ground are bound to acquire ticks than those scavenging in bushes and trees (Mitra et al. 2010). The frequency of ticks on birds fluctuates between location, years, season, and distinctive species of birds. The frequency of ticks on various birds depends chiefly on the level of the ground feeding (Hasle 2013).

19.2 Epizootiology and Mode of Transmission

Tick-borne pathogens mostly spread biologically, which implies that these pathogens affect their tick vector, where they multiply as well as develop before their transmission to another vertebrate host (Hornok et al. 2014; Stich et al. 2008). Typically, they have a life cycle of four stages: egg, larva, nymph, and adult, and the life cycle from egg to egg can take up to 2 years to complete. In every cycle, at least one blood feeding is required from vertebrates by the larva and nymph to mount into the next stage of the life cycle (Franke et al. 2010). Tick-borne pathogens in the adult female ticks might have the capacity to access another host, on the off chance that they go over the ovaries of the female ticks in the next generation. The propagation of tick-borne pathogens includes tick's vector and hosts from the vertebrate. Peoples living in normal foci of tick-borne pathogens with zoonotic potential can end up infected. *Ixodes persulcatus* and *Ixodes ricinus* are the most important vectors for tick-borne pathogens in Asia and Europe, respectively (Geller et al. 2013). As most of the tick-borne pathogens can survive transstadially, that is, from the larval stage to nymphs, or from nymphs to adults, the birds can carry ticks infected by such pathogens as tick-borne encephalitis virus (Waldenström et al. 2007), *Anaplasma marginale* (Stich et al. 1989), *Babesia divergens* (Bonnet et al. 2007), and *Anaplasma phagocytophilum* (Franke et al. 2010).

19.3 Tick-Borne Diseases

Tick-borne diseases (TBDs) are significant to both medical and veterinary clinical fields. The range of TBDs influencing domestic animals and humans has expanded recently; numerous vital zoonotic TBDs, for example, babesiosis, anaplasmosis, ehrlichiosis, and Lyme borreliosis, are progressively having more consideration from doctors and veterinarians. The list of potential tick-borne pathogens keeps on expanding with the advancement in the molecular biology, as new strains, species, or genetic variants of microbes are being identified in ticks worldwide (Duh et al. 2010; Pacheco et al. 2011; Subramanian et al. 2012; Dantas-Torres et al. 2012). The known species of ticks are grouped into three families: Argasidae, Ixodidae, and Nuttalliellidae (Apanaskevich et al. 2011; Guglielmone et al. 2010; Dantas-Torres et al. 2012). A portion of these agents, for example, *Rickettsia slovaca*, *R. parkeri*, and *R. massiliae* were recognized in ticks, years before these were related to human sickness diseases (Paddock 2009). While others like flaviviruses (for example, Kyasanur forest disease virus, Omsk hemorrhagic fever virus, and Powassan encephalitis virus) have been associated with human disease in new geological areas (Piesman and Eisen 2008; Dobler 2010).

The majority of the cases of human parasitism are identified with hard ticks. However, soft ticks' infestations are also reported, which may prompt extreme medical complications in the persons bitten by them. For example, many *Ornithodoros* spp. are related to the epidemiology of TBDs (like relapsing fever) affecting humans (Estrada-Peña and Jongejan 1999; Charrel et al. 2007); likewise, the pigeon tick (Weckesser et al. 2010). Ticks belonging to the *Ornithodoros* are perceived to cause systemic illness, and local lesions called tick toxicosis. Recently Yadav et al. (2018) identified and characterized a tick-borne virus named Wad Medani virus (WMV) isolated from the tick *Hyalomma marginatum* from Maharashtra State, India, using the next-generation sequencing.

19.3.1 Crimean Congo Hemorrhagic Fever (CCHF)

CCHF is one of the dangerous hemorrhagic fevers which are endemic in Asia, Africa, the Middle East, and Eastern Europe. CCHF is a tick-borne zoonotic viral infection brought about by the CCHF virus (CCHFV). CCHFV belonging to the genus *Nairovirus* (family *Bunyaviridae*) is an RNA virus. It is transferred to people by tick bites or through contact with blood or tissues from tainted ticks, animals, or humans. Humans are tainted by the bite of the ticks or by pounding an infected tick of the *Hyalomma* spp., on exposed skin. The contamination can likewise be obtained by permucosal and percutaneous path by contact of creature blood or tissues and by drinking unpasteurized milk. The likelihood of airborne transmission is supposed in a few cases in Russia; however, no evident proof exists (Watts et al. 1988). Human-to-human transmission is conceivable and is essential in the nosocomial setup when mucous membranes or skin is presented to the blood and liquids from the patients with discharge. Moreover, the possible horizontal transmission has been accounted for from a mother to a child.

CCHFV has an extensive geographic transmission as any other tick-borne infection, including about 30 nations from eastern China through Asia, the Middle East, and southeastern Europe to Africa (Ergönül 2006; Hoogstraal 1979). Amid the previous decade, the infection has risen in new territories of Europe, Africa, the Middle East, and Asia and has expanded in illness endemic regions (Leblebicioglu 2010). India was constantly under the potential risk of CCHF viral disease until a flare-up hit in some parts of Gujarat. As of 2021, the status of CCHFV in India is still obscure. A large number of wild and domestic animals may also get CCHFV infection. The infection has been usually found in smaller wild animal species, for example, hedgehogs and bunnies, which harbor the tick vectors that are at juvenile stages (Hoogstraal 1979; Watts et al. 1988). These little animals are supposed to act as amplifying hosts and keep up the virus in nature. In any case, there exists no proof of the virus causing infection in animals. CCHFV has been found in around 31 species of ticks in seven genera of the Ixodidae family. Amid the different genera of Ixodidae, the most proficient and basic vectors for CCHF are the members of the genus *Hyalomma*. Other ixodid ticks from the genera *Boophilus*, *Dermacentor*, *Rhipicephalus*, and *Ixodes* may likewise transmit the virus. These vectors have both transovarial and transstadial transmission of infection, in this way adding to the course of the disease in nature by staying affected all through their formative stages, furthermore passing to the next generation. Ticks belonging to the *H. marginatum* are common in many parts of the Eurasian and African landmasses. The juvenile ticks feed for the most part on birds, and to a lesser degree, on small mammals, though the grown-ups effectively look for bigger well-evolved animals, including bunnies, wild and trained ungulates, or humans (Hoogstraal 1979).

Humans are the primary host of CCHFV where disease signs are clearly visible. The CCHF infection has four different stages: incubation period, pre-hemorrhagic stage, hemorrhagic stage, and convalescent stage. The incubation time for CCHF infection varies between 3 and 7 days. The mean span is, to a great extent, impacted by the route of contamination, viral burden, and source of infected blood or tissue from livestock. The base viral load required for the transmission of infection is 1–10 organisms (Franz et al. 1997). The sickness starts with the pre-hemorrhagic stage described by nonspecific prodromal indications amid which it impersonates other viral ailments. The main symptoms are high fever, myalgia, migraine, sickness, stomach torment, and non-bloody diarrhea. This is joined by relative bradycardia, hypotension, tachypnea, pharyngitis, conjunctivitis, and cutaneous flushing or rash (Hoogstraal 1979). The pre-hemorrhagic stage goes on for 4–5 days, and in a more significant part of the patients, it advances to the hemorrhagic stage. The hemorrhagic stage is commonly short and has a fast course with indications of dynamic drain and diathesis. These incorporate petechiae, epistaxis, conjunctival discharge, hematemesis, melena, and hemoptysis. Some patients may likewise have hepatosplenomegaly (Bakir et al. 2005; Ergonul et al. 2004; Ozkurt et al. 2006). The infection is deadly in 40–60% of the cases (Sannikova et al. 2007; Doganci et al.

2008). In the survivors, the healing time frame starts 10–20 days after the beginning of the disease. During this stage, patients may have a weak heartbeat, loss of hearing, tachycardia, and loss of hair and memory. However, these delayed consequences have been accounted for just in few outbreaks (Hoogstraal 1979; Schwarz et al. 1997; Karti et al. 2004). Migratory birds are long-distance transporters of ticks and have been accounted for containing different human pathogens (Elfving et al. 2010). Pre-adult ticks can remain attached to bird hosts amid movement, after that separating at breeding or stopover locales, where mammalian hosts can conceivably set up new foci (Hoogstraal 1979). Concerning findings in Spain (Estrada-Peña et al. 2012), one could speculate that new cycles of CCHFV transmission could be started through viremic or nonviremic co-feeding systems, including, for instance, transstadially infected adult *H. rufipes* ticks and susceptible *H. lusitanicum* ticks that are benefiting from a same mammalian host.

19.3.2 Dermanyssus gallinae

Dermanyssus gallinae, also known as the Poultry Red Mite, is the most crippling blood-feeding arthropod attacking birds and mammals. *D. gallinae* is a hematophagous parasite that infects many bird species. From an economic perspective, it is among the most vital ectoparasites impacting egg-laying hens in numerous parts of the world (Wang et al. 2018; Chauve 1998). Extreme pervasions of *D. gallinae* in egg-laying units can result in restlessness, irritation, anemia, diminished egg generation and quality, and even demise of birds (Kirkwood 1967).

Moreover, *D. gallinae* invasions have caused dermatitis in poultry laborers and occupants nearby poultry houses (Rosen et al. 2002). Adult and nymph mites feed on the hens for short time frames (1.5–2.0 h) at night; however, they invest the majority of their time hiding in cracks of roosting and nesting sites. Their area of the host and the resulting migratory behavior have prompted troubles in the control of *D. gallinae*.

The first source of concerns related to red mite invasion is the very high and expanding predominance of this sickness in Europe. A current epidemiological review reports that 83% of the European ranches are infected by *D. gallinae*. This frequency achieves 94% in The Netherlands, Germany, and Belgium (Mul 2013). Poultry red mite invasion influences all product types, from the backyard or organic farms, to more intensive, enriched cage or barn systems (Sparagano et al. 2014). Apart from the high prevalence of the disease, another concern is the seriousness of the impacts caused by *D. gallinae* parasitism on the birds' well-being and welfare. The first clinical sign in infected animals is subacute anemia because of rehashed mite bites. A laying hen can lose over 3% of its blood volume consistently (van Emous 2005). In extraordinary cases, *D. gallinae* invasion might be heavy to the point that hens may die from severe anemia (Cosoroaba 2001; Wojcik et al. 2000; Pilarczyk et al. 2004). *D. gallinae* is additionally a possible vector of viral and bacterial disease agents with *Borrelia anserine* (Pritchard et al. 2015), *Erysipelothrix rhusiopathiae* (Chirico et al. 2003), *Salmonella enterica* (Hamidi et al. 2011), *Pasteurella multocida*, eastern equine encephalomyelitis virus (Sparagano et al. 2014), and Fowl poxvirus (Durden et al. 1992).

The mites in the production house initiate high stress in the birds. Stress is prompted by skin irritation and pain related to rehashed mite bites supported by the high parasite load regular of red mite pervasions, with densities running from 25,000 to 500,000 mites for every hen (Mul 2013; van Emous 2005; Kilpinen et al. 2005). Also, mite pervasions initiate forceful feather pecking and savage conduct, expanded water and feed intake, and reduction of general animal health (van Emous 2005; Chauve 1998; Kilpinen et al. 2005; Mul 2009). *D. gallinae* is associated with the transmission of various poultry pathogens, including zoonotic pathogens like *Salmonella enteritidis* (Valiente et al. 2007; Valiente et al. 2009, 2010; Mul 2009), which causes one of the most widespread zoonoses around the world, non-typhoidal salmonellosis. This malady has the most noteworthy worldwide human death rate of any zoonotic illness, with most cases being of foodborne origin, and poultry items being among the most widely recognized source of the infection (Valiente et al. 2009; Majowicz et al. 2010). *Borrelia burgdorferi*, which causes avian influenza A virus and Lyme ailment, is recently added to the rundown of zoonotic pathogens conceivably transmitted by *D. gallinae* (George et al. 2015; Sommer et al. 2016).

19.4 Diagnosis, Control, and Treatment

Tick-borne diseases (TBDs) are the most widely recognized vector-borne diseases (Adams et al. 2015). Recently, as of 2017, 19 protozoans, bacterial, and viral agents have been associated with TBDs (TBDs 2017). *Borrelia burgdorferi*, the causal agent of Lyme ailment, records for an expected 300,000 yearly instances of TBDs (Burgdorfer et al. 1982; Nelson et al. 2015; Hinckley et al. 2014). The most commonly used serologic tests for TBD diagnoses are enzyme-linked immunosorbent assay (ELISA), indirect immunofluorescent assay (IFA), and western blot (Connally et al. 2016; Theel 2016; Wormser et al. 2006). The analytic exactness of these tests might be influenced by different intrinsic restrictions that can hinder actual diagnosis. For Lyme's illness, the prescribed technique for determination is a two-tiered testing calculation comprising of an ELISA pursued by supplemental western blots. Even though this strategy is specific and sensitive in dispersed disease, its utility in the early period of the malady is restricted. It precisely identifies <40% of patients with initial sickness and can result in up to 28% of IgM western blots giving false-positive results (Aguero-Rosenfeld et al. 2005; Seriburi et al. 2012). This technique is additionally labor-intensive and utilizes subjective criteria for the elucidation of consequences that may affect reproducibility (Fallon et al. 2014; Ang et al. 2011). IFA is suggested for analysis of tick-borne diseases with Babesia, Anaplasma, Ehrlichia, and Rickettsia. The precision of IFAs can change broadly among testing research facilities principally because of the absence of standardized antigenic targets, cross-reactivity, and the subjective establishment of positive thresholds (Biggs et al. 2016). For other tick-borne diseases, specific serologic tests are still not accessible, as in the case of Heartland virus or Powassan virus, they are just performed in specific laboratories (Pastula et al. 2014; Piantadosi et al. 2016). An extra test is the absence of measures that, at the same time, test for an introduction to the full range of tick-borne agents. This is an essential impediment in TBD diagnosis in light of observational studies that demonstrate ticks can harbor more than one human pathogen (Aliota et al. 2014; Prusinski et al. 2014; Tokarz et al. 2017).

An improvement in the methodology for settling the inherent lack of specificity related with protein tests is the utilization of synthetic peptides or recombinant antigens that show immunodominant epitopes (Dessau et al. 2015; Signorino et al. 2014; Arnaboldi et al. 2013; Nayak et al. 2016; Molins et al. 2017). This methodology increases assay specificity by taking out nonspecific and possibly cross-responsive epitopes. Focusing on a combination of such antigens increases assay sensitivity and has been appeared to enhance the analysis of early Lyme disease (Embers et al. 2016). The TBD-Serochip, another procedure, offers a significant improvement in target determination by utilizing a broad range of linear peptides that recognize critical specific immunodominant epitopes (Tokarz et al. 2018). Microscopy and PCR assay of blood samples are employed to analyze active HGA and babesiosis.

In many different multiple trials, the adequacy of oral doxycycline, cefuroxime axetil, and amoxicillin for treating Lyme sickness has been set up. In the parenteral antibiotic treatment process, ceftriaxone is the suggested antibiotic. Numerous trials for treating erythema migrans have appeared for a 10-day course of oral doxycycline and a 14-day course for treating early neurologic Lyme sickness in mobile patients. Proof demonstrates that for HGA a 10-day course of oral doxycycline is convincing, and a 7- to 10-day course of azithromycin in addition to atovaquone is viable for mild babesiosis. From the reports of numerous cases, in addition to quinine, a 7- to 10-day course of clindamycin is regularly used to treat severe babesiosis. An ongoing report supports at least a month and a half of antibiotics for extremely immunocompromised patients with babesiosis, without any parasites identified on blood smear for at least the last 2 weeks of treatment (Sanchez et al. 2016).

There is no particular antiviral treatment for tick-borne encephalitis (TBE). Patients when in doubt need hospitalization and strong consideration depending on the seriousness of signs/symptoms, and more often involves the use of analgesics, antipyretics, antiemetics, upkeep of electrolyte and water balance, and if necessary administration of anticonvulsive agents. In patients with neuromuscular loss of motion prompting respiratory failure, intubation, and ventilatory help is important. In an extensive study in Germany, 12% of patients were treated in the emergency unit, and in 5% ventilation was required (Kaiser 1999). Among the TBE patients hospitalized at the University Medical Center Ljubljana, in Slovenia, during the years 2000–2004, 6.9% were admitted in ICU, and 22.5% of them required mechanical ventilation (Logar et al. 2006; Bogovic and Strle 2015).

19.5 Role of Birds as Vectors of Fluke Worms (*Trichobilharzia szidati*)

Parasitic flatworms belonging to the family Schistosomatidae have complex biology and life history and usually are critical pathogens. The life cycles include aquatic snails as intermediate hosts, where asexual propagation occurs, and vertebrates, for example, birds and mammals as definitive hosts, where sexual proliferation takes place (Warren 1982; Horák et al. 2002; Colley et al. 2014). Individuals from the *Schistosoma* genus infect people and different mammals as final hosts. In humans, *Schistosoma mansoni*, *S. haematobium*, and *S. japonicum* are the three dominating species and are huge pathogens in their very own right. They influence around 230 million individuals globally and cause hepato-intestinal or urogenital diseases, jointly known as schistosomiasis (Rollinson et al. 2009; Leontovyč et al. 2019).

Bird schistosomes belonging to the genus *Trichobilharzia* have a two-host life cycle, just as the human schistosomes. Species from *Trichobilharzia* commonly utilizes pulmonate snails as middle host and fowls as the final hosts. Snails develop infection by the entrance of ciliated larvae. After asexual reproduction, cercariae are formed (Horák et al. 2002). They are discharged into the water and infect the final hosts by entering their skin. At that point, the larvae change to schistosomula and move through the circulatory or nervous systems of the host body to the target organ where egg-laying happens. Schistosomes from the genus *Trichobilharzia* infect birds as their characteristic last host. However, they can likewise infect mammals as unplanned (anomalous) host. Movement and development are finished just inside the natural host. However, the initial process of migration to the host body additionally happens inside the unusual hosts (Horák et al. 2002), and different mammals (hamsters, mice, rabbits, guinea pigs, rhesus monkeys) have been utilized as lab models (Olivier 1953). From the skin, schistosomula of instinctive species of *Trichobilharzia* enter the veins of the host where they begin vascular migration (Bourns et al. 1973). The primary spot of accumulation of these larvae is the lungs of the birds just as in mammalian hosts (Haas and Pietsch 1991). The parasites go through the lungs of birds, and afterward, they proceed with relocation using veins to the spot of final destination (Haas and Pietsch 1991; Horák and Kolářová 2000).

In the lungs, schistosomules can cause severe parasitic pneumonia, followed by lymphatic lesions and an invasion of macrophages, eosinophils, and heterophils, and additionally death of the animal in extreme cases (Chanová et al. 2007). In the host, *T. szidati* schistosomules dodge the immune reaction of the host and ingest blood to secure nutrients (Kašný et al. 2011). Yet, the molecular basis for nutrient obtaining and their survival is unclear. Most of the molecular biological studies generally concentrated on human schistosomes (Gobert et al. 2014; Rinaldi et al. 2015; Hahnel et al. 2018). Still, there are no genomic sequences available for *Trichobilharzia* species (except *T. regenti* (parasite.wormbase.org)), and also the information on transcriptomic and proteomics is scarce.

References

Adams D et al (2015) Summary of notifiable infectious diseases and conditions—United States, 2013. MMWR Morb Mortal Wkly Rep 62:1–122. https://doi.org/10.15585/mmwr.mm6253a1

Aguero-Rosenfeld ME, Wang G, Schwartz I, Wormser GP (2005) Diagnosis of Lyme borreliosis. Clin Microbiol Rev 18:484–509. https://doi.org/10.1128/CMR.18.3.484-509.2005

Aliota MT et al (2014) The prevalence of zoonotic tick-borne pathogens in *Ixodes scapularis* collected in the Hudson Valley, New York State. Vector Borne Zoonotic Dis 14:245–250. https://doi.org/10.1089/vbz.2013.1475

Ang CW, Notermans DW, Hommes M, Simoons-Smit AM, Herremans T (2011) Large differences between test strategies for the detection of anti-Borrelia antibodies are revealed by comparing eight ELISAs and five immunoblots. Eur J Clin Microbiol Infect Dis 30:1027–1032. https://doi.org/10.1007/s10096-011-1157-6

Apanaskevich DA et al (2011) A new species of Ixodes (Acari: Ixodidae) from South African mammals. J Parasitol 97:389–398

Arnaboldi PM et al (2013) Outer surface protein C peptide derived from *Borrelia burgdorferi sensu stricto* as a target for serodiagnosis of early Lyme disease. Clin Vaccine Immunol 20:474–481. https://doi.org/10.1128/CVI.00608-12

Bakir M, Ugurlu M, Dokuzoguz B, Bodur H, Tasyaran MA, Vahaboglu H et al (2005) Crimean-Congo haemorrhagic lever outbreak in Middle Anatolia: a multicentre study of clinical features and outcome measures. J Med Microbiol 54:385–389

Biggs HM et al (2016) Diagnosis and management of tickborne rickettsial diseases: Rocky Mountain spotted fever and other spotted fever group rickettsioses, ehrlichioses, and anaplasmosis—United States. MMWR Recomm Rep 65:1–44. https://doi.org/10.15585/mmwr.rr6502a1

Bogovic P, Strle F (2015) Tick-borne encephalitis: a review of epidemiology, clinical characteristics, and management. World J Clin Cases 3(5):430–441

Bonnet S, Jouglin M, Malandrin L, Becker C, Agoulon A, L'Hostis M, Chauvin A (2007) Transstadial and transovarial persistence of *Babesia divergens* DNA in *Ixodes ricinus* ticks fed on infected blood in a new skin-feeding technique. Parasitology 134:197–207

Bourns TKR, Ellis JC, Rau ME (1973) Migration and development of *Trichobilharzia ocellata* (Trematoda: Schistosomatidae) in its duck hosts. Can J Zool 51:1021–1030

Burgdorfer W et al (1982) Lyme disease—a tick-borne spirochetosis? Science 216:1317–1319

Chanová M, Vuong S, Horák P (2007) *Trichobilharzia szidati*: the lung phase of migration within avian and mammalian hosts. Parasitol Res 100:1243–1247

Charrel RN et al (2007) Alkhurma hemorrhagic fever virus in *Ornithodoros savignyi* ticks. Emerg Infect Dis 13:153–155

Chauve C (1998) The poultry red mite *Dermanyssus gallinae* (De Geer, 1778): current situation and future prospects for control. Vet Parasitol 79:239–245

Chirico J, Eriksson H, Fossum O, Jansson D (2003) The poultry red mite, *Dermanyssus gallinae*, a potential vector of *Erysipelothrix rhusiopathiae* causing erysipelas in hens. Med Vet Entomol 17:232–234

Colley DG, Bustinduy AL, Secor WE, King CH (2014) Human schistosomiasis. Lancet 383:2253–2264

Connally NP et al (2016) Testing practices and volume of non-Lyme tickborne diseases in the United States. Ticks Tick Borne Dis 7:193–198. https://doi.org/10.1016/j.ttbdis.2015.10.005

Cosoroaba I (2001) Massive *Dermanyssus gallinae* invasion in battery-husbandry raised fowls. Revue Méd Vét 152:89–96

Dantas-Torres F et al (2012) Description of a new species of bat-associated argasid tick (Acari: Argasidae) from Brazil. J Parasitol 98:36–45

Dessau RB, Moller JK, Kolmos B, Henningsson AJ (2015) Multiplex assay (Mikrogen recomBead) for detection of serum IgG and IgM antibodies to 13 recombinant antigens of *Borrelia*

burgdorferi sensu lato in patients with neuroborreliosis: the more the better? J Med Microbiol 64:224–231. https://doi.org/10.1099/jmm.0.000009
Dobler G (2010) Zoonotic tick-borne flaviviruses. Vet Microbiol 140:221–228
Doganci L, Ceyhan M, Tasdeler NF, Sarikayalar H, Tulek N (2008) Crimean-Congo hemorrhagic fever and diffuse alveolar haemorrhage. Trop Dr 38:252–254
Duh D et al (2010) *Rickettsia hoogstraalii* sp. nov., isolated from hard- and soft-bodied ticks. Int J Syst Evol Microbiol 60:977–984
Durden LA, Linthicum KJ, Turell MJ (1992) Mechanical transmission of Venezuelan equine encephalomyelitis virus by hematophagous mites (Acari). J Med Entomol 29:118–121
Elfving K, Olsen B, Bergström S, Waldenström J, Lundkvist Å, Sjöstedt A et al (2010) Dissemination of spotted fever rickettsia agents in Europe by migrating birds. PLoS One 5:e8572. https://doi.org/10.1371/journal.pone.0008572
Embers ME et al (2016) Five-antigen fluorescent bead-based assay for diagnosis of Lyme disease. Clin Vaccine Immunol 23:294–303. https://doi.org/10.1128/CVI.00685-15
Ergönül O (2006) Crimean-Congo haemorrhagic fever. Lancet Infect Dis 6:203–214
Ergonul O, Celikbas A, Dokuzoguz B, Eren S, Baykam N, Esener H (2004) Characteristics of patients with Crimean-Congo hemorrhagic fever in a recent outbreak in turkey and impact of oral ribavirin therapy. Clin Infect Dis 39:284–287
Estrada-Peña A, Jongejan F (1999) Ticks feeding on humans: a review of records on human-biting Ixodoidea with special reference to pathogen transmission. Exp Appl Acarol 23:685–715
Estrada-Peña A, Palomar AN, Santibáñez P, Sánchez N, Habela MA, Portillo A et al (2012) Crimean-Congo hemorrhagic fever virus in ticks, southwestern Europe, 2010. Emerg Infect Dis 18:179–180. https://doi.org/10.3201/eid1801.111040
Fallon BA, Pavlicova M, Coffino SW, Brenner CA (2014) Comparison of Lyme disease serologic test results from 4 laboratories in patients with persistent symptoms after antibiotic treatment. Clin Infect Dis 59:1705–1710
Franke J, Fritzsch J, Tomaso H, Straube E, Dorn W, Hildebrandt A (2010) Coexistence of pathogens in host-seeking and feeding ticks within a single natural habitat in Central Germany. Appl Environ Microbiol 76:6829–6836
Franz DR, Jahrling PB, Friedlander AM, McClain DJ, Hoover DL, Bryne WR et al (1997) Clinical recognition and management of patients exposed to biological warfare agents. JAMA 278:399–411
Geller J, Nazarova L, Katargina O, Leivits A, Järvekülg L, Golovljova I (2013) Tick-borne pathogens in ticks feeding on migratory passerines in Western part of Estonia. Vector-Borne Zoonotic Dis 13:443–448
George DR, Finn RD, Graham KM, Mul MF, Maurer V, Valiente Moro C, Sparagano OA (2015) Of mites and men: should the poultry red mite *Dermanyssus gallinae* be of wider concern for medical science? In: 1st COST Conference and Management Committee Meeting Proceedings, pp 27–28
Gobert GN, You H, McManus DP (2014) Gaining biological perspectives from schistosome genomes. Mol Biochem Parasitol 196:21–28
Guglielmone AA et al (2010) The Argasidae, Ixodidae and Nuttalliellidae (Acari: Ixodida) of the world: a list of valid species names. Zootaxa 2528:1–28
Haas W, Pietsch U (1991) Migration of *Trichobilharzia ocellata schistosomula* in the duck and in the abnormal murine host. Parasitol Res 77:642–644
Hahnel S et al (2018) Tissue-specific transcriptome analyses provide new insights into GPCR signalling in adult *Schistosoma mansoni*. PLoS Pathog 14:e1006718
Hamidi A, Sherifi K, Muji S, Behluli B, Latifi F, Robaj A, Postoli R, Hess C, Hess M, Sparagano O (2011) *Dermanyssus gallinae* in layer farms in Kosovo: a high risk for Salmonella prevalence. Parasit Vectors 4:136
Hasle G (2013) Transport of ixodid ticks and tick-borne pathogens by migratory birds. Front Cell Infect Microbiol 3:48

Hinckley AF et al (2014) Lyme disease testing by large commercial laboratories in the United States. Clin Infect Dis 59:676–681. https://doi.org/10.1093/cid/ciu397

Hoogstraal H (1979) The epidemiology of tick-borne Crimean-Congo hemorrhagic fever in Asia, Europe, and Africa. J Med Entomol 15:307–417

Horák P, Kolářová L (2000) Survival of bird schistosomes in mammalian lungs. Int J Parasitol 30: 65–68

Horak I, Camicas JL, Keirans J (2002) The Argasidae, Ixodidae and Nuttalliellidae (Acari: Ixodida): a world list of valid tick names. Exp Appl Acarol 28:27–54

Horák P, Kolářová L, Adema CM (2002) Biology of the schistosome genus *Trichobilharzia*. Adv Parasitol 52:155–233

Hornok S, Kováts D, Csörgő T, Meli ML, Gönczi E, Hadnagy Z, Takács N, Farkas R, Hofmann-Lehmann R (2014) Birds as potential reservoirs of tick-borne pathogens: first evidence of bacteraemia with *Rickettsia helvetica*. Parasit Vectors 7:128

Kaiser R (1999) The clinical and epidemiological profile of tick-borne encephalitis in southern Germany 1994-98: a prospective study of 656 patients. Brain 122(Pt 11):2067–2078

Karti SS, Odabasi Z, Korten V, Yilmaz M, Sonmez M, Caylan R et al (2004) Crimean-Congo hemorrhagic fever in Turkey. Emerg Infect Dis 10:1379–1384

Kašný M et al (2011) Cathepsins B1 and B2 of *Trichobilharzia* spp., bird schistosomes causing cercarial dermatitis. Adv Exp Med Biol 712:136–154

Kilpinen O, Roepstorff A, Permin A, Nørgaard-Nielsen G, Lawson LG et al (2005) Influence of *Dermanyssus gallinae* and *Ascaridia galli* infections on behaviour and health of laying hens (*Gallus gallus domesticus*). Br Poult Sci 46:26–34

Kirkwood AC (1967) Anaemia in poultry infested with the red mite *Dermanyssus gallinae*. Vet Rec 80:514–516

Kuo CC, Lin YF, Yao CT, Shih HC, Chung LH, Liao HC, Hsu YC, Wang HC (2017) Tick-borne pathogens in ticks collected from birds in Taiwan. Parasit Vectors 10:587

Leblebicioglu H (2010) Crimean-Congo haemorrhagic fever in Eurasia. Int J Antimicrob Agents 36 (Suppl 1):S43–S46

Leontovyč R, Young ND, Korhonen PK, Hall RS, Bulantová J, Jeřábková V, Kašný M, Gasser RB, Horák P (2019) Molecular evidence for distinct modes of nutrient acquisition between visceral and neurotropic schistosomes of birds. Sci Rep 9:1347

Logar M, Bogovic P, Cerar D, Avsic-Zupanc T, Strle F (2006) Tick-borne encephalitis in Slovenia from 2000 to 2004: comparison of the course in adult and elderly patients. Wien Klin Wochenschr 118:702–707

Loss SR, Noden BH, Hamer GL, Hamer SAA (2016) Quantitative synthesis of the role of birds in carrying ticks and tick-borne pathogens in North America. Oecologia 182:947–959

Majowicz SE, Musto J, Scallon E, Angulo FJ, Kirk M, O'Brien SJ et al (2010) The global burden on non-typhoidal *Salmonella gastroenteritis*. Clin Infect Dis 50:882–889

Mitra SS, Buckley PA, Buckley FG, Ginsberg HS (2010) Highly variable acquisition rates of *Ixodes scapularis* (Acari: Ixodidae) by birds on an Atlantic Barrier Island. J Med Entomol 47:1019–1027

Molins CR, Delorey MJ, Replogle A, Sexton C, Schriefer ME (2017) Evaluation of bioMerieux's dissociated Vidas Lyme IgM II and IgG II as a first-tier diagnostic assay for Lyme disease. J Clin Microbiol 55:1698–1706. https://doi.org/10.1128/JCM.02407-16

Mul M (2009) Control methods for *Dermanyssus gallinae* in systems for laying hens: results of an international seminar. World Poult Sci J 65:589–599

Mul M (2013) Fact sheet: the poultry red mite, *Dermanyssus gallinae* (De Geer, 1778). A small pest that packs a big punch. https://www.researchgate.net/publication/258553789_Fact_sheet_Poultry_Red_Mite_in_Europe

Nayak S et al (2016) Microfluidics-based point-of-care test for serodiagnosis of Lyme disease. Sci Rep 6:35069. https://doi.org/10.1038/srep35069

Nelson CA et al (2015) Incidence of clinician-diagnosed Lyme disease, United States, 2005-2010. Emerg Infect Dis 21:1625–1631. https://doi.org/10.3201/eid2109.150417

Olivier L (1953) Observations on the migration of avian schistosomes in mammals previously unexposed to cercariae. J Parasitol 39(Suppl 3):237–243

Ozkurt Z, Kiki I, Erol S, Erdem F, Yilmaz N, Parlak M et al (2006) Crimean-Congo hemorrhagic fever in Eastern Turkey: clinical features, risk factors and efficacy of ribavirin therapy. J Infect 52:207–215

Pacheco RC et al (2011) *Rickettsia monteiroi* sp. nov., infecting the tick *Amblyomma incisum* in Brazil. Appl Environ Microbiol 77:5207–5211

Paddock CD (2009) The science and fiction of emerging rickettsioses. Ann N Y Acad Sci 1166: 133–143

Pastula DM et al (2014) Notes from the field: heartland virus disease—United States, 2012-2013. MMWR Morb Mortal Wkly Rep 63:270–271

Piantadosi A et al (2016) Emerging cases of powassan virus encephalitis in New England: clinical presentation, imaging, and review of the literature. Clin Infect Dis 62:707–713. https://doi.org/10.1093/cid/civ1005

Piesman J, Eisen L (2008) Prevention of tick-borne diseases. Annu Rev Entomol 53:323–343

Pilarczyk B, Balicka-Ramisz A, Ramisz A, Pająk B (2004) Wpływ inwazji *Dermanyssus gallinae* na zdrowotność i produkcyjność kur niosek. Med Weter 60:874–876

Pritchard J, Kuster T, Sparagano O, Tomley F (2015) Understanding the biology and control of the poultry red mite *Dermanyssus gallinae*: a review. Avian Pathol 44:143–153

Prusinski MA et al (2014) Prevalence of *Borrelia burgdorferi* (Spirochaetales: Spirochaetaceae), *Anaplasma phagocytophilum* (Rickettsiales: Anaplasmataceae), and *Babesia microti* (Piroplasmida: Babesiidae) in *Ixodes scapularis* (Acari: Ixodidae) collected from recreational lands in the Hudson Valley region, New York State. J Med Entomol 51:226–236

Rinaldi G et al (2015) New research tools for urogenital schistosomiasis. J Infect Dis 211:861–869

Rollinson D et al (2009) Genetic diversity of schistosomes and snails: implications for control. Parasitology 136:1801–1811

Rosen S, Yeruham I, Braverman Y (2002) Dermatitis in humans associated with the mites *Pyemotes tritici, Dermanyssus gallinae, Ornithonyssus bacoti* and *Androlaelaps casalis* in Israel. Med Vet Entomol 16:442–444

Sanchez E, Vannier E, Wormser GP, Hu LT (2016) Diagnosis, treatment, and prevention of Lyme disease, human granulocytic anaplasmosis, and babesiosis: a review. JAMA 315:1767–1777

Sannikova IV, Pacechnikov VD, Maleev VV (2007) Respiratory lesions in Crimean-Congo hemorrhagic fever. Ter Arkh 79:20–23

Schwarz TF, Nsanze H, Ameen AM (1997) Clinical features of Crimean-Congo haemorrhagic fever in the United Arab Emirates. Infection 25:364–367

Seriburi V, Ndukwe N, Chang Z, Cox ME, Wormser GP (2012) High frequency of false positive IgM immunoblots for *Borrelia burgdorferi* in clinical practice. Clin Microbiol Infect 18:1236–1240

Signorino G, Arnaboldi PM, Petzke MM, Dattwyler RJ (2014) Identification of OppA2 linear epitopes as serodiagnostic markers for Lyme disease. Clin Vaccine Immunol 21:704–711. https://doi.org/10.1128/CVI.00792-13

Sommer D, Heffels-Redmann U, Köhler K, Lierz M, Kaleta EF (2016) Role of the poultry red mite (*Dermanyssus gallinae*) in the transmission of avian influenza a virus. Tierärztliche Praxis Grosstiere 44:26–33

Sparagano OA, George DR, Harrington DW, Giangaspero A (2014) Significance and control of the poultry red mite *Dermanyssus gallinae*. Annu Rev Entomol 59:447–466. https://doi.org/10.1146/annurev-ento-011613-162101

Stich RW, Kocan KM, Palmer GH, Ewing SA, Hair JA, Barron SJ (1989) Transstadial and attempted transovarial transmission of *Anaplasma marginale* by *Dermacentor variabilis*. Am J Vet Res 50(8):1377–1380

Stich RW, Schaefer JJ, Bremer WG, Needham GR, Jittapalapong S (2008) Host surveys, ixodid tick biology and transmission scenarios as related to the tick-borne pathogen, *Ehrlichia canis*. Vet Parasitol 158:256–273

Subramanian G et al (2012) *Diplorickettsia massiliensis* as a human pathogen. Eur J Clin Microbiol Infect Dis 31:365–369

Theel ES (2016) The past, present, and (possible) future of serologic testing for Lyme disease. J Clin Microbiol 54:1191–1196. https://doi.org/10.1128/JCM.03394-15

Tickborne diseases of the United States (2017). https://www.cdc.gov/ticks/diseases/. Accessed 10 Aug 2017

Tokarz R et al (2017) Detection of *Anaplasma phagocytophilum*, *Babesia microti*, *Borrelia burgdorferi*, *Borrelia miyamotoi*, and Powassan virus in ticks by a multiplex real-time reverse transcription-PCR assay. mSphere 2. https://doi.org/10.1128/mSphere.00151-17

Tokarz R, Mishra N, Tagliafierro T, Sameroff S, Caciula A, Chauhan L, Patel J, Sullivan E, Gucwa A, Fallon B, Golightly M, Molins C, Schriefer M, Marques A, Briese T, Lipkin WI (2018) A multiplex serologic platform for diagnosis of tick-borne diseases. Sci Rep 8:3158

Valiente MC, Chauve C, Zenner L (2007) Experimental infection of *Salmonella enteritidis* by the poultry red mite, *Dermanyssus gallinae*. Vet Parasitol 146:329–336

Valiente MC, de Luna CJ, Tod A, Guy JH, Sparagano OA, Zenner L (2009) The poultry red mite (*Dermanyssus gallinae*): a potential vector of pathogenic agents. Exp Appl Acarol 48:93–104

Valiente MC, Fravalo P, Amelot M, Chauve C, Salvat G, Zenner L (2010) Experimental studies on the potential role of the poultry red mite *Dermanyssus gallinae*, as a vector of *Salmonella serotype enteritidis*. In: Sabelis MW, Bruin J (eds) Trends Acarology: Proceedings of the 12th International Congress. Springer, pp 521–525

van Emous R (2005) Wage war against the red mite. Poultry Int 44:26–33

Waldenström J, Lundkvist A, Falk KI et al (2007) Migrating birds and tickborne encephalitis virus. Emerg Infect Dis 13(8):1215–1218

Wang C, Ma Y, Huang Y, Xu J, Cai J, Pan B (2018) An efficient rearing system rapidly producing large quantities of poultry red mites, *Dermanyssus gallinae* (Acari: Dermanyssidae), under laboratory conditions. Vet Parasitol 258:38–45

Warren KS (1982) Schistosomiasis: host-pathogen biology. Rev Infect Dis 4:771–775

Watts DM, Ksiazeck TG, Linthicum KJ, Hoogstraal H (1988) Crimean-Congo hemorrhagic fever. In: Monath TP (ed) The arboviruses: epidemiology and ecology. CRC Press, Florida, pp 177–222

Weckesser S et al (2010) Anaphylactic reactions to bites of the pigeon tick *Argas reflexus*. Eur J Dermatol 20:244–245

Wojcik AR, Greygon-Franckiewicz B, Zbikowska E, Wasielewski L (2000) [Invasion of *Dermanyssus gallinae* (De Geer, 1778) in poultry farms in the Torun region.] Wiad Parazytol 46:511–551. (In Polish)

Wormser GP et al (2006) The clinical assessment, treatment, and prevention of lyme disease, human granulocytic anaplasmosis, and babesiosis: clinical practice guidelines by the Infectious Diseases Society of America. Clin Infect Dis 43:1089–1134. https://doi.org/10.1086/508667

Yadav PD, Shete AM, Nyayanit DA, Albarino CG, Jain S, Guerrero LW, Kumar S, Patil DY, Nichol ST, Mourya DT (2018) Identification and characterization of novel mosquito-borne (Kammavanpettai virus) and tick-borne (Wad Medani) reoviruses isolated in India. J Gen Virol 99:991–1000

Chapter 20
Mycotic Diseases (Aspergillosis)

Abstract Fungal diseases are consistently ignored throughout the years regardless of their disturbing effect on human well-being. It is evaluated those fungal contaminations kill more than 1.5 million individuals each year. Recent global estimates have found 3,000,000 cases of chronic pulmonary aspergillosis in a year. Aspergillosis is among the most vital and common diseases in pet birds, caused by *Aspergillus*, mainly by *A. fumigatus*. It is widely distributed in the birds which are exposed to *Aspergillus* spores. Aspergillosis can be of acute or chronic form. The acute form develops quickly in a couple of days, and the birds may die due to lung infection and respiratory problems. In the chronic form, the bird shows slow development of clinical symptoms, loss of weight, and formation of nodules in the internal organs. The leading cause of about 90% of the pulmonary mycosis in humans is the *A. fumigatus*. The disease is more severe in immunosuppressed persons having neoplastic treatment. The antifungal agents that have been used in the treatment of birds are clotrimazole, itraconazole, enilconazole, terbinafine, and amphotericin B. The primary factor in controlling the avian aspergillosis is environmental control. Efforts in minimizing environmental factors responsible for the growth of *Aspergillus* around the birds may help in reducing the development of aspergillosis.

Keywords Fungal diseases · Aspergillosis · *A. fumigatus* · Birds · Pulmonary mycosis · Immunosuppression

Fungi are common in the environment, and some are even viewed as typical occupants of the skin, gastrointestinal tract, and different mucous membrane surfaces. As a rule, birds can avert disease if their immune system is unblemished and completely operational. However, in cases where the immune system is compromised it can lead to the advancement of serious complications (Sorci 2013). The fungal infections in avian species can be managed by early recognition of the ability of the fungi to cause disease, to provide proper antifungal medications for the severity of infection, and to monitor the response of the patient to the therapy continually. The medicinal, veterinary, and environmental significance of mycotic diseases have expanded significantly during the last few decades. The current

Y. S. Malik et al., *Role of Birds in Transmitting Zoonotic Pathogens*, Livestock Diseases and Management, https://doi.org/10.1007/978-981-16-4554-9_20

problem of mycotic ailments in humans surpasses a few million cases around the world (Rodrigues and Albuquerque 2018; Brown et al. 2012). The reasons for the present circumstance are intricate. However, they, for the most part, mirror conjunction of human exercises that have brought about restorative advancement, travel, and trade joined with the upheaval of the HIV pestilence (Casadevall 2018). This section aims to provide brief information on the several fungal diseases in companion avian species.

Fungal infections contrast from bacterial diseases in that they will, in general, be incessant and kill the host gradually. Fungal diseases are additionally progressively obstinate to treatment with the end goal that most prominent mycoses require treatment courses enduring months or more. Compared to bacterial and viral infection, leading human fungal diseases are rarely transferable, and this has prompted diminished enthusiasm by public health authorities in observation. Hence, moderately little data on the frequency and predominance of mycoses is available. Fungal epidemics, when they happen, will, in general, reflect the increased occurrence of hosts with compromised immunity and susceptible to diseases (e.g., cryptococcosis pestilences in patients with HIV infection (Park et al. 2009), exposures to a vast inoculum, for example, flare-ups of histoplasmosis following tree cuttings in endemic regions (Gustafson et al. 1981), and iatrogenic causes, for example, the episode of *Exserohilum rostratum* following polluted steroid solutions (Andes and Casadevall 2013)).

Intrusive fungal infections are a developing general medical issue. The prevalent mycoses in North America are specifically blastomycosis, coccidioidomycosis, and histoplasmosis, which are caused by thermally dimorphic organisms and can cause infection in immunocompetent or immunocompromised hosts, frequently bringing about serious ailment and demise (Pfaller and Diekema 2010). Infection is usually procured through the inward breath of contagious spores, which results in a respiratory sickness, even though the clinical symptoms can extend from asymptomatic to serious distributed disease (Pfaller and Diekema 2010; Benedict et al. 2012; Thompson III 2011; Wheat 2006). Most diseases with *Blastomyces*, *Coccidioides*, and *Histoplasma* spp. happen sporadically in regions where these fungi are geologically limited, additionally alluded to as mycosis-endemic regions (Pfaller and Diekema 2010). Eumycetoma and Chromoblastomycosis, for example, cause crippling diseases influencing skin, subcutaneous tissues, and fundamental bones (Lamoth and Kontoyiannis 2018; Ahmed et al. 2017). Vitally, the study of disease transmission of fungal maladies is dynamic and challenging to anticipate. Recently, *Candida auris*, a multidrug-resistant pathogen, has risen as a serious risk to humans, with a few strains resistant to every principal class of antifungal drugs (Lamoth and Kontoyiannis 2018). The city of Rio de Janeiro, Brazil, faced the most significant pandemic of sporotrichosis ever (Gremiao et al. 2017). Paracoccidioidomycosis is as yet a standout among the essential foundational mycoses in Latin America and the main reason for mortality in non-immunosuppressed people in Brazil (Queiroz-Telles et al. 2017; Gonzalez and Hernandez 2016).

The treatment for most intrusive fungal diseases remains insufficient despite the best accessible antifungal treatments. For instance, aspergillosis and cryptococcosis have high mortality even though they are treated with the best treatment and are frequently persistent in hosts with compromised immunity. The chronicity of some diseases requires extended treatment, which increases the risk of resistance to antifungal drugs. These sicknesses are gradually connected with antifungal medication opposition. New pathogenic fungi have shown up a high level of resistance from existing antifungal agents, for example, *C. auris*. Despite destruction brought about by fungal diseases in vulnerable people and different biological systems, the fungal kingdom, in general, tends to be disregarded concerning bacteria, viruses, and parasites. Recently, there has been some ongoing advancement in mycotic diseases (Gow et al. 2018).

Fungal diseases are consistently ignored throughout the years regardless of their disturbing effect on human well-being. Latest estimates (Bongomin et al. 2017) propose a worldwide yearly event of around 3,000,000 instances of chronic pulmonary aspergillosis, 200,000 cryptococcal cases of meningitis, 700,000 instances of invasive candidiasis, 250,000 cases of intrusive aspergillosis, 100,000 instances of scattered histoplasmosis, 500,000 instances of *Pneumocystis jirovecii* pneumonia, more than 10,000,000 instances of fungal asthma, and 1,000,000 instances of fungal keratitis. It is evaluated that fungal contaminations kill more than 1.5 million individuals each year (GAFFI 2018). Seeing the wide prevalence of various filamentous fungi, especially the *Aspergillus* species in the environment, despite their positive uses, the negative impacts of these fungi are something to be dealt with intense care. The presumption that contaminations brought about by these fungi seldom occur must not be encouraged because whether we concur or not these filamentous fungi cause a disturbing number of impending diseases that should not be ignored (Egbuta et al. 2017).

20.1 Aspergillosis

20.1.1 Historical Background

The term "aspergillosis" usually refers to the infection caused by the genus *Aspergillus* that includes around 600 fungal species. Aspergillosis is also known as brooder pneumonia, mycotic pneumonia, mycosis, as well as cytomycosis. Aspergillosis is a noncontagious disease caused by inhalation of *Aspergillus* spores. The fungus had been first reported by Micheli in 1729 as fungal conidiophore and called them as Aspergillus based on their morphology (Thom and Raper 1945). In birds, aspergillosis had been reported in the 1800s in the captive flamingo by Owen in 1832 (Austwick 1968). Aspergillosis is among the most vital and common diseases in pet birds, caused by *Aspergillus*, mainly by *A. fumigatus*. Infection with other genera like *Penicillium* and *Mucor* are uncommon, yet the disease of these fungi is also known as "aspergillosis." Single birds or flocks of various avian species might be influenced and experience the acute or chronic types of this localized or systemic disease (Arne et al. 2011; Beernaert et al. 2010; Converse 2008; Tell 2005). Thus, the acute ailment may evolve due to overexposure to the fungal spores, while chronic cases are ordinarily brought about by the fungal and other predisposing elements, for example, immunosuppression, bacterial contaminations, vitamin A deficiency, and hereditary factors (Fisher et al. 2012; Redig 2007; Redig 1993). Aspergillosis outbreaks have been reported in 180 wood ducks and 1000 Mallards in Illinois and died due to the consumption of rotten silage (Pearson 1969).

20.1.2 Epizootiology

Aspergillosis is a fatal disease in different regions. It is widely distributed in the birds that are exposed to *Aspergillus* spores (O'Meara and Witter 1971). Currently several studies had shown successful isolation of the fungus or its spore from Antarctica (Wicklow 1968; Cong et al. 2017). *Aspergillus* species can survive under various ecological conditions as they have versatile features to meet their requirements and make the species a universal contagious pathogen for a wide scope of hosts, including humans and different animals. *Aspergillus fumigatus* conidia are scattered more effectively, noticeable all around than those of other fungi (Kwon-Chung and Sugui 2013). The air current can make conidia scatter because of their momentous hydrophobicity, and the conidia are shielded from UV because of the melanin in their cell wall (Kwon-Chung and Sugui 2013; Brakhage and Liebmann 2005).

Aspergillus fumigatus is present in a broad range of natural conditions with the ideal temperature of 37 °C (thriving in the range of 12–65 °C), and the pH for growth extends in the range of 2.1–8.8 (Kwon-Chung and Sugui 2013). Thermotolerance encourages the fungus' development, not just in rotting organic matter, but also in the mammals and birds' respiratory tract. *Aspergillus fumigatus* generally dwells in compost, a dynamic place that experiences extensive changes in temperature as well as high microbial activity. The capacity to flourish in these natural surroundings requires a considerable level of thermotolerance that has been guessed to add to virulence (Tekaia and Latge 2005). Moreover, the presence of various glycosyl hydrolases (Tekaia and Latge 2005) in the *A. fumigatus* genome validates the capacity of the fungus to degrade polysaccharides from plant cell walls and obtain nitrogen sources made accessible by the degradation of proteinaceous substances (Seyedmousavi et al. 2015; Abad et al. 2010).

Aspergillus is ubiquitous and requires moisture and a temperature of 23–26 °C for growth and sporulation. The leading cause of infection is the *Aspergillus* spore inhalation. At least six million spores are required to cause a lethal infection in 1-day-old chicken (Austwick 1968). While in wild birds, only a few spores are required to cause infection (Hillgarth and Kear 1979). The oral route is an unusual way of infection. In an experiment, chickens were orally given wheat-containing *A. fumigatus,* and *A. flavus* show only typical mycotoxin signs and not the respiratory signs as in *Aspergillus* infection (Chute 1965).

Aspergillosis can be of acute or chronic form. The acute form develops quickly in a couple of days, and the birds may die due to lung infection and respiratory problems. In the chronic form, the bird shows slow development of clinical symptoms, loss of weight, and formation of nodules in the internal organs. The chronic form is generally related to an individual bird rather than a group. Stress is also an important factor in the severity of the disease (Munir et al. 2017). The two factors or theories explain the development of disease in the birds. Firstly, the infectious dose having high spore content may surpass the natural immunity of the bird. Secondly, the birds are the carriers of the spore, and under any stress condition, develop the disease and get sick (Friend and Trainer 1969). Lead intoxication also acts as a predisposing factor for aspergillosis as reported in several wild bird species like the mallard ducks, North America Geese (*Branta Canadensis*), the Swans (*Cygnus buccinators*), and the Andean Condor (*Vultur gryphus*) (Bair et al. 1988; Locke et al. 1969). Outbreaks of aspergillosis have been reported in places having different stress factors like spore concentration, improper ventilation, overcrowding, and mold-contaminated food (Atkinson and Brojer 1998).

20.1.3 Mode of Transmission

Aspergillus belongs to the family Trichocomaceae of the class Eurotiomycetes, and phylum Ascomycota as per CABI Bioscience, in 2004. The three well-characterized species are *A. fumigates, A. niger* (produce mycotoxins), and *A. flavus* (produce the aflatoxins) (Chute 1965). *A. flavus*, *A. nidulans*, and *A. terreusare* are additionally viewed as pathogenic in avian species. *A. fumigatus* is the main species infecting poultry birds, while *A. flavus* is less reported in poultry. Both the species are lacking sexual stage in their life cycle. Even in wild birds, aspergillosis has been reported either as sporadic or acute epizootics. Most birds are vulnerable to disease, particularly young ones having compromised immune systems. The important source of aspergillosis infection is through contaminated water (amid showering), inhalation of airborne conidia, and nosocomial infections can be through hospital fabrics and plastics. Congestion, poor sanitation, poor ventilation, poor nourishment, age, contact to respiratory toxins, simultaneous infection, and damp/dry dusty conditions might encourage presentation to a large number of spores and, eventually, the advancement of the disease. Numerous wild avian species might be influenced together with raptors (red-tailed hawks, goshawks, and gyrfalcons), galliformes (quail, fowl, and turkeys), waterfowl (plunging and shorebirds), and also penguins. Among companion birds, there is, by all accounts, a higher pervasiveness of contamination in mynah, blue-fronted Amazon parrots, and African gray parrots. Among all these waterfowls, raptors and gulls are the most affected bird's species. Aspergillosis has been reported in 13 orders, including 35 bird species (O'Meara and Witter 1971).

Respiratory illness brought about by mycotic disease, particularly by *Aspergillus* spp., is regularly detected in caged pet birds and is viewed as a typical reason for death in avian patients. For these contagious agents, a high predisposition is found in psittacine, waterfowl, mynahs, and birds of prey in captivity. Aside from the structural particularities of birds, predisposing factors, for example, inappropriate supervision conditions (such as low humidity for tropic psittacines, stress, insufficient food, and high stocking density), are related to fungal exposure (Bailey 2008). The predisposing factors should be considered when treatment protocols for a diseased bird are developed and are usually done by enhancing the management, by decreasing the fungal spores in the surrounding, and by enhancing food hygienic conditions and quality. Moreover, in all cases, stress factors should be avoided (Krautwald-Junghanns et al. 2015).

20.1.4 Pathogenesis, Clinical Signs, and Pathology

Aspergillus spp. are saprophytic and present in soil abundantly releasing conidia into the environment. However, the risk of exposure depends upon humidity, temperature, and wind (Panackal et al. 2010). If conidia are inhaled, a healthy individual rarely develops the disease, while in immunocompromised individuals *Aspergillus* can develop an infection with high morbidity and mortality (Challa 2018). In birds, inhalation is the main route of *A. fumigatus* infection (Oglesbee 1997), as spores are too small they reach lungs and air sac through nasal cavity or trachea (Fedde 1998). The primary infection sites are the air sacs as the inhaled air first reaches the thoracic and abdominal air sacs before reaching lung's epithelial cells (Nardoni et al. 2006). In the lungs, spores enter the atria and parts of the infundibula, where they are engulfed by the phagocytic cells (Maina 2002). When the number of spores is high or the birds are immunocompromised, then spores may form loosely attached plaques. These plaques may overgrow in the respiratory tract and obstruct the bronchi or trachea (Oglesbee 1997). Sometime, sporulation also occurs in the air sacs (Nardoni et al. 2006; Cacciuttolo et al. 2009) and produce fruiting bodies which can penetrate air sacs to the adjacent organs causing serositis and superficial necrosis (Tsai et al. 1992).

It is mostly assumed that the scarcity of avian respiratory macrophages leads to the improper avian respiratory, immune system causing birds to be susceptible to the respiratory pathogen (Toth and Siegel 1986). However, the susceptibility of birds to *Aspergillus* may be due to different anatomical, physiological, and respiratory immune system characteristics compared with mammals (Toth 2000; Maina 2002; Tell 2005). The aspergillosis does not have a specific clinical sign, so making diagnosis difficult (Dahlhausen et al. 2004). Diagnosis usually depends on the history of stress/environmental factors/an immunosuppressive condition, clinical presentation, serology, hematology and biochemistry, endoscopy, radiographic changes, and culture of the fungus (Jones and Orosz 2000). Clinical sign in birds is associated with the site of infection. In birds, symptoms include the decline in activity and food intake, change in vocalization, weight loss, and difficulty in breathing. Once the symptoms are observed the disease progress rapidly, and treatment is challenging. Hematology and plasma biochemistry gives an indicative diagnosis of aspergillosis (Jones and Orosz 2000). Leukocytosis, monocytosis, heterophilia, and lymphopenia are reported in aspergillosis cases (Forbes et al. 1992). Acute infections show increased β-globulins, while chronic infection shows increased β-globulins and γ-globulin fractions. In immunocompromised birds, hypoproteinemia is observed (Ivey 2000; Cray et al. 2009) and WBC count may be normal (Flammer and Orosz 2008). However, changes in protein electrophoresis are nonspecific but can be used as an indicative diagnosis of aspergillosis (Ivey 2000; Cray et al. 2009).

Serological tests are used for early and definite diagnosis of aspergillosis (Peden and Rhoades 1992). However, in acute and immunosuppressed birds, antibody production is delayed and low, which may give false-negative results (Brown and Redig 1994). In these cases, detection of antigen is better (Cray et al. 2006), while in chronic cases where antigens are low, antibodies detection is more useful (Jones and Orosz 2000). Serological tests like agar gel immunodiffusion, counter-immunoelectrophoresis, and enzyme-linked immunosorbent assays are used for aspergillosis diagnosis in birds. These tests are considered positive when supported by other evidence (Cray et al. 2006, 2009). Cytology and histopathology detect the presence of the organism, and further culture identification provides a definite diagnosis (Dahlhausen et al. 2004). Only the isolation of *Aspergillus* organisms does not confirm infection as they are ubiquitous and can be contaminants (Jensen et al. 1997; Flammer and Orosz 2008). However, only abundant culture from any part or organ should be considered as diagnostic (Jensen et al. 1997). A few studies used PCR, including real-time PCR and supported this assay for diagnosis of avian aspergillosis (Dahlhausen et al. 2004; Cray et al. 2009).

20.1.5 Public Health and Animal Health Concerns

The leading cause of about 90% of the pulmonary mycosis in humans is the *A. fumigatus*. The disease is more severe in immunosuppressed persons having neoplastic treatment (Latgé 2001). Though the aspergillosis is not a contagious disease, continuous exposure to the same source of infection may cause more severe disease and even leads to infection of other organs. The re-exposure to the *Aspergillus* spp. may be fatal in the organism having leucopenia or tuberculosis or under cancer treatment (Rüchel and Reichard 1999). Precaution should be taken when handling infectious fecal matter or any other material in the area having a history of aspergillosis. The wild birds generally do not transfer *Aspergillus* directly to domestic animals. However, there is a possibility of infection when wild birds share the habitat with domestic animals. Aspergillosis in wild birds may cause heavy losses in birds, and in some cases, epizootics may develop (Bowes 1990).

The high number of *Aspergillus* conidia in the environment is an important risk factor for developing the obtrusive disease in hematological and immunocompromised patients (Cavallo et al. 2013). Subsequently, risk factors related to the event of *Aspergillus* conidia in extremely colonized environments, like bird farms, should be precisely observed. An investigation by Cafarchia et al. (2014) described the

Aspergillus species in the hen farms, their number in the environment, disease transmission, and the existence of symptoms related to birds and workers. Albeit a high number of airborne *Aspergillus* conidia usually did not cause infection in birds, a critical relationship was seen between the existence of the fungi and human colonization (Seyedmousavi et al. 2015; Cafarchia et al. 2014), showing a potential danger of contamination for human health. The *Aspergillus* species is widely distributed in the environment and has negative impacts, so should deal with care as they are causing several diseases that should not be ignored (Egbuta et al. 2017).

20.1.6 Diagnosis, Treatment, and Control Measures

The diagnosis of aspergillosis can turn out to be a troublesome test for the avian veterinarian. Despite a significant number of the advances made in diagnostic capacity, the determination is not as simple as it appears. At present, final analysis for aspergillosis depends on combined proof from a few analytic tests (Savelieff et al. 2018; Jones and Orosz 2000). Potential sources of contagious disease, inclining elements, and the management conditions might be distinguished amid anamnesis. Suspect cases can be affirmed by clinical examination, radiology, or registered tomography. Also, results from a biochemical blood test, total blood count, acute stage protein examination, and protein electrophoresis might be utilized as subordinate tests to decide a patient's condition. *Aspergillus* antigen, *Aspergillus*-specific antibody, and toxins identifications from blood tests might help in defining a conclusion, but still not considered trustworthy for all species of birds. *Aspergillus* antigen, toxin, and *Aspergillus*-specific antibody detections from blood samples may aid in establishing a diagnosis but are currently not regarded as reliable for all avian species (Richardson and Page 2018). In most of the cases of aspergillosis, pathological and histopathological injuries are indicative, and molecular biological examinations are accessible to distinguish the contagious agents from a biopsy or necropsy samples. A complete finding depends on parasitic identification with histopathological proof showing the infectious agent inside the influenced tissue. Nonetheless, endoscopic or laparoscopic tests submitted for parasitic culture and consequent identification are still viewed as substantial for clinical analysis of aspergillosis in birds (Fischer and Lierz 2015).

In veterinary medicine, aspergillosis treatment is challenging. As most of the antifungal agents used are fungistatic, so long-term therapy is required for the treatment of aspergillosis. However, very few reports are available for the treatment of aspergillosis, but these reports are exception rather than rule (Tell et al. 2019). The antifungal agents that have been used in the treatment of birds are clotrimazole, itraconazole, enilconazole, terbinafine, and amphotericin B. Treatment of the disease depends on the seriousness and the site of infection. The treatment of aspergillosis is mostly done by tropical therapy after removing granulomatous tracheal lesions by suction (Westerhof 1995). In the case of birds, the removal of granulomatous lesions is difficult as they are located within the respiratory system (Hernandez-Divers

2002). So mostly, antifungal therapy is used for aspergillosis. However, topical therapy can be applied through nasal or air sac flushing, nebulization, or surgical irrigation of the abdominal cavities (Oglesbee 1997; Abrams et al. 2001), while systemic drugs can be applied intravenously and orally.

Treatment of aspergillosis is generally long-term with continuous monitoring of the patient's response by serological testing to the medication. In most of the avian species, itraconazole is given once daily at a dose of 5–10 mg/kg orally. For the African dark parrots (*Psittacus erithacus*), itraconazole is judiciously used, and the dose is reduced to 2.5–5.0 mg/kg. Another antifungal Terbinafine hydrochloride is also used at a dose of 10–15 mg/kg orally every 12–24 h. Similarly, amphotericin B is frequently used in the treatment of aspergillosis in birds and is given intravenously and intratracheally by catheter, or can be applied directly to granulomatous injury in the coelomic cavity. If single antifungal therapy does not work, then a combination of fluconazole or itraconazole with amphotericin B is given. The primary factor in controlling the avian aspergillosis is environmental control. Efforts in minimizing environmental factors responsible for the growth of *Aspergillus* around the birds may help in reducing the development of aspergillosis (Tell et al. 2019). Protective immunity obtained by vaccination is considered in preventing and treating avian aspergillosis. However, no successful attempts have been obtained (Beernaert et al. 2010). Research on the virulence factors and the role of immunity in avian aspergillosis may help in developing a more effective vaccine.

References

Abad A, Fernandez-Molina JV, Bikandi J et al (2010) What makes *Aspergillus fumigatus* a successful pathogen? Genes and molecules involved in invasive aspergillosis. Rev Iberoam Micol 27(4):155182

Abrams GA, Paul-Murphy J, Ramer JC, Murphy CJ (2001) *Aspergillus* blepharitis and dermatitis in a peregrine falcon-gyrfalcon hybrid (*Falco peregrinus*× *Falco rusticolus*). J Avian Med Surg 15(2):114–120

Ahmed AA, van de Sande W, Fahal AH (2017) Mycetoma laboratory diagnosis: review article. PLoS Negl Trop Dis 11:e0005638. PMID:28837657

Andes D, Casadevall A (2013) Insights into fungal pathogenesis from the iatrogenic epidemic of *Exserohilum rostratum* fungal meningitis. Fungal Genet Biol 61:143–145

Arne P, Thierry S, Wang D et al (2011) *Aspergillus fumigatus* in poultry. Int J Microbiol 746356. https://doi.org/10.1155/2011/746356

Atkinson R, Brojer C (1998) Unusual presentations of aspergillosis in wild birds. In: Proc Association of Avian Veterinarians, pp 177–181

Austwick PKC (1968) Effects of adjustment to the environment of fungal form. In: Ainsworth GC, Sussman AS (eds) The fungi. Academic Press, New York, pp 419–442

Bailey T (2008) Aspergillosis, raptors: respiratory problems. In: Chitty J, Lierz M (eds) BSAVA manual of raptors, pigeons and passerine birds, 1st edn. Replika Press Pvt. Ltd., India, pp 230–233

Bair WC, Simpson SG, Windingstad RM (1988) Acute aspergillosis in mallards at Oahe seep near Pierre, South Dakota. Prairie Naturalist 20(3):153–156

Beernaert LA, Pasmans F, Van Waeyenberghe L, Haesebrouck F, Martel A (2010) *Aspergillus* infections in birds: a review. Avian Pathol 39(5):325–331

Benedict K, Roy M, Chiller T, Davis JP (2012) Epidemiologic and ecologic features of blastomycosis: a review. Curr Fungal Infect Rep 6:327–335. https://doi.org/10.1007/s12281-012-0110-1

Bongomin F, Gago S, Oladele RO, Denning DW (2017) Global and multi-national prevalence of fungal diseases-estimate precision. J Fungi (Basel) 3:pii: E57

Bowes VA (1990) British Columbia. An outbreak of aspergillosis in wild waterfowl. Can Vet J 31(4):303

Brakhage AA, Liebmann B (2005) *Aspergillus fumigatus* conidial pigment and cAMP signal transduction: significance for virulence Med Mycol 43(Suppl 1):S75S82

Brown PA, Redig PT (1994) *Aspergillus* ELISA: a tool for detection and management. In: Proc Annu Conf Assoc Avian Vet, pp 295–300

Brown GD, Denning DW, Gow NA, Levitz SM, Netea MG, White TC (2012) Hidden killers: human fungal infections. Sci Transl Med 4(165):165rv13. https://doi.org/10.1126/scitranslmed.3004404

Cacciuttolo E, Rossi G, Nardoni S, Legrottaglie R, Mani P (2009) Anatomopathological aspects of avian aspergillosis. Vet Res Commun 33(6):521–527

Cafarchia C, Camarda A, Iatta R, Danesi P, Favuzzi V, Di Paola G, Pugliese N, Caroli A, Montagna MT, Otranto D (2014) Environmental contamination by *Aspergillus* spp. in laying hen farms and associated health risks for farm workers. J Med Microbiol 63:464–470

Casadevall A (2018) Fungal diseases in the 21st century: the near and far horizons. Pathog Immun 3(2):183–196

Cavallo M, Andreoni S, Martinotti MG et al (2013) Monitoring environmental Aspergillus spp. contamination and meteorological factors in a haematological unit. Mycopathologia 176(5–6):387394

Challa S (2018) Pathogenesis and pathology of invasive aspergillosis. Curr Fungal Infect Rep 12(1):23–32

Chute HL (1965) Diseases caused by fungi. In: Diseases of poultry, 5th edn. The Iowa State University Press, Ames, pp 494–502

Cong B, Wang N, Liu S, Liu F, Yin X, Shen J (2017) Isolation, characterization and transcriptome analysis of a novel Antarctic *Aspergillus sydowii* strain MS-19 as a potential lignocellulosic enzyme source. BMC Microbiol 17(1):129

Converse KA (2008) Aspergillosis. In: Thomas NJ, Hunter DB, Atkinson CT (eds) Infectious diseases of wild birds. Blackwell Publishing Professional, Ames, pp 360–374

Cray C, Watson T, Rodriguez M, Arheart K (2006) Assessment of aspergillosis diagnostics. In: Proc. Annu. Conf. Assoc. Av. Vet, pp 59–61

Cray C, Reavill D, Romagnano A, Van Sant F, Champagne D, Stevenson R, Rolfe V, Griffin C, Clubb S (2009) Galactomannan assay and plasma protein electrophoresis findings in psittacine birds with aspergillosis. J Avian Med Surg 23(2):125–135

Dahlhausen B, Abbott R, VanOverloop P (2004) Rapid detection of pathogenic Aspergillus species in avian samples by real-time PCR assay: a preliminary report. In: Proc Annu Conf Assoc Avian Vet, pp 37–40

Egbuta MA, Mwanza M, Babalola OO (2017) Health risks associated with exposure to filamentous fungi. Int J Environ Res Public Health 14(7):719. Published 2017 Jul 4. https://doi.org/10.3390/ijerph14070719

Fedde MR (1998) Relationship of structure and function of the avian respiratory system to disease susceptibility. Poult Sci 77(8):1130–1138

Fischer D, Lierz M (2015) Diagnostic procedures and available techniques for the diagnosis of Aspergillosis in birds. J Exotic Pet Med 24:283–295

Fisher MC, Henk DA, Briggs CJ, Brownstein JS, Madoff LC, McCraw SL, Gurr SJ (2012) Emerging fungal threats to animal, plant and ecosystem health. Nature 484(7393):186–194. https://doi.org/10.1038/nature10947. PubMed: 22498624

Flammer K, Orosz S (2008) Avian mycoses: managing these difficult diseases. In: Proceedings of the 29th Annual Conference & Expo of the Association of the Avian Veterinarians with the Association of the European College of Avian Medicine and Surgery, p 153

Forbes NA, Simpson GN, Goudswaard MF (1992) Diagnosis of avian aspergillosis and treatment with itraconazole. Vet Rec 130(23):519–520

Friend M, Trainer DO (1969) Aspergillosis in captive herring gulls. Bull Wildl Dis Assoc 5(3):271–275

GAFFI (2018) Global fund for fungal infections. http://www.gaffi.org. Accessed 1 Feb 2018

Gonzalez A, Hernandez O (2016) New insights into a complex fungal pathogen: the case of *Paracoccidioides* spp. Yeast 33:113–128. PMID:26683539

Gow NAR, Amin T, McArdle K, Brown AJP, Brown GD, Warris A, The Wtsa-Mmfi Consortium (2018) Strategic research funding: a success story for medical mycology. Trends Microbiol 26(10):811–813. https://doi.org/10.1016/j.tim.2018.05.014

Gremiao ID, Miranda LH, Reis EG, Rodrigues AM, Pereira SA (2017) Zoonotic epidemic of sporotrichosis: cat to human transmission. PLoS Pathog 13:e1006077. PMID:28103311

Gustafson TL, Kaufman L, Weeks R, Ajello L, Hutcheson RH Jr, Wiener SL, Lambe DW Jr, Sayvetz TA, Schaffner W (1981) Outbreak of acute pulmonary histoplasmosis in members of a wagon train. Am J Med 71(5):759–765

Hernandez-Divers SJ (2002) Endosurgical debridement and diode laser ablation of lung and air sac granulomas in psittacine birds. J Avian Med Surg 16(2):138–145

Hillgarth N, Kear J (1979) Diseases of seaducks in captivity. Wild 30(30):135–141

Ivey ES (2000) Serologic and plasma protein electrophoretic findings in 7 psittacine birds with aspergillosis. J Avian Med Surg 14(2):103–106

Jensen HE, Christensen JP, Bisgaard M, Nielsen OL (1997) Immunohistochemistry for the diagnosis of aspergillosis in Turkey poults. Avian Pathol 26(1):5–18

Jones MP, Orosz SE (2000) The diagnosis of aspergillosis in birds. Semin Avian Exotic Pet Med 9(2):52–58

Krautwald-Junghanns M-E, Vorbrüggen S, Böhme J (2015) Aspergillosis in birds: an overview of treatment options and regimens. J Exotic Pet Med 24(3):296–307. ISSN 1557-5063

Kwon-Chung KJ, Sugui JA (2013) *Aspergillus fumigatus*-what makes the species a ubiquitous human fungal pathogen? PLoS Pathog 9:12.e1003743

Lamoth F, Kontoyiannis DP (2018) The *Candida auris* alert: facts and perspectives. J Infect Dis 217:516–520. PMID:29390110

Latgé JP (2001) The pathobiology of *Aspergillus fumigatus*. Trends Microbiol 9(8):382–389

Locke LN, Frickie DN, Bagley GE, Young LT (1969) Lead poisoning and aspergillosis in an Andean condor. J Am Vet Med Assoc 155(7):1052–1056

Maina JN (2002) Some recent advances on the study and understanding of the functional design of the avian lung: morphological and morphometric perspectives. Biol Rev 77(1):97–152

Munir MT, Rehman ZU, Shah MA, Umar S (2017) Interactions of *Aspergillus fumigatus* with the respiratory system in poultry. Worlds Poult Sci J 73(2):321–336

Nardoni S, Ceccherelli R, Rossi G, Mancianti F (2006) Aspergillosis in Larus cachinnans micaellis: survey of eight cases. Mycopathologia 161(5):317–321

O'Meara DC, Witter JF (1971) Aspergillosis. In: Infectious and parasitic diseases of wild birds, pp 153–162

Oglesbee BL (1997) Mycotic diseases. In: Altman RB (ed) Avian medicine and surgery, 1st edn. W.B. Saunders Company, Philadelphia, PA, pp 323–361

Panackal AA, Li H, Kontoyiannis DP, Mori M, Perego CA, Boeckh M, Marr KA (2010) Geoclimatic influences on invasive aspergillosis after hematopoietic stem cell transplantation. Clin Infect Dis 50(12):1588–1597

Park BJ, Wannemuehler KA, Marston BJ, Govender N, Pappas PG, Chiller TM (2009) Estimation of the current global burden of cryptococcal meningitis among persons living with HIV/AIDS. AIDS 23(4):525–530

Pearson GL (1969) Aspergillosis in wintering mallards. J Wildl Dis 5(4):404–405

Peden WM, Rhoades KR (1992) Pathogenicity differences of multiple isolates of *Aspergillus fumigatus* in turkeys. Avian Dis 36:537–542

Pfaller MA, Diekema DJ (2010) Epidemiology of invasive mycoses in North America. Crit Rev Microbiol 36:1–53. https://doi.org/10.3109/10408410903241444

Queiroz-Telles F, Fahal AH, Falci DR, Caceres DH, Chiller T et al (2017) Neglected endemic mycoses. Lancet Infect Dis 17:e367–e377. PMID:28774696

Redig PT (1993) Avian Aspergillosis. In: Fowler ME (ed) Zoo and wild animal medicine current therapy, vol 3. WB Saunders, Philadelphia, pp 178–181

Redig PT (2007) Fungal diseases—aspergillosis. In: Samour JH (ed) Avian medicine. Mosby, Amsterdam, pp 373–387

Richardson M, Page I (2018) Role of serological tests in the diagnosis of mold infections. Curr Fungal Infect Rep 12:127–136

Rodrigues ML, Albuquerque PC (2018) Searching for a change: the need for increased support for public health and research on fungal diseases. PLoS Negl Trop Dis 12(6):e0006479. https://doi.org/10.1371/journal.pntd.0006479

Rüchel R, Reichard U (1999) Pathogenesis and clinical presentation of aspergillosis. Contrib Microbiol 2:21–43

Savelieff MG, Pappalardo L, Azmanis P (2018) The current status of avian aspergillosis diagnoses: veterinary practice to novel research avenues. Vet Clin Pathol 47:342–362

Seyedmousavi S, Guillot J, Arne P, de Hoog GS, Mouton JW, Melchers WJ, Verweij PE (2015) *Aspergillus* and aspergilloses in wild and domestic animals: a global health concern with parallels to human disease. Med Mycol 53:765–797

Sorci G (2013) Immunity, resistance and tolerance in bird–parasite interactions. Parasite Immunol 35:350–361. https://doi.org/10.1111/pim.12047

Tekaia F, Latge JP (2005) *Aspergillus fumigatus*: saprophyte or pathogen? Curr Opin Microbiol 8(4):385392

Tell LA (2005) Aspergillosis in mammals and birds: impact on veterinary medicine. Med Mycol 43 (Suppl 1):S71–S73

Tell LA, Burco JD, Woods L, Clemons KV (2019) Aspergillosis in birds and mammals: considerations for veterinary medicine. In: Recent developments in fungal diseases of laboratory animals. Springer, Cham, pp 49–72

Thom C, Raper KB (1945) A manual of the aspergilli. LWW

Thompson GR III (2011) Pulmonary coccidioidomycosis. Semin Respir Crit Care Med 32:754–763. https://doi.org/10.1055/s-0031-1295723

Toth TE (2000) Nonspecific cellular defense of the avian respiratory system: a review. Dev Comp Immunol 24(2–3):121–139

Toth TE, Siegel PB (1986) Cellular defense of the avian respiratory tract: paucity of free-residing macrophages in the normal chicken. Avian Dis 1:67–75

Tsai SS, Park JH, Hirai K, Itakura C (1992) Aspergillosis and candidiasis in psittacine and passeriforme birds with particular reference to nasal lesions. Avian Pathol 21(4):699–709

Westerhof I (1995) Treatment of tracheal obstruction in psittacine birds using a suction technique: a retrospective study of 19 birds. J Avian Med Surg 1:45–49

Wheat LJ (2006) Histoplasmosis: a review for clinicians from non-endemic areas. Mycoses 49: 274–282. https://doi.org/10.1111/j.1439-0507.2006.01253.x

Wicklow DT (1968) *Aspergillus fumigatus* Fresenius isolated from ornithogenic soil collected at Hallett Station, Antarctica. Can J Microbiol 14(6):717–719

Chapter 21
Sporadic Fungal Infections

Abstract We here discuss some of the sporadic fungal infections of birds, viz., Dermatophytosis (caused by *Microsporum* and *Trichophyton*), Dactylariosis (caused by *Dactylaria gallopava*), Histoplasmosis (*Histoplasma capsulatum*), Cryptococcosis (caused by *Cryptococcus neoformans*), Candidiasis (caused by *Candida* spp.), Rhodotoruliasis (caused by *Rhodotorula* spp.), and Mucormycosis (caused by *Mucor* spp., *Absidia* spp., and *Rhizopus* spp.). Zoonotic transmission of these fungal pathogens from birds often involves aerosolization of spores from contaminated soil associated with fecal droppings of birds, or spores can also be carried as contaminant on avian species. In the present discourse, an overview of abovementioned infections has been given emphasizing zoonotic episodes, diseases in birds, epidemiology, diseases in human, and control strategies.

Keywords Sporadic fungal infections · Birds · Zoonoses · Dermatophytosis · Cryptococcosis · Histoplasmosis · Candidiasis · Mcormycosis

21.1 Dermatophytosis (Favus)

Dermatophytosis is the fungus causing skin infection and possesses the ability of human-to-human transfer. It is also known as ringworm, dermatomycosis, as well as favus. The main causal agent is the *Microsporum gallinae*. Dermatophytosis is caused in several birds like turkey, chickens, ducks, quail, and canary (Charlton et al. 2008) including backyard, hobby, or even the game chickens. The symptoms include white scaly or crusty lesions on the comb of chickens and also on the skin of the head and neck, which becomes unfathered. The skin becomes thick due to hyperkeratosis serocellular crusts. Diagnostics of dermatophytosis involves scraping of skin in 10% potassium hydroxide (KOH) and viewing the fragments of hyphae under a microscope. Other diagnostics include culturing of the sample on Sabaroud's dextrose agar. If animal or bird is infected with the fungus, it should be quarantined from others to prevent the spread of the disease. People handling these infected animals should take extra precautions to prevent zoonotic transfer of the disease by adopting good hygienic and biosecurity practices.

Y. S. Malik et al., *Role of Birds in Transmitting Zoonotic Pathogens*, Livestock Diseases and Management, https://doi.org/10.1007/978-981-16-4554-9_21

Dermatophytes are filamentous fungi that can attack keratinized tissue (skin, hair, nails) of humans and other animals (Hube et al. 2015). The causative agents of dermatophytosis are grouped into three genera: *Microsporum*, *Trichophyton,* and *Epidermophyton* (Deuteromycetes). Based on their habitat dermatophytes are grouped as Anthropophilic dermatophytes (parasitic forms that infect humans), Zoophilic dermatophytes (parasitic forms that infect animals, and also humans), and Geophilic dermatophytes (saprophytic fungi related with keratinous materials in soil) (Didehdar et al. 2016; Abd Elmegeed et al. 2015; Tampieri 2004). Dermatophytes have fungal species that infect humans, just like those that likewise infect different animals or develop in the environment. The *Trichophyton rubrum* is a common species involved in the skin infection of immunocompetent persons (Persinoti et al. 2018).

The dermatophyte fungi contain around 30 types of keratinophilic molds, causing infections of the skin, nail, and hair of animals and feathers of birds (Borman et al. 2007). *Trichophyton violaceum* and *T. rubrum* are a common cause of human dermatophyte infection, the first one is found on the scalp, while the last is limited to the nail and glabrous skin. The two species are phylogenetically similar but are different phenotypically (Zhan et al. 2018). Keratinolytic microorganisms can likewise be imperative in the birds because of their probable role in feather degradation (Goodenough and Stallwood 2010). Organism induced feather degradation can prevent flight and mate fascination, and also thermoregulation (Burtt Jr and Ichida 1999, 2004; Shawkey et al. 2007), in spite of the fact that the degree and extent of such impacts are at present unclear (Gunderson 2008). Keratinolytic microorganisms earlier linked with birds include *Microsporum gallinae* (Hubálek 2000) *Chrysosporium tropicum*, and *Bacillus licheniformis* (Riffel and Brandelli 2006). Keratin waste hydrolysis by microorganisms is thought of as a biotechnological substitute for recycling and valorization. Keratinases, which act as proteases having keratinolytic activity, act synergistically with other keratinolytic proteins to decompose the complex supramolecular association of keratin (Călin et al. 2017; Jin et al. 2017).

Microsporum gallinae is a zoophilic dermatophyte and is related to cocks specifically. Even though *M. gallinae* sporadically causes ringworm in dogs, cats, and monkeys (Khosravi and Mahmoudi 2003; Grunder et al. 2005), human infection is uncommon. The first human instance of *M. gallinae* disease was described by Sabouraud (1910) and Georg and Torres (1956)). The identification of the *M. gallinae* isolates from the patient is confirmed by the morphological appearance and the DNA sequence analysis. Further DNA sequence analysis of the *M. gallinae* confirmed the zoophilic nature by describing that the closeness of the sequences with the other zoophilic fungi, *A. grubyi,* and *M. vanbreuseghemii.* In the human *M. gallinae* infections, three patients were found exposed or came in contact with cocks or pigeons (del Palacio et al. 1992; Poblete-Gutierrez et al. 2006).

Further, Miyasato et al. (2011) also found a patient who had a breed of fighting cocks over many years. Although they did not find *M. gallinae* during field investigation, the breeding cocks were supposed to be the source of the *M. gallinae* infection. Similarly, a homing pigeon (*Columba liviadomestica*) was eluded with the incidence of ulceration, bleeding, feather loss and scabs, and outside layer in the head. Mycological assessment demonstrated concurrent contaminations by *Malassezia* and *Microsporum gallinae*. It was reasoned that keratinophilic organisms, particularly dermatophytes in skin and quills of birds, could cause extreme infections and maybe a potential source for the spread of disease in humans and animals (Eidi et al. 2017).

Chickens from California's Central Valley and different Oriental breeds (Shamo and Aseel) were seen to have an unusual skin condition and feather loss. The presence of white plaques on the comb, face, and additionally on-ear lobes was followed by feather loss beginning at the caudal base of the comb and advancing down the neck. Even though the cockerels were influenced first, the condition spread to the hens matched with those roosters (Bradley et al. 1993). *Microsporum gallinae*, identified by the histological and mycological tests, was found as the causative agent of favus (avian dermatophytosis). Chickens (*Gallus domesticus*) and their nursing surroundings contain dermatophytes and related species. The dominant dermatophyte species identified with chickens are *Microsporum gallinae* and *Arthroderma simii*; and are known causative agents of mycotic zoonoses (Yamaguchi et al. 2014; Chermette et al. 2008). The former species was isolated from a human (Miyasato et al. 2011) and a fighting cockerel (Murata et al. 2013) in Japan. Human infection of *Arthroderma simii* is rare. However, fungal species spread worldwide (Chermette et al. 2008).

Moreover, keratinophilic fungi like *Chrysosporium* species have also been isolated from chickens and their habitats in Egypt (Abdel-Gawad and Moharram 1989; el-Naghy et al. 1998), India (Dixit and Kushwaha 1991; Saxena et al. 2004), Iran (Khosravi and Mahmoudi 2003), Germany (Grunder et al. 2005), Bahrain (Mandeel et al. 2011), and Saint Kitts and Nevis (Gungnani et al. 2012). Sharma and Sumbali (2017) studied the feathers of two migratory birds, that is, bar-headed geese and common teal, for the diversity of dermatophytes and other keratinophilic fungi visiting Gharana wetland in Jammu, India. From the feathers of bar-headed geese and common teal, 33 keratinophilic fungal species belonging to 17 different genera were recovered. Two species of dermatophytes belonging to *Microsporum* were *M. gypseum* and *M. canis* (Sharma and Sumbali 2017). Depending on the geophilic, zoophilic, or anthropophilic starting point of the fungus and on the immunological status of the patient, symptomatology can broadly contrast. Identification methods used for distinguishing species have been changed since the last two decades with more emphasis on molecular methods for species identification. It ends up evident that the calculation utilized for the identification of related species needs to use both phenotypic and genomic techniques (Hayette and Sacheli 2017).

21.2 Dactylariosis

It is one of the other sporadic fungal infections of birds. Dactylariosis is usually caused by one of the thermophilic fungi called *Dactylaria gallopava*. It has the same host as the favus, including chickens, turkey poults, and quails. Although it is one of the sporadic diseases, when it affects certain flocks, particularly the younger chicken's flocks, it leads to high mortality rates due to fatal encephalitis. The clinical signs are consistent with the nervous manifestations in the form of the loss of balance, tremors, and paralysis. This disease is more associated with the litter contamination as well as the egg incubators (Waldrip et al. 1974).

D. gallopava grows well at 25–35 °C while the optimum temperature is 45 °C and low pH (<5) conditions. Birds are infected by inhaling spores, which spread in the wake of getting discharged into the air (Abbas et al. 2017; Waldrip et al. 1974), but the disease is produced by angioinvasion and hematogenous spread to CNS (Dhama et al. 2013). Birds between starting 1 and 5 weeks of age are vulnerable. Mortality amid infection flare-up extends between 3 and 20%, primarily because of neurological sickness (Waldrip et al. 1974; Shane et al. 1985). Torticollis and in-coordination because of paresis are seen in tainted poults.

In some cases, visible injuries result in visual impairment. In uncommon instances, pneumonic granuloma causes dyspnea, as viewed in aspergillosis (Sonne et al. 2012; Dhama et al. 2013). *Verruconis gallopava* is a natural dematiaceous fungus that is perceived progressively as a reason for human illness, particularly for immunocompromised people. Contamination can run from superficial and limited lesions to the pneumonic association and central nervous system abscesses. Optimal treatment is still unknown. However, posaconazole was found to successfully treat pulmonary infection by *V. gallopava* in a cardiac patient (Moran et al. 2019).

Ochroconis gallopava (also known as *Dactylaria gallopava* and *Dactylaria constricta* var. *gallopava*) is seemingly the most critical clinical individual from the Ochroconiales. A thermostable organism of the family Ochroconiaceae, *O. gallopava* is usually found in warm, acidic conditions, including rotting flora and manure (Waldrip et al. 1974). Sporadic and epizootic *O. gallopava* contaminations are very much described in poultry; be that as it may, in nondomestic species, reports are constrained to sporadic sickness in snowy owls (*Nyctea scandiaca*) and gray-winged trumpeters (*Psophia crepitans*) and a single flare-up in Japanese quail chicks (Karesh et al. 1987; Salkin et al. 1990; Shane et al. 1985). While Ossiboff et al. (2015) reported encephalitis and ophthalmitis because of ochroconiasis in tinamou (*Eudromia elegans*) chicks. *D. gallopava* can cause encephalitis in chickens and other birds, but contact with birds has not been related to human disease (Salkin et al. 1990; Ranck Jr et al. 1974). *Ochroconis gallopava* cause pulmonary infection in an immunocompetent patient with adrenal inefficiency on chronic, low dose steroids (Whitehurst et al. 2018). In another recent report, a fruitful treatment strategy against *O. gallopava* endophthalmitis was accounted using a combination of intravitreal amphotericin B and voriconazole infusions, as well as IV posaconazole and voriconazole, which brought about an amazing visual result (Kim et al. 2018).

Although dematiaceous fungi are found all over the world in rotting vegetation and soil but are common in mainly tropical and subtropical regions, the patients infected with *R. mackenziei* are mostly found from Middle Eastern countries, including Syria, Saudi Arabia, or Kuwait (Sutton et al. 1998; Revankar et al. 2004; Kantarcioglu and de Hoog 2004), yet cases have also been accounted from Afghanistan (Cristini et al. 2010), India (Badali et al. 2010) and Pakistan (Jabeen et al. 2011). Cerebral phaeohyphomycosis is much of the time a lethal sickness caused by genuinely neurotropic dematiaceous fungi. Albeit uncommon, this infection happens particularly among immunocompetent patients, and the clinical side effects are frequently misdiagnosed as a cerebral tumor or bacterial brain sore. The proper conclusion and treatment of cerebral infections by melanized fungi are challenging if caused by fungi of an unidentified ecological niche. A recent incidence of cerebral phaeohyphomycosis in Iran was the first culture-confirmed case due to *Rhinocladiella mackenziei* (Mohammadi et al. 2018).

21.3 Histoplasmosis

Histoplasmosis is caused by a dimorphic fungus *Histoplasma capsulatum* and exists in soils contaminated with bat or bird droppings as spore-forming fungi. Around 90% population of the Ohio and Mississippi river valleys in the United States are exposed to the microconidia of *Histoplasma* through their lifespans, but only about 1% will show indicative infections (Diaz 2018; Kauffman 2007; Manos et al. 1956). While this percent can reach up to 50% during outbreaks of histoplasmosis (Kauffman 2007; Manos et al. 1956), previously assumed to be restricted to limited geographical areas of the Americas, autochthonous and imported instances of histoplasmosis are accounted worldwide (Armstrong et al. 2018; Bahr et al. 2015). Other endemic areas for histoplasmosis are Asia, Australia, Central and South America, and sub-Saharan Africa (Bahr et al. 2015; Colombo et al. 2011; Chakrabarti and Slavin 2011; McLeod et al. 2011; Baddley et al. 2011). Benedict and Mody (2016) described 105 outbreaks (characterized as more than one case), including 2850 cases in 26 continental states and Puerto Rico from 1938 to 2013. The most well-known introduction sources were chicken coops and buildings experiencing destruction or renovation. In most of the cases (around 77%), birds and bats and their droppings were found (Benedict and Mody 2016).

Generally, areas with void structures, chicken coops, ranches, and tennis complexes have high-hazard for contracting histoplasmosis. These structures and settings all will, in general, attract groups of resting or perching birds, for example, seagulls briefly protecting from harsh oceans on the grounds of seaside and lakeside. Previously histoplasmosis outbreaks were more regularly connected with chicken coop and farm exposures. Current outbursts are frequent in nature and work environment where birds and bats and their droppings are existent, and the environment is disturbed by uncovering or construction (Benedict and Mody 2016).

The disruption of contaminated soil aerosolized the infective microconidia of *H. capsulatum,* which can change into budding yeast when breathed to the alveoli level of the lungs. The majority of the immunocompetent individuals with positive histplasmin skin tests will stay asymptomatic, although infected, as the pulmonary macrophages overwhelm microconidia and carry them to lymph nodes for destruction. About 20% of exposed persons, after inhaling a large spore inoculum, will show temporary severe influenza-like symptoms after 3–21 days. A few will have intense pneumonia that may advance to chronic granulomatous pneumonia enduring months, particularly in immunocompromised people. Intense and unending diseases may advance quickly to scattered contaminations, particularly in immunocompromised people with diabetes, HIV disease, and AIDS (Diaz 2018; Kauffman 2007). Immunocompetent persons with asymptomatic histoplasmosis do not require any treatment. In many cases, of immunocompetent patients will likewise require no antifungal treatment in intense histoplasmosis pneumonia except if side effects, particularly fever and discomfort, endure for over a month (Wheat et al. 2000). In such cases, oral itraconazole, for 3–6 months, should be given to avoid the development of chronic pulmonary disease. In both immunocompetent and immunocompromised patients, the chronic pulmonary disease should be cured with oral itraconazole (Diaz 2018).

21.4 Cryptococcosis

Cryptococcosis causing fungi infect animals worldwide, also a large number of mammals (including humans) and, seldom, birds, reptiles, and amphibians. It is caused by genus *Cryptococcus*, from basidiomycetous yeast, specifically, *Cryptococcus neoformans* and *Cryptococcus gattii* species complexes (Singh et al. 2018). The two species, *C. neoformans* and *C. gattii* as specified by the present classification (Kwon-Chung and Varma 2006; Kwon-Chung et al. 2002), are with the serotypes A, D, and AD for the first one, and B and C for the last one. *Cryptococcus neoformans* as of now comprises of two varieties: *C. neoformans* var. *grubii* (serotype A) (Franzot et al. 1999) and *C. neoformans* var. *neoformans* (serotype D) (Kwon-Chung et al. 2011).

The ecological repository of *C. neoformans*is normally identified with bird feces, especially pigeon droppings. Though they have also been found in rotting trees, wood and plant trash, conduits, and soil, all normally sullied with bird feces (Singh et al. 2018; El-Hariri et al. 2015, 2016; Saleh et al. 2011). *Cryptococcus neoformans* is found in natural habitats, for example, soil, and develops in birds' excreta, particularly that of pigeons. Diseases in birds are not common. *C. neoformans* are found in the feces of canaries (26%), bearer pigeons (18%), budgerigars (2%), and psittacine birds (1%), aside from domestic poultry (Singh and Dash 2008; Saremi et al. 2004). *Cryptococcus* species is related to the trachea of fowls, isolated from chickens of a poultry-processing unit (Laubscher et al. 2000). *Cryptococcus laurentii* was found to cause feather loss in a glossy starling (*Lamprotornis chalybaeus*). The bird showed irregular feather loss, particularly around the back and beak area, and gray crusts piercing firmly to the underlying skin. The feathers appear greasy and disseminate a moldy odor.

In an Australian zoological park, a red-tailed black cockatoo (*Calyptorhynchus banksii*) was diagnosed with cryptococcosis, caused by *Cryptococcus gattii* VGI, living in a large coop with other different avian species, so a systematic investigation was conducted to know about the environmental sources of infection. A tree with a broad hollow was recognized as the probable source of contamination depends on the influential culture of *C. gattii* VGI, phylogenetic analysis, and multi-locus sequencing of the environmental and disease-related isolates. This prompted the cautious expulsion of the tree to diminish the danger of future instances of cryptococcosis in this aviary (Schmertmann et al. 2018).

Birds never become ill, since their inside temperature is higher than the highest temperature required for the development of the fungus. In people, the disease happens through the inward breath. The disease mostly occurs in persons with low immunity. The infection initiates chronic or acute cryptococcosis meningitis, which shows cerebral pain, disquietude, and in the last stages, a firm neck and fever. It may even prompt the death of the patient. The disease can likewise influence the lungs and skin (Malik et al. 2003). Pulmonary cryptococcosis is a significant intrusive disease in immunocompromised patients. However, it is also progressively observed in immunocompetent patients. The primary human pathogens are *C. neoformans* and *C. gattii*, which have wide dispersion. It can impersonate pulmonary tuberculosis, lung malignancy, bacterial pneumonia, and other pneumonic mycoses (Setianingrum et al. 2018).

An investigation from Uganda found that 11% of the patients with HIV infection in the hospital had pulmonary cryptococcosis as an optional infection (Yoo et al. 2011). In a recent study from Thailand, 13% of HIV patients having pneumonia carry cryptococcal antigen in their serum, even though 33% of this populace never had cryptococcal meningitis or its history (Harris et al. 2012). Mostly the respiratory diseases in these patients were cryptococcal. A survey from China uncovered that 60% of the patients of pulmonary cryptococcosis were immunocompetent non-HIV patients (Liu et al. 2016). In another study, it was found that in immunocompetent patients, 67% of them can disseminate pulmonary cryptococcosis, into the central nervous system causing cryptococcal meningitis (Suwatanapongched et al. 2013). Patients with pneumonic cryptococcosis associated with cryptococcal meningitis indicated less response to the treatment and poor clinical result in comparison with cryptococcal meningitis alone (Cao et al. 2018).

C. neoformans and *C. gattii* share numerous virulence factors that permit infection in the human (Kronstad et al. 2011). These primary factors are polysaccharide capsule, melanin pigment, growth at 37 °C, and extracellular enzymes (Zaragoza et al. 2009; Nosanchuk and Casadevall 2006; Rodrigues et al. 2008). The polysaccharide reduces the macrophage functions may be by disturbing the T-cell responses in the lung (Levitz and Specht 2006; Vecchiarelli et al. 1994) while the melanin reduces the oxidative killing of the immune cells against *Cryptococcus* with redox buffer mechanism (Nosanchuk and Casadevall 2006). The superoxide dismutase supports the survival of *Cryptococcus* at body temperature, as the null mutant of this gene is both avirulent and loses the capacity to develop at 37 °C. The considerable dissimilarity between *C. neoformans* and *C. gattii* is the reduced capacity of *C. gattii* to produce inside macrophages (Voelz et al. 2014). So, the pulmonary cryptococcosis is one of the regular diseases in both immunocompromised and immunocompetent people. So, there is an incredible enthusiasm for the investigation of environmental niches, genetic diversity, and pathogenesis of this disease, given the outbreak of *C. gattii* in a few regions (Setianingrum et al. 2018).

Cryptococcosis has been accounted for in a few avian animal varieties including little passerines and partner psittacine, and some bird species, for example, nondomesticated pigeons that act as carriers of the fungi. Avian and fascinating pet experts ought to likewise know about its zoonotic potential. Nosanchuk et al. (2000) depicted the probable spread of *C. neoformans* from a cockatoo to a 72-year-old elderly person getting immunosuppressive prescriptions after renal transplantation. Isolates of *C. neoformans* firmly recommend that the patient's disease came about because of presentation to aerosolized cockatoo excreta. Clinical indications of Cryptococcus diseases might be non-explicit and include weakness, exhaustion, diarrhea, despondency, dyspnea, anorexia, weight reduction, paralysis, and visual deficiency. Neurological symptoms may happen with CNS association. Moderate to extreme dyspnea might be seen with the inclusion of both the lower or upper respiratory tract. The antemortem finding of Cryptococcosis is troublesome (Abassi et al. 2015). Complete identification of Cryptococcosis needs the exhibition of the oval to round shape yeast having a mucopolysaccharide capsule on histologic or cytologic examination of tracheal discharge and also by isolation of the culture. A coagulated, mucoid exudate might be found inside the respiratory tract, long bones, coelomic cavity, sinuses, and necropsy. In any avian species, the CNS signs with coagulated mucoid exudates are considered exceedingly doubtful for Cryptococcosis (Maziarz and Perfect 2016). Avian veterinarians should utilize extraordinary alert when dealing with patients associated with having Cryptococcosis. Zoonotic diseases may happen through breathing in residue from dried feces of pigeons, starlings, and different species. In birds, itraconazole, Amphotericin B, and ketoconazole have been prescribed as conceivable treatment choices; although, fluconazole might be a superior decision for diseases of the CNS. Anticipation for effective treatment is incredibly weak.

21.5 Candidiasis

Candidiasis is a critical infrequent opportunistic fungal disease in poultry. It has additionally been accounted for to be a disease or intestinal contamination in various types of wild birds that are being brought up in confinement (Milton 1999). Outbreaks occur where appropriate management protocols are not pursued, as the disease is sporadic. *Candida albicans* is the leading cause of candidiasis. *C. albicans* is a typical environmental fungus and an opportunistic pathogen which are typically found in the bird's crop. Candidiasis has been seen in chickens, turkeys, pigeons, guinea fowl, geese, quail, parrots, and birds (Moretti et al. 2000). The contaminated drinking water or food is a conventional means of disease transmission. The source of candida exposure to birds is the contaminated environments, like litter from poultry farms, and game bird rearing facilities, areas contaminated with bird discharge, and human waste (Mugale et al. 2015). The infected food, beak, and tongue injuries expose the birds to the candidiasis. The young birds with crop stasis are more susceptible (Bauck 1994; Kunkle 2003). *Candida* is the second most crucial isolated genus of yeast, having great significance for animals and humans (Brilhante et al. 2013). Although the most commonly isolated species is *Candida albicans*, other species are also considered as emerging pathogens because of their high prevalence and their ability to cause an infection like *C. guilliermondii* was the most frequent isolate (Reis et al. 2018).

Candidiasis, also known as the "thrush," occurs when surface colonization of the gastrointestinal tract advances to profound tissue attack. The outcome is smooth development and colonization of the gastrointestinal mucosa by the fungi, and if it remains unhindered, *Candida* sp. may develop to systemic letting for dispersion to occur. Clinical indications of candidiasis differ contingent on the area of disease (Bondaryk et al. 2013). Local contaminations inside the oropharynx may cause trouble or hesitance to swallow sustenance and halitosis. Oropharyngeal conditions ordinarily result in pseudomembranes of necrotic trash covering the mucous layers or catarrhal inflammation, which give the mucous layers a "Turkish towel" form. Infection inside the crop shows vomiting, spewing forth, anorexia, delayed crop emptying, and discernable thickening of the crop. Ventricular and proventricular diseases may cause diarrhea, weight reduction, vomiting, and general disquietude. *Candida* spp. may likewise colonize the respiratory tract prompting dyspnea, visual or nasal release, and sniffling. Less ordinarily, *Candida* spp. may also taint skin, liver, bone marrow, visual tissues, pericardial tissues, and the CNS. *Candida parasilosis* is accounted for causing fundamental contamination of the liver and bone marrow just as splenic deterioration (Tan et al. 2016). Candidiasis diagnosis is generally detected by the budding present in the *Candida* sp. (3–6 μm) with Grams', new methylene blue, lactophenol cotton blue, or Diff-Quik stains of the feces, crop contents, or spewed/retched material or lesions. Cellophane tape and skin scrapings tests may likewise be performed to help in the finding of associated yeast diseases with the skin.

Caged, as well as free-living birds, may act as a source of microbial dispersion, essentially through their feces, which act as a rich medium, perfect for fungal growth (Tampieri 2006). Generally, birds are susceptible to fewer pathogens than other animals; they can go about as hosts or repositories and scatter pathogens in the environment (Contreras et al. 2016). However, less is known about the dispersal role of the specific birds. The pathogenic yeast to humans is generally found in the gastrointestinal tract of the birds, which act as bearers and disseminators of microbes to the environment (Reis et al. 2018; Brilhante et al. 2013). Candidiasis is an opportunistic parasitic disease that generally influences the upper alimentary tract in chickens (Dykstra et al. 2013). *Candida albicans*, the essential pathogen, is also a part of the healthy microbial flora of the alimentary tract (Dykstra et al. 2013). In a recent study, simultaneous fowlpox and candidiasis diseases happened in a chicken group (Ogasawara et al. 2016). Four perished chickens (one Nagoya breed and three white silkie chickens) were inspected for analysis. White curd-like plaques were seen in the crop during necropsy. Parasitic components that were positive for *Candida albicans* with immunohistochemistry were dispersed all through the tongue, choanal mucosa, throat, and crop. PCR and genetic sequencing distinguished the *C. albicans* gene in the crop, and fowlpox gene in the bursa of Fabricius. These outcomes demonstrate that the outbreak was an uncommon introduction of fowlpox in suddenly infected chickens, with unexpected pox injuries in the BF (Ogasawara et al. 2016).

Candida glabrata is among the most frequently recognized yeast types of the human gut mycobiota. This species has additionally risen as a noteworthy organism of human mucosal, systemic, and circulation system yeast diseases, second just to *C. albicans* (Al-Yasiri et al. 2016; Silva et al. 2012; Falagas et al. 2010). *C. glabrata* contaminations are portrayed by a high (around up to 70%) casualty rate, particularly in immunocompromised patients (Krcmery and Barnes 2002). Nonetheless, the source of *C. glabrata* has not been described well. Aside from humans, this yeast is also a usual commensal of numerous types of pet birds, for example, parakeets, lovebirds, cockatiels, and cockatoos or common migratory birds, for instance, *Sylvia communis* (whitethroat) and the *Muscicapa striata* (spotted flycatcher) (Francesca et al. 2014; Brilhante et al. 2010; Cafarchia et al. 2006; Mancianti et al. 2002). A few pathogenic microbes appeared to be transmitted over long distances among people and birds, for example, aquatic bird-borne flu virus (Fouchier et al. 2003). Several studies have emphasized the threat of spreading different antibiotic-resistant microorganisms using contaminated feces of birds (Horton et al. 2011; Lord et al. 2010; Bonnedahl et al. 2009).

Yellow-legged gulls are remarkably more synanthropic compared to other sea birds. This is a noteworthy reason that gull populaces have developed associatively with human-made conditions, including human reject locales, along with the Mediterranean littoral (Neves et al. 2006). In addition, beach sand may likewise also act as a repository for *C. glabrata* (Shah et al. 2011). The birds turn into a repository as the yeast matures into a part of the gut microbiota. The yellow-legged gull can generally fly long distances to fodder from landfills along the Mediterranean littoral. In this way, the urbanized areas with garbage dumps might be a prospective source of clinically imperative yeast transferred by gulls. Because of the high movement of gulls, they encourage the dispersal of the yeast by discharging their droppings over a broad region of the marine and earthy environment.

Humans might be tainted with *C. glabrata* directly through the droppings of the bird in the environment or by food contaminated with bird droppings. Thus yellow-legged gulls have an essential role in the diffusion of *C. glabrata* attained from an anthropic environment as suggested by the genetic homogeny of human and bird isolates. By this speculation, yellow-legged gulls spread and transmit potential human pathogens in different environments. Further, Al-Yasiri et al. (2016) have demonstrated that yellow-legged gulls act as a carrier, reservoir, and disseminator of *C. glabrata* as they showed the presence of genetically homogeneous *C. glabrata* isolate resistant to fluconazole in both humans and gulls in Marseille.

Liu et al. (2018) isolated and identified *C. albicans* from live birds with supposed candidiasis from the poultry in the southern Hebei Province (China). They are determined by differential medium culture, morphology, and rDNA internal transcribed spacer sequencing. The distinct *C. albicans* strains were examined by ABC genotyping and multilocus sequence typing. In light of this investigation and others, they suggested that *C. albicans* isolates from poultry were generally independent but not different from the isolates of a human. The strains with the equivalent or firmly related genotypes, however, recovered from the birds, and humans may have exchanged and evolved between the two kinds of the host (Liu et al. 2018).

21.6 Rhodotoruliasis

Rhodotorula is typical environmental yeast found in soil, air, lakes, drain, sea water, and organic product juice. *Rhodotorula* is in the Sporidiobolaceae family, which colonizes humans, plants, and different mammals. The *Rhodotorula* has eight species, and *R. glutinis*, *R.mucilaginosa*, and *R. minuta* are the recognized source of disease in humans (Granero et al. 2017; Larone 1995). The colonies of *Rhodotorula* are pink to red and blastoconidia are unicellular lacking hyphae and pseudohyphae. *Rhodotorula* has been isolated from various biological systems and environments. *Rhodotorula* spp. has been perceived as rising yeast pathogens in people over the most recent two decades. While no instances of *Rhodotorula* disease were accounted before 1985, the number of contaminations expanded after that time may be as a result of the more extensive utilization of medicines and central venous catheters (Kim et al. 2013; Miceli et al. 2011).

Rhodotorula mucilaginosa usually infects the skin and is infrequently found in raptors (particularly falcons). The yeast commonly causes oily, yellowish-dark colored outside layer overlying broken and discolored zones of skin in the axillary region of the wings or between the body divider and the thigh musculature. Without treatment, sores may progress toward becoming hyperkeratotic, bringing about proliferative, horny developments on the affected skin (Giovannini et al. 2014). Diagnosis of the disease depends on physical check-up, cytology, histopathology, or culture of infected tissues. Treatment needs expulsion of horny developments if present and therapy with a topical antifungal cream for several weeks. Topical or systemic antibiotics ought to be considered to avoid secondary bacterial contamination from rising. Furthermore, the development of the infected regions ought to likewise be limited to enable the injuries to recuperate.

Several findings of the pathogenicity of *Rhodotorula* spp. have been found in animals, among them, one reported infection caused by *R. mucilaginosa* in chickens skin and another lung disease in sheep (Aruo 1980; Monga and Garg 1980).

Rhodotorula was accounted for causing epididymitis, skin sores in sea lion, and dermatitis in the cat having mastitis and crusted injuries (Alvarez-Perez et al. 2010; Bourdeau et al. 1992; Kadota et al. 1995; Costa et al. 1993). Interestingly, *Rhodotorula* is also found in pools where ocean animals are retained in confinement (Buck 1980). Duarte et al. (2001) have demonstrated the presence of *R. mucilaginosa* in the ear canal of five (11.1%) out of 45 cattle with external parasitic otitis (Duarte et al. 2001). A few authors have demonstrated the *Rhodotorula* as a colonizing fungus in the cloaca and oropharynx of ostriches, in the cloaca and fecal samples of pigeons and wild birds, the genital tract of female camels, in cats, and rhesus monkeys (Melville et al. 2004; Costa et al. 2010; Shokri et al. 2010; Brotto et al. 2005; Amaral et al. 1998). Recently, the respiratory tract in a dog was reported with mixed infection, and authors claimed it as the first case of dogs' respiratory tract infection caused by *Rhodotorula mucilaginosa* and *Trichosporon jirovecii* (Biegańska et al. 2018).

Formerly considered nonpathogenic, *Rhodotorula* species have risen as opportunistic pathogens with the capacity to colonize and taint vulnerable patients. Ongoing investigations have shown that the occurrence of fungemia caused by *Rhodotorula* was somewhere in the range of 0.5 and 2.3% in the United States (Wirth and Goldani 2012; Lunardi et al. 2006; Louria et al. 1960) and Europe (Duboc de Almeida et al. 2008). Most instances of disease with *Rhodotorula fungemia* are related to catheters in the patient's having hematologic malignancies (Potenza et al. 2018; Lunardi et al. 2006). Taking into account that *Rhodotorula* is a universal and saprophytic fungus, its isolation from nonsterile human destinations, particularly from the mucous layers, has regularly been of faulty clinical essentialness. Restricted contaminations without fungemia including onychomycosis, endophthalmitis, prosthetic joint diseases, meningitis, and peritonitis (for the most part connected with consistent peritoneal dialysis) have been accounted for in immunocompromised and immunocompetent patients.

García-Suárez et al. (2011) investigated 29 instances of Rhodotorula fungemia in patients with hematological conditions. This examination demonstrated that 100% of patients who had fungemia by *Rhodotorula* had some type of central venous access, for example, a Hickman catheter. Tuon and Costa (2008) conducted the first methodical survey of contaminations caused by *Rhodotorula* in 128 patients (Tuon and Costa 2008) in all the papers about *Rhodotorula* diseases until January 2006 (Lunardi et al. 2006). The most commonly recognized *Rhodotorula* species was *R. mucilaginosa*, followed by *R. glutinis* and *R. minuta* (Moț et al. 2017). Nearly 40% of the patients are found in immunosuppression, and the most widely recognized hidden condition related to *Rhodotorula* contamination was the utilization of central venous catheters. In another report, Spiliopoulou et al. (2012) depicted a patient who had *Rhodotorula fungemia* in an emergency unit. They found that *Rhodotorula* is resistant to fluconazole and echinocandins. Then again, in vitro susceptibility studies uncovered that *Rhodotorula* is commonly susceptible to amphotericin B and flucytosine (Spader et al. 2019).

21.7 Mucormycosis (*Mucor* spp., *Absidia* spp., and *Rhizopus* spp.)

Mucormycosis is an opportunistic saprobic disease caused by fungi in the order Mucorales (Hoffmann et al. 2013). The frequently distinct species in the order Mucorales are in the genera *Rhizopus*, *Mucor*, *Rhizomucor*, *Lichtheimia* (in the past *Absidia*), *Apophysomyces*, *Cunninghamella*, and *Saksenaea*. The Mucorales are naturally found in soil, and they are generally isolated from rotting natural material. The organisms are regularly likewise found in the indoor and open-air, in nourishment stuff, and dust (Hoffmann et al. 2013). Mucormycosis in animals and humans are comparable in the disease transmission, entryway of passage, limitation, and development of lesions (Seyedmousavi et al. 2018; Reynaldi et al. 2017; Isidoro-Ayza et al. 2014; Muir and Raidal 2012; Cunha et al. 2011).

The pathogenic individuals from the Mucorales are omnipresent in indoor habitats. Inside the domesticated environment of animals, however, contamination perpetually is built up just when the typical balance among animals and the agent are disturbed. By other fungal diseases in animals, for instance, candidiasis and aspergillosis, inclining factors are not identified with the animal species but rather to the tainted animals (Seyedmousavi et al. 2018; Reynaldi et al. 2017; Muir and Raidal 2012). General inclining factors preferring mucormycosis in humans also apply for animals; that is, contaminations are found in hosts that are immunocompromised or generally crippled because of a metabolic issue. However, exposure to mucoralean fungi or disruption of the bacterial microbiota in the forestomach may cause disease generally in healthy animals (Jensen et al. 1992).

The saprophytic fungi in the mucor group of organisms cause disease of variable forms relying on the organ framework affected. Clinical indications of the infection might be the aftereffect of air sacculitis, enteritis, myocarditis, osteomyelitis, and nephritis. Mycotic contaminations because of *Mucor* spp. have been accounted for in African grey parrots. An antemortem determination is not common, and no active treatment has been described. Intense necrosis of the ventriculus is an exceptionally extraordinary sore in birds. Muir and Raidal (2012) explained a deadly instance of intense necrotizing ventriculitis in a female *Eclectus roratus* (eclectus parrot) caused by *Rhizopus microsporus* var *chinensis*. The bird was intensely dull and dormant, was regurgitating, and had brilliant green droppings, suggesting intense heavy metal poisoning. The bird was treated with liquids and chelation treatment, however, died within 12 h. Necropsy, cytology, histopathology, and culture results showed dual fungal attack of the ventriculus related to hemorrhage, transmural necrosis, intense irritation, and more *Rhizopus microsporus* var. *chinensis* and less *Candida krusei* were found.

Rhizopus is the most widely recognized *Mucorales* genus, and *Rhizopus microsporus* var. *rhizopodiformis* is the second most common species in *Rhizopus* (Roden et al. 2005). The mode of transmission of the disease is assumed to be using an inward breath of spores from the environment, deposition of spores on wounds, and ingestion of polluted food (Ribes et al. 2000). Major factors for the development of mucormycosis are diabetes mellitus, neutropenia, and corticosteroid treatment (Ribes et al. 2000). Human mucormycosis is depicted in paranasal sinuses, lungs, skin, cerebrum, and the gastrointestinal tract and is frequently spread (Kimura et al. 2012).

Rhizopus arrhizus was the predominant agent of mucormycosis identified in India, followed by *Rhizopus microsporus*, *Apophysomyces variabilis*, and *Rhizopus homothallicus* (Prakash et al. 2018). Hazard factors are diabetic ketoacidosis, immunosuppressive treatment, neutropenia, protein-calorie malnutrition, serious injury, and iron over-burden with or without deferoxamine (Kwon-Chung and Bennett 1992; Sugar 1992; Rinaldi 1989). Clinical symptoms are pulmonary, cutaneous, rhinocerebral, gastrointestinal, central nervous system, scattered, and various disorders. The most commonly found etiologic agent among humans is *Rhizopus oryzae* (arrhizus). Schmitz et al. (2018) reported *Mucor ramosissimus* as the causal agent of cutaneous zygomycosis (mucormycosis) in a patient with aplastic anemia who was having immunosuppressive treatment. The seclusion of Mucorales is not successful in a significant number of cases because of the delicate hyphae of the fungus. Once in a while, tissue materials are sent in formalin to a histopathology research center as it were. The molecular techniques have enhanced the detection of fungi, even in paraffin-implanted tissue (Rickerts et al. 2006; Bialek et al. 2005). However, *Rhizopus arrhizus* is the most prevalent causal agent worldwide (Chakrabarti et al. 2009; Skiada et al. 2011). Regardless of the increase in attention to the disease, the early detection of mucormycosis stays subtle because of trouble in getting the sample from deep tissues and the lack of a biomarker (Kontoyiannis et al. 2012). Recently, polymerase chain response for early detection of mucormycosis has been assessed with excellent outcomes; yet, no standard kit is available for routine testing (Millon et al. 2016).

References

Abassi M, Boulware DR, Rhein J (2015) Cryptococcal meningitis: diagnosis and management update. Curr Trop Med Rep 2(2):90–99

Abbas MS, Yassein SN, Khalaf JM (2017) Isolation and identification of some important mycological isolates from dropping of birds in Baghdad. J Entomol Zool Stud 5(3):671–673

Abd Elmegeed AS, Ouf SA, Moussa TA, Eltahlawi SM (2015) Dermatophytes and other associated fungi in patients attending to some hospitals in Egypt. Braz J Microbiol 46:799–805

Abdel-Gawad KM, Moharram AM (1989) Keratinophilic fungi from the duck nails in Egypt. J Basic Microbiol 29:259–263

Alvarez-Perez S, Mateos A, Dominguez L, Martinez-Nevado E, Blanco JL, Garcia ME (2010) Isolation of *Rhodotorula mucilaginosa* from skin lesions in a Southern sea lion (*Otaria flavescens*): a case report. Veterinarni Medicina 55(6):297–301

Al-Yasiri MH, Normand A-C, L'Ollivier C, Lachaud L, Bourgeois N, Rebaudet S, Piarroux R, Mauffrey J-F, Ranque S (2016) Opportunistic fungal pathogen *Candida glabrata* circulates between humans and yellow-legged gulls. Sci Rep 6:36157. Published online 2016 Oct 26. https://doi.org/10.1038/srep36157

Amaral RC, Ibanez JF, Mamizuka EM, Gambale W, de Paula CR, Larsson CE (1998) Microbiota indıgena do meato acustico externo de gatos hıgidos. Ciencia Rural, Santa Maria 28(3):4441–4445

Armstrong PA, Jackson BR, Haselow D (2018) Multistate epidemiology of histoplasmosis, United States, 2011-2014. Emerg Infect Dis 24(3):425–431

Aruo SK (1980) Necrotizing cutaneous rhodotorulosis in chickens in Uganda. Avian Dis 24(4):1038–1043

Badali H, Chander J, Bansal S et al (2010) First autochthonous case of *Rhinocladiella mackenziei* cerebral abscess outside the Middle East. J Clin Microbiol 48:646

Baddley JW, Winthrop KL, Patkar NM (2011) Geographic distribution of endemic fungal infections among older persons. Emerg Infect Dis 17(9):1664–1669

Bahr NC, Antinori S, Wheat LJ, Sarosi GA (2015) Histoplasmosis infections worldwide: thinking outside the Ohio River Valley. Curr Trop Med Rep 2:70–80

Bauck L (1994) Mycoses. In: Ritchie BW, Harrison GJ, Harrison LR (eds) Avian medicine: principles and application, 1st edn. Wingers Publishing, Lake Worth, p 1005

Benedict K, Mody RK (2016) Epidemiology of histoplasmosis outbreaks, United States, 1938-2013. Emerg Infect Dis 22(3):370–378

Bialek R, Konrad F, Kern J et al (2005) PCR based identification and discrimination of agents of mucormycosis and aspergillosis in paraffin wax embedded tissue. J Clin Pathol 58:1180–1184

Biegańska MJ, Rzewuska M, Dąbrowska I et al (2018) Mixed infection of respiratory tract in a dog caused by Rhodotorula mucilaginosa and Trichosporon jirovecii: a case report. Mycopathologia 183:637. https://doi.org/10.1007/s11046-017-0227-4

Bondaryk M, Kurzątkowski W, Staniszewska M (2013) Antifungal agents commonly used in the superficial and mucosal candidiasis treatment: mode of action and resistance development. Postepy Dermatol Alergol 30(5):293–301

Bonnedahl J et al (2009) Dissemination of *Escherichia coli* with CTX-M type ESBL between humans and yellow-legged gulls in the south of France. PLoS One 4:e5958

Borman AM, Campbell CK, Fraser M, Johnson EM (2007) Analysis of the dermatophyte species isolated in the British Isles between 1980 and 2005 and review of worldwide dermatophyte trends over the last three decades. Med Mycol 45:131–141

Bourdeau P, Hubert B, Magnol JP (1992) Suspicion de dermatomycose `a *Rhodotorula mucilaginosa* chez un chat infect´e par le FeLV et le FIV. Recueil de M´edecine Veterinaire 168(2):91–96

Bradley FA, Bickford AA, Walker RL (1993) Diagnosis of favus (avian dermatophytosis) in Oriental breed chickens. Avian Dis 37:1147–1150

Brilhante RSN et al (2010) Characterization of the gastrointestinal yeast microbiota of cockatiels (*Nymphicus hollandicus*): a potential hazard to human health. J Med Microbiol 59:718–723

Brilhante RSN, de Alencar LP, Cordeiro RA et al (2013) Detection of Candida species resistant to azoles in the microbiota of rheas (*Rhea americana*): possible implications for human and animal health. J Med Microbiol 62:889–895

Brotto TL, Andrade MCR, Goncalves MAB, Gimenis F, Pina A (2005) Identification of fungi microflora in the ear conducts of rhesus macaques (*Macaca mulatta*) kept in captivity. Braz J Vet Res Anim Sci 42(6):459–464

Buck JD (1980) Occurrence of human-associated yeasts in the feces and pool waters of captive bottlenosed dolphins (*Tursiops truncatus*). J Wildl Dis 16(1):141–149

Burtt EH Jr, Ichida JM (1999) Occurrence of feather-degrading bacilli in the plumage of birds. Auk 166:364–372

Burtt EH Jr, Ichida JM (2004) Gloger's rule, feather-degrading bacteria, and color variation among song sparrows. Condor 106:681–686

Cafarchia C, Camarda A, Romito D et al (2006) Occurrence of yeasts in cloacae of migratory birds. Mycopathologia 161:229–234

Călin M, Constantinescu-Aruxandei D, Alexandrescu E, Răut I, Doni MB, Arsene M-L, Oancea F, Jecu L, Lazăr V (2017) Degradation of keratin substrates by keratinolytic fungi. Electron J Biotechnol 28:101–112

Cao W, Jian C, Zhang H, Xu S (2018) Comparison of clinical features and prognostic factors of cryptococcal meningitis caused by *Cryptococcus neoformans* in patients with and without pulmonary nodules. Mycopathologia:1–8

Chakrabarti A, Slavin MA (2011) Endemic fungal infections in the Asia-Pacific region. Med Mycol 49(4):337–344

Chakrabarti A, Chatterjee SS, Das A et al (2009) Invasive zygomycosis in India: experience in a tertiary care hospital. Postgrad Med J 85:573–581

Charlton BR, Chin RP, Barnes HJ (2008) Fungal infections. In: Saif YM (ed) Diseases of poultry, 12th edn. Blackwell Publishing, Ames, pp 989–1001

Chermette R, Ferreiro L, Guillot J (2008) Dermatophytoses in animals. Mycopathologia 166:385–405

Colombo AL, Tobon A, Restrepo A, Queiroz Telles F, Nucci M (2011) Epidemiology of endemic fungal infections in Latin America. Med Mycol 49(8):785–798

Contreras A, Gomez-Martın A, Paterna A et al (2016) Epidemiological role of birds in the transmission and maintenance of zoonoses. Rev Sci Tech l'OIE 35:845–862

Costa EO, Gandra CR, Pires MF, Coutinho SD, Castilho W, Teixeira CM (1993) Survey of bovine mycotic mastitis in dairy herds in the State of Sao Paulo, Brazil. Mycopathologia 124(1):13–17

Costa AKF, Sidrim JJC, Cordeiro RA, Brilhante RSN, Monteiro AJ, Rocha MFG (2010) Urban pigeons (*Columba livia*) as a potential source of pathogenic yeasts: a focus on antifungal susceptibility of Cryptococcus strains in Northeast Brazil. Mycopathologia 169(3):207–213

Cristini A, Garcia-Hermoso D, Celard M et al (2010) Cerebral phaeohyphomycosis caused by *Rhinocladiella mackenziei* in a woman native to Afghanistan. J Clin Microbiol 48:3451

Cunha SC, Aguero C, Damico CB et al (2011) Duodenal perforation caused by *Rhizomucor* species in a cat. J Feline Med Surg 13:205–207

del Palacio A, Pereiro-Miguens M, Gimeno C et al (1992) Widespread dermatophytosis due to *Microsporum (Trichophyton) gallinae* in a patient with AIDS-a case report from Spain. Clin Exp Dermatol 17:449–453

Dhama K, Chakraborty S, Verma AK, Tiwari R, Barathidasan R, Kumar A, Singh SD (2013) Fungal/mycotic diseases of poultry-diagnosis, treatment and control: a review. Pak J Biol Sci 16:1626–1640

Diaz JH (2018) Environmental and wilderness-related risk factors for histoplasmosis: more than bats in caves. Wilderness Environ Med 29:531–540

Didehdar M, Shokohi T, Khansarinejad B, Sefidgar SAA, Abastabar M, Haghani I, Amirrajab N, Mondanizadeh M (2016) Characterization of clinically important dermatophytes in North of Iran using PCR-RFLP on ITS region. J Mycol Med 26:345–350

Dixit AK, Kushwaha RK (1991) Occurrence of keratinophilic fungi on Indian birds. Folia Microbiol (Praha) 36:383–386

Duarte ER, Resende JCP, Rosa CA, Hamdan JS (2001) Prevalence of yeasts and mycelial fungi in bovine parasitic otitis in the state of Minas Gerais, Brazil. J Vet Med 48(8):631–635

Duboc de Almeida GM, Costa SF, Melhem M et al (2008) *Rhodotorula* spp. isolated from blood cultures: clinical and microbiological aspects. Med Mycol 46(6):547–556

Dykstra MJ, Charlton BR, Chin RP, Barnes HJ (2013) Fungal infections. In: Swayne DE (ed) Diseases of poultry, 13th edn. Wiley-Blackwell, Ames, IA, pp 1077–1096

Eidi S, Nourani H, Naghibi A (2017) Necrotic ulcerative dermatitis due to simultaneous infections of *Malassezia* and *Microsporum gallinae* in a pigeon (*Columba livia domestica*). Iran J Vet Sci Technol 8(2):20–24. https://doi.org/10.22067/veterinary.v8i2.59131

El-Hariri M, Hamza D, Elhelw R, Refai M (2015) Lovebirds and cockatiels risk reservoir of *Cryptococcus neoformans*, a potential hazard to human health. J Vet Sci Med Diagn 4(4)

El-Hariri M, Hamza D, Elhelw R, Refai M (2016) Eucalyptus tree: a potential source of *Cryptococcus neoformans* in Egyptian environment. Int J Microbiol 2016:4080725, 5 pages. https://doi.org/10.1155/2016/4080725

el-Naghy MA, el-Ktatny MS, Fadl-Allah EM, Nazeer WW (1998) Degradation of chicken feathers by *Chrysosporium georgiae*. Mycopathologia 143:77–84

Falagas ME, Roussos N, Vardakas KZ (2010) Relative frequency of albicans and the various non-albicans *Candida* spp among candidemia isolates from inpatients in various parts of the world: a systematic review. Int J Infect Dis 14:e954–e966

Fouchier RAM, Osterhaus ADME, Brown IH (2003) Animal influenza virus surveillance. Vaccine 21:1754–1757

Francesca N et al (2014) Yeasts vectored by migratory birds collected in the Mediterranean island of Ustica and description of *Phaffomyces usticensis* f.a. sp. nov., a new species related to the cactus ecoclade. FEMS Yeast Res 14:910–921

Franzot SP, Salkin IF, Casadevall A (1999) *Cryptococcus neoformans* var. *grubii*: separate varietal status for *Cryptococcus neoformans* serotype A isolates. J Clin Microbiol 37(3):838–8407

García-Suárez J, Gómez-Herruz P, Cuadros JA, Burgaleta C (2011) Epidemiology and outcome of *Rhodotorula* infection in haematological patients. Mycoses 54(4):318–324

Georg LK, Torres G (1956) A human case of trichophyton gallinae infection; disease contracted from chickens. AMA Arch Derm 74:191–197

Giovannini J, Lee R, Zhang SX, Jun AS, Bower KS (2014) Rhodotorula keratitis: a rarely encountered ocular pathogen. Case Rep Ophthalmol 5(3):302–310

Goodenough AE, Stallwood B (2010) Intraspecific variation and interspecific differences in the bacterial and fungal assemblages of blue tit (*Cyanistes caeruleus*) and great tit (*Parus major*) nests. Microb Ecol 59:221–232

Granero V, Peyronel E, Mensa N, Liuzzi N, Costa C, Cavallo MR (2017) An unusual case of *Rhodotorula mucilaginosa* fungaemia in a cancer patient. Microbiol Med 32(3). https://doi.org/10.4081/mm.2017.6827

Grunder S, Mayser P, Redmann T, Kaleta EF (2005) Mycological examinations on the fungal flora of the chicken comb. Mycoses 48:114–119

Gunderson AR (2008) Feather-degrading bacteria: a new frontier in avian and host–parasite research? Auk 125:972–979

Gungnani HC, Sharma S, Gupta B (2012) Keratinophilic fungi recovered from feathers of different species of birds in St Kitts and Nevis. West Indian Med J 61(9):912–915

Harris JR, Lindsley MD, Henchaichon S et al (2012) High prevalence of cryptococcal infection among HIV-infected patients hospitalized with pneumonia in Thailand. Clin Infect Dis 54:43–50

Hayette MP, Sacheli R (2017) Unusual species of dermatophytes: rarely identified or new? Mycopathologia 182:203. https://doi.org/10.1007/s11046-016-0066-8

Hoffmann K, Pawlowska J, Walther G et al (2013) The family structure of the Mucorales: a synoptic revision based on comprehensive multigene-genealogies. Persoonia 30:57–76

Horton RA et al (2011) Fecal carriage and shedding density of CTX-M extended-spectrum {beta}-lactamase-producing *Escherichia coli* in cattle, chickens, and pigs: implications for environmental contamination and food production. Appl Environ Microbiol 77:3715–3719

Hubálek Z (2000) Keratinophilic fungi associated with free-living mammals and birds. In: Kushwaha RKS, Guarro J (eds) Biology of dermatophytes and other keratinophilic fungi. Revista Iberoamericanade Micología, São Paulo, pp 93–103

Hube B et al (2015) Dermatomycoses and inflammation: the adaptive balance between growth damage, and survival. J Mycol Med 25:e44–e58

Isidoro-Ayza M, Perez L, Cabanes FJ et al (2014) Central nervous system mucormycosis caused by *Cunninghamella bertholletiae* in a bottlenose dolphin (*Tursiops truncatus*). J Wildl Dis 50:634–638

Jabeen K, Farooqi J, Zafar A et al (2011) *Rhinocladiella mackenziei* as an emerging cause of cerebral phaeohyphomycosis in Pakistan: a case series. Clin Infect Dis 52:213

Jensen HE, Aalbaek B, Basse A et al (1992) The occurrence of fungi in bovine tissues in relation to portals of entry and environmental factors. J Comp Pathol 107:127–140

Jin H-S, Park SY, Kim K, Lee Y-J, Nam G-W, Kang NJ et al (2017) Development of a keratinase activity assay using recombinant chicken feather keratin substrates. PLoS One 12:e0172712

Kadota K, Uchida K, Nagatomo T et al (1995) Granulomatous epididymitis related to *Rhodotorula glutinis* infection in a dog. Vet Pathol 32(6):716–718

Kantarcioglu AS, de Hoog GS (2004) Infections of the central nervous system by melanized fungi: a review of cases presented between 1999 and 2004. Mycoses 47:4

Karesh WB, Russell R, Gribble D (1987) Dactylaria gallopava encephalitis in two grey-winged trumpeters (*Psophia crepitans*). Avian Dis 31(3):685–688

Kauffman CA (2007) Histoplasmosis: a clinical and laboratory update. Clin Microbiol Rev 20(1):115–132

Khosravi AR, Mahmoudi M (2003) Dermatophytes isolated from domestic animals in Iran. Mycoses 46:222–225

Kim HA, Hyun M, Ryu S-Y (2013) Catheter-associated *Rhodotorula mucilaginosa* fungemia in an immunocompetent host. Infect Chemother 45(3):339–342

Kim EL, Patel SR, George MS, Ameri H (2018) Ochroconis gallopava endophthalmitis successfully treated with intravitreal voriconazole and amphotericin B. Retinal Cases Brief Rep 12

Kimura M, Nishimura K, Enoki E et al (2012) Chlamydospores of *Rhizopus microsporus* var. *rhizopodiformis* in Tissue of Pulmonary Mucormycosis. Mycopathologia 174:441. https://doi.org/10.1007/s11046-012-9560-9

Kontoyiannis DP, Lewis RE, Lortholary O et al (2012) Future directions in mucormycosis research. Clin Infect Dis 54:79–85

Krcmery V, Barnes AJ (2002) Non-*albicans Candida* spp. causing fungaemia: pathogenicity and antifungal resistance. J Hosp Infect 50:243–260

Kronstad JW, Attarian R, Cadieux B et al (2011) Expanding fungal pathogenesis: cryptococcus breaks out of the opportunistic box. Nat Rev Microbiol 9:193–203

Kunkle RA (2003) Fungal infections. In: Saif YM, Barnes HJ, Glisson JR, Fadly AM, McDougald LR, Swayne DE (eds) Diseases of poultry, 11th edn. Iowa State Press, Ames, p 901

Kwon-Chung KJ, Bennett JE (1992) Medical mycology. Lea & Febiger, Philadelphia, pp 524–559

Kwon-Chung KJ, Varma A (2006) Do major species concepts support one, two or more species within *Cryptococcus neoformans*? FEMS Yeast Res 6(4):574–587

Kwon-Chung KJ, Boekhout T, Fell JW, Diaz M (2002) Proposal to conserve the name *Cryptococcus gattii* against *C. hondurianus* and *C bacillisporus* (Basidiomycota, Hymenomycetes, Tremellomycetidae). Taxon 51(4):804–806

Kwon-Chung KJ, Boekhout T, Wickes BL, Fell JW (2011) Systematics of the genus Cryptococcus and its type species C. neoformans. In: Heitman J, Kozel TR, Kwon-Chung KJ, Perfect JR, Casadevall A (eds) Cryptococcus: from human pathogen to model yeast. ASM Press, pp 3–16. (Chapter 1)

Larone DH (1995) Medically important fungi—a guide to identification, 3rd edn. American Society for Microbiology, Washington, DC

Laubscher WDF, Viljoen BC, Albertyn J (2000) The yeasts flora occurring in the trachea of broiler chicken. Food Technol Biotechnol 38:77–80

Levitz SM, Specht CA (2006) The molecular basis for the immunogenicity of *Cryptococcus neoformans* mannoproteins. FEMS Yeast Res 6:513–524

Liu K, Ding H, Xu B et al (2016) Clinical analysis of non-AIDS patients pathologically diagnosed with pulmonary cryptococcosis. J Thorac Dis 8:2813–2821

Liu J, Liu H, Yan J, Liu N, Zhang H, Zhao C, Liu Y (2018) Molecular typing and genetic relatedness of 72 clinical Candida albicans isolates from poultry. Vet Microbiol 214:36–43

Lord ATK, Mohandas K, Somanath S, Ambu S (2010) Multidrug resistant yeasts in synanthropic wild birds. Ann Clin Microbiol Antimicrob 9:1–5

Louria DB, Greenberg SM, Molander DW (1960) Fungemia caused by certain nonpathogenic strains of the family Cryptococcaceae. N Engl J Med 263:1281–1284

Lunardi LW, Aquino VR, Zimerman RA, Goldani LZ (2006) Epidemiology and outcome of *Rhodotorula* fungemia in a tertiary care hospital. Clin Infect Dis 43(6):e60–e63

Malik R, Krockenberger MB, Cross G, Doneley R, Madill DN, Black D, Mcwhirter P, Rozenwax A, Rose K, Alley M, Forshaw D, Russel-Brown I, Johnstone AC, Martin P, O'Brien CR, Love DN (2003) Avian cryptococcosis. Med Mycol 41:115–124

Mancianti F, Nardoni S, Ceccherelli R (2002) Occurrence of yeasts in psittacines droppings from captive birds in Italy. Mycopathologia 153:121–124

Mandeel Q, Nardoni S, Mancianti F (2011) Keratinophilic fungi on feathers of common clinically healthy birds in Bahrain. Mycoses 54:71–77

Manos NE, Ferebee SH, Kerschbaum WF (1956) Geographic variation in the prevalence of histoplasmin sensitivity. Dis Chest 29(6):649–668

Maziarz EK, Perfect JR (2016) Cryptococcosis. Infect Dis Clin North Am 30(1):179–206

McLeod DS, Mortimer RH, Perry-Keene DA (2011) Histoplasmosis in Australia: report of 16 cases and literature review. Medicine 90(1):61–68

Melville PA, Cogliati B, Mangiaterra MBBCD et al (2004) Determinacao da microbiota presente na cloaca e orofaringe de avestruzes (Struthio camelus) clinicamente sadios. Ciencia Rural 34(6):1871–1876

Miceli MH, Dıaz JA, Lee SA (2011) Emerging opportunistic yeast infections. Lancet Infect Dis 11(2):142–151

Millon L, Herbrecht R, Grenouillet F et al (2016) Early diagnosis and monitoring of mucormycosis by detection of circulating DNA in serum: retrospective analysis of 44 cases collected through the French Surveillance Network of Invasive Fungal Infections (RESSIF). Clin Microbiol Infect 22:810.e1–810.e8

Milton F (1999) Candidiasis. In: Milton F, Franson JC (eds) Field manual of wildlife diseases, 1st edn. U.S. Department of interior. U. S. Geological Survey, pp 135–136

Miyasato H, Yamaguchi S, Taira K, Hosokawa A, Kayo S, Sano A, Uezato H, Takahashi K (2011) Tinea corporis caused by *Microsporum gallinae*. The first clinical case in Japan. J Dermatol 38: 473–478

Mohammadi R, Mohammadi A, Ashtari F et al (2018) Cerebral phaeohyphomycosis due to *Rhinocladiella mackenziei* in Persian Gulf region: a case and review. Mycoses 61:261–265. https://doi.org/10.1111/myc.12734

Monga DP, Garg DN (1980) Ovine pulmonary infection caused by *Rhodotorula rubra*. Mykosen 23(4):208–211

Moran C, Delafield NL, Kenny G et al (2019) A case of *Verruconis gallopava* infection in a heart transplant recipient successfully treated with posaconazole. Transpl Infect Dis 21(2):e13044

Moretti AD, Piergili FL, Bọncio P, Pasquali ED, Rossi (2000) Isolation of *Candida rugosa* from turkeys. J Vet Med B 47:433–439

Moț AC, Pârvu M, Pârvu AE, Roşca-Casian O, Dina NE, Leopold N, Silaghi-Dumitrescu R, Mircea C (2017) Reversible naftifine-induced carotenoid depigmentation in *Rhodotorula mucilaginosa* (A. Jörg.) F.C. Harrison causing onychomycosis. Sci Rep 7:11125

Mugale M, Bhat AA, Gavhane DS, Bhat SA (2015) Outbreaks of thrush in pigeons in Punjab State of India. Comp Clin Path 24(3):635–638. Published online 2014 Jul 8. https://doi.org/10.1007/s00580-014-1958-y

Muir M, Raidal S (2012) Necrotising ventriculitis due to combined infection with *Rhizopus microsporus* var. *chinensis* and *Candida krusei* in an eclectus parrot (*Eclectus roratus*). Austr Vet J 90:277–280

Murata M, Takahashi H, Takahashi S, Takahashi Y, Chibana H, Murata Y, Sugiyama K, Kaneshima T, Yamaguchi S, Miyasato H, Murakami M, Kano R, Hasegawa A, Uezato H, Hosokawa A, Sano A (2013) Isolation of *Microsporum gallinae* from a fighting cock (*Gallus gallus domesticus*) in Japan. Med Mycol 51:144–149

Neves VC, Murdoch N, Furness RW (2006) Population status and diet of the yellow-legged Gull in the Azores. Life Mar Sci 23A:59–73

Nosanchuk JD, Casadevall A (2006) Impact of melanin on microbial virulence and clinical resistance to antimicrobial compounds. Antimicrob Agents Chemother 50:3519–3528

Nosanchuk JD, Shoham S, Fries BC, Shapiro DS, Levitz SM, Casadevall A (2000) Evidence of zoonotic transmission of *Cryptococcus neoformans* from a pet cockatoo to an immunocompromised patient. Ann Intern Med **132**:205–208

Ogasawara F, Yamamoto Y, Sato Y, Fukunari K, Murata K, Yaegashi G, Goto M, Murakami R (2016) Concurrent Fowlpox and Candidiasis diseases in backyard chickens with unusual pox lesions in the bursa of Fabricius. Avian Dis 60:705–708

Ossiboff RJ, Clancy MM, Terio KA, Conley KJ, McAloose D (2015) Cerebral and ocular ochroconiasis in 2 elegant crested tinamou (*Eudromia elegans*) chicks. Vet Pathol 52:716–719

Persinoti GF, Martinez DA, Li W, Dögen A, Billmyre RB, Averette A, Goldberg JM, Shea T, Young S, Zeng Q, Oliver BG, Barton R, Metin B, Hilmioğlu-Polat S, Ilkit M, Gräser Y, Martinez-Rossi NM, White TC, Heitman J, Cuomo CA (2018) Whole-genome analysis illustrates global clonal population structure of the ubiquitous dermatophyte pathogen *Trichophyton rubrum*. Genetics 208:1657

Poblete-Gutierrez P, Abuzahra F, Becker F, Krause H, Merk HF, Frank J (2006) Onychomycosis in a diabetic patient due to *Trichophyton gallinae*. Mycoses 49:254–257

Potenza L, Chitasombat MN, Klimko N et al (2018) Rhodotorula infection in haematological patient: risk factors and outcome. Mycoses 62(3):223–229. https://doi.org/10.1111/myc.12875

Prakash H, Ghosh AK, Rudramurthy SM, Singh P, Xess I, Savio J, Pamidimukkala U, Jillwin J, Varma S, Das A, Panda NK, Singh S, Bal A, Chakrabarti A (2018) A prospective multicenter study on mucormycosis in India: epidemiology, diagnosis, and treatment. Med Mycol

Ranck FM Jr, Georg LK, Wallace DH (1974) Dactylariosis—a newly recognized fungus disease of chickens. Avian Dis 18:4

Reis EJC, Buscariolo F, Siqueira JPZ, Castilho EM, Almeida MTG (2018) *Agapornis* sp. pet birds: source of dissemination of azole-resistant yeasts. Med Mycol 57(2):133–150. https://doi.org/10.1093/mmy/myy061

Revankar SG, Sutton DA, Rinaldi MG (2004) Primary central nervous system phaeohyphomycosis: a review of 101 cases. Clin Infect Dis 38:206

Reynaldi FJ, Giacoboni G, Cordoba SB et al (2017) Mucormicosys due to *Saksenaea vasiformis* in a dog. Med Mycol Case Rep 16:4–7

Ribes JA, Vanover-Sams CL, Baker DJ (2000) Zygomycetes in human disease. Clin Microbiol Rev 13:236–301

Rickerts V, Just-Nuling G, Konrad F et al (2006) Diagnosis of invasive aspergillosis and mucormycosis in immunocompromised patients by semi-nested PCR assay of tissue samples. Eur J Clin Microbiol Infect Dis 25:8–13

Riffel A, Brandelli A (2006) Keratinolytic bacteria isolated from feather waste. Braz J Microbiol 37: 395–399

Rinaldi MG (1989) Zygomycosis. Infect Dis Clin N Am 3:19–41

Roden MM, Zaoutis TE, Buchanan WL, Knudsen TA, Sarkisova TA, Schaufele RL et al (2005) Epidemiology and outcome of zygomycosis: a review of 929 reported cases. Clin Infect Dis 41: 634–653

Rodrigues ML, Nakayasu ES, Oliveira DL et al (2008) Extracellular vesicles produced by *Cryptococcus neoformans* contain protein components associated with virulence. Eukaryot Cell 7: 58–67

Sabouraud R (1910) Maladies de Cuir Chevelu. III. Les Maladies Cryptogamiques. Les Teignes. Masson etCie, Paris

Saleh H, Moawad AA, ElHariri M, Refai MK (2011) Prevalence of yeasts in human, animals and soil sample in El-Fayoum Governorate in Egypt. Int J Microbiol 2(3):233–239

Salkin IF, Dixon DM, Kemna ME et al (1990) Fatal encephalitis caused by *Dactylaria constricta* var. *gallopava* in a snowy owl chick (*Nyctea scandiaca*). J Clin Microbiol 28(12):2845–2847

Saremi F, Go JL, Zee CS (2004) Imaging in CNS cryptococcosis. http://emedicine.medscape.com/article/339576-overview

Saxena P, Kumar A, Shrivastava JN (2004) Diversity of keratinophilic mycoflora in the soil of Agra (India). Folia Microbiol (Praha) 49:430–434

Schmertmann LJ, Bodley K, Meyer W, Malik R, Krockenberger MB (2018) Multi-locus sequence typing as a tool to investigate environmental sources of infection for cryptococcosis in captive birds. Med Mycol

Schmitz A, Korbel R, Rinder M (2018) Zygomycosis due to Mucor racemosus associated with a circovirus infection in a Blue Tit (*Cyanistes caeruleus*). J Comp Pathol 165:67–71

Setianingrum F, Rautemaa-Richardson R, Denning DW (2018) Pulmonary cryptococcosis: a review of pathobiology and clinical aspects. Med Mycol 57(2):133–150

Seyedmousavi S, Bosco SMG, de Hoog S, Ebel F, Elad D, Gomes RR, Jacobsen ID, Jensen HE, Martel A, Mignon B, Pasmans F, Pieckova E, Rodrigues AM, Singh K, Vicente VA, Wibbelt G, Wiederhold NP, Guillot J (2018) Fungal infections in animals: a patchwork of different situations. Med Mycol 56:165–187

Shah AH et al (2011) Indicator microbes correlate with pathogenic bacteria, yeasts and helminthes in sand at a subtropical recreational beach site. J Appl Microbiol 110:1571–1583

Shane SM, Markovits J, Snider TG et al (1985) Encephalitis attributed to dactylariosis in Japanese quail chicks (*Coturnix coturnix japonica*). Avian Dis 29(3):822–828

Sharma S, Sumbali G (2017) Diversity and distribution of dermatophytes and other keratinophilic fungi in the soil of gharana wetland, India—an important stop over site for migratory birds. Int J Pharm Biosci 8(2):956–963

Shawkey MD, Pillai SR, Hill GE, Siefferman LM, Roberts SR (2007) Bacteria as an agent for change in structural plumage color: correlational and experimental evidence. Am Nat 169: S112–S121

Shokri H, Khosravi A, Sharifzadeh A, Tootian Z (2010) Isolation and identification of yeast flora from genital tract in healthy female camels (*Camelus dromedarius*). Vet Microbiol 144(1–2):183–186

Silva S et al (2012) *Candida glabrata*, *Candida parapsilosis* and *Candida tropicalis*: biology, epidemiology, pathogenicity and antifungal resistance. FEMS Microbiol Rev 36:288–305

Singh SD, Dash BB (2008) Spontaneous lesions of cryptococcosis in White Leghorn chicken. Indian J Vet Pathol 32:68–69

Singh K, Ilkit M, Shokohi T, Tolooe A, Malik R, Seyedmousavi S (2018) Cryptococcosis: emergence of *Cryptococcus gattii* in animals and zoonotic potential. In: Seyedmousavi S, de Hoog G, Guillot J, Verweij P (eds) Emerging and epizootic fungal infections in animals. Springer, Cham

Skiada A, Pagano L, Groll A et al (2011) Zygomycosis in Europe: analysis of 230cases accrued by the registry of the European Confederation of Medical Mycology (ECMM) Working Group on Zygomycosis between 2005 and 2007. Clin Microbio Infect 17:1859–1867

Sonne C, Alstrup AKO, Therkildsen OR (2012) A review of the factors causing paralysis in wild birds: Implications for the paralytic syndrome observed in the Baltic Sea. Sci Total Environ 416: 32–39

Spader TB, Ramírez-Castrillón M, Valente P et al (2019) In vitro interactions of amphotericin b combined with non-antifungal agents against *Rhodotorula mucilaginosa* strains. Mycopathologia. https://doi.org/10.1007/s11046-019-0317-6

Spiliopoulou A, Anastassiou ED, Christofidou M (2012) *Rhodotorula* fungemia in an intensive care unit patient and review of published cases. Mycopathologia 174(4):301–309

Sugar AM (1992) Mucormycosis. Clin Infect Dis 14(Suppl):S126–S129

Sutton DA, Slifkin M, Yakulis R, Rinaldi MG (1998) U.S. case report of cerebral phaeohyphomycosis caused by *Ramichloridium obovoideum* (*R. mackenziei*): criteria for identification, therapy, and review of other known dematiaceous neurotropic taxa. J Clin Microbiol 36:708

Suwatanapongched T, Sangsatra W, Boonsarngsuk V, Incharoen P (2013) Clinical and radiologic manifestations of pulmonary cryptococcosis in immunocompetent patients and their outcomes after treatment. Diagnostic Interv Radiol 19:438–446

Tampieri MP (2004) Update on the diagnosis of dermatomycosis. Parassitologia 46:183–186

Tampieri MP (2006) Mycetes and urban areas. Parassitologia 48:121–124

Tan X, Chen R, Zhu S et al (2016) Candida albicans airway colonization facilitates subsequent *Acinetobacter baumannii* pneumonia in a rat model. Antimicrob Agents Chemother 60(6):3348–3354. Published 2016 May 23. https://doi.org/10.1128/AAC.02180-15

Tuon FF, Costa SF (2008) *Rhodotorula* infection. A systematic review of 128 cases from literature. Revista Iberoamericana de Micologia 25(3):135–140

Vecchiarelli A, Pietrella D, Dottorini M et al (1994) Encapsulation of *Cryptococcus neoformans* regulates fungicidal activity and the antigen presentation process in human alveolar macrophages. Clin Exp Immunol 98:217–223

Voelz K, Johnston SA, Smith LM, Hall RA, Idnurm A, May RC (2014) "Division of labour" in response to host oxidative burst drives a fatal *Cryptococcus gattii* outbreak. Nat Commun 5: 5194

Waldrip DW, Padhye AA, Ajello L et al (1974) Isolation of *Dactylaria gallopava* from broiler-house litter. Avian Dis 18(3):445–451

Wheat J, Sarosi G, McKinsey D (2000) Practice guidelines for the management of patients with histoplasmosis. Clin Infect Dis 30(4):588–595

Whitehurst C, Patel C, Boethel C (2018) Ochroconis gallopava infection in a patient with bronchiectasis on low dose steroids. C58. Fungal Case Rep A5386

Wirth F, Goldani LZ (2012) Epidemiology of *Rhodotorula*: an emerging pathogen. Interdiscip Perspect Infect Dis 2012:465717, 7 pages. https://doi.org/10.1155/2012/465717

Yamaguchi S, Sano A, Hiruma M, Murata M, Kaneshima T, Murata Y, Takahashi H, Takahashi S, Takahashi Y, Chibana H, Touyama H, Ha NT, Nakazato Y, Uehara Y, Hirakawa M, Imura Y, Terashima Y, Kawamoto Y, Takahashi K, Sugiyama K, Hiruma M, Murakami M, Hosokawa A, Uezato H (2014) Isolation of dermatophytes and related species from domestic fowl (*Gallus gallus domesticus*). Mycopathologia 178:135–143

Yoo SD, Worodria W, Med M et al (2011) The prevalence and clinical course of HIV-associated pulmonary cryptococcosis in Uganda. J Acquir Immune Defic Syndr 54:269–274

Zaragoza O, Rodrigues ML, De Jesus M et al (2009) The capsule of the fungal pathogen *Cryptococcus neoformans*. Adv Appl Microbiol 68:1–64

Zhan P, Dukik K, Li D, Sun J, Stielow JB, van den Ende BG, Brankovics B, Menken SBJ, Mei H, Bao W, Lv G, Liu W, de Hoog GS (2018) Phylogeny of dermatophytes with genomic character evaluation of clinically distinct *Trichophyton rubrum* and *T. violaceum*. Stud Mycol 89:153–175